Praise from the Experts

"The genesis of the Ramírez work is the legendary *Experimental Statistics, NBS Handbook 91* assembled by Mary Natrella of the National Bureau of Standards (now National Institute of Standards and Technology). The authors have skillfully blended one of the finest traditional statistical works with the contemporary software capability of JMP. The result is a powerful, yet user-friendly resource the practicing engineer/scientist can rely upon to solve the immediate problem at hand.

"The authors are seasoned industrial statisticians responding to the needs of frontline engineers and scientists. Unlike traditional textbooks, each chapter focuses upon a real life technical problem rather than a statistical technique. The book is rich with many examples across both industry and discipline.

"For example, both young and seasoned investigators will enjoy and appreciate the dynamic JMP analysis of data from the first published paper of a young scientist named Einstein.

"The book will also serve as a valuable supplement to traditional engineering/scientific statistical textbooks at both the undergraduate and graduate level. The authors deftly dovetail both graphical and computational analysis and in the process clarify and quantify the industrial challenge under investigation.

"We look forward to utilizing the book in our next offering of engineering statistics.

"This book deserves a place on the bookshelf of every practicing engineer and scientist."

James C. Ford Ph.D.
College of Engineering
University of Delaware

"*Analyzing and Interpreting Continuous Data Using JMP: A Step by Step Guide*, by José G. Ramírez, Ph.D., and Brenda S. Ramírez, M.S., does a wonderful job blending the practical side of industrial data with a rigorous statistical handling in an intuitive and graphical framework utilizing JMP. Industrial engineers/scientists interested in proper handling of statistical comparisons related to experimentation and calibration will find its step-by-step framework complete and well documented via detailed process steps, examples, and visuals. This work provides a unique connection back to *Experimental Statistics, NBS Handbook 91* (now NIST SP 958). Lastly, the book is very much a ready-for-use reference guide for common statistical problems in engineering and science."

Tim Rey
Leader, Data Mining and Modeling
The Dow Chemical Company

"*Analyzing and Interpreting Continuous Data Using JMP: A Step-by-Step Guide*, by José G. Ramírez and Brenda S. Ramírez, is not just an introductory guide to statistical analysis; it is above all an excellent tool for quality and reliability engineers and for Six Sigma practitioners. The authors have made data analysis and interpretation very easy to do by using examples that are encountered on a daily basis in all industries. While it is very instructive and covers just about all the subjects applicable to all aspects of management, especially quality and reliability, the book keeps mathematical reasoning to a strict minimum. The many examples and the graphs that accompany each chapter make the book easier to understand by a wider audience. This is just an excellent book for beginners and a reference for practitioners!"

Issa Bass
Senior Six Sigma Consultant with Manor House and Associates

SAS Publishing

Analyzing and Interpreting Continuous Data Using JMP®

A Step-by-Step Guide

José G. Ramírez, Ph.D.

Brenda S. Ramírez, M.S.

THE POWER TO KNOW®

The correct bibliographic citation for this manual is as follows: Ramírez, José G., and Brenda S. Ramírez. 2009. *Analyzing and Interpreting Continuous Data Using JMP®: A Step-by-Step Guide.* Cary, NC: SAS Institute Inc.

Analyzing and Interpreting Continuous Data Using JMP®: A Step-by-Step Guide

ISBN 978-1-59994-488-3

SAS Institute Inc., SAS Campus Drive, Cary, North Carolina 27513.

1st printing, August 2009
2nd printing, October 2010

To our daughter Oriana Sofía
Our joyful "Golden Wisdom"

To Brittany, Lindsay, Harrison, Jadon, and Nehum
Future engineers and scientists

To Professor G.E.P. Box
Advisor, Mentor, Friend ("Pel")

Contents

Foreword

John Tukey predicted that computers would revolutionize the practice of statistics. But I think most of us are surprised by the extent to which the practice of statistics has been "democratized" by the computer. It is very probable that more statistical methods are used by non-statisticians than by those of us who think of ourselves as statisticians. Consequently, there is an increasing need for books that are application-focused, treat software as an integral part of the process of using statistical methods to help solve a particular problem, and offer concise, authoritative advice. José G. Ramírez, Ph.D., and Brenda S. Ramírez, M.S., have written such a book.

Ramírez and Ramírez focus on continuous data, and use JMP software as the analytical engine. The authors have provided much more than a software manual, offering straightforward explanations and practical insight into fundamental concepts such as formulating null and alternative hypotheses, interpretation of *p*-values, confidence intervals, and interpreting ANOVA and regression results. A seven-step problem-solving procedure is emphasized throughout the book.

The book is organized by application, not by method. This is a refreshing and appealing approach that invites the reader to "dive in" and learn the appropriate techniques to solve your problem and to quickly understand how the techniques you need are implemented in JMP. After reading Chapters 1 and 2, the reader can easily move into any of the following five chapters: "Characterizing the Measured Performance of a Material, Process, or Product"; "Comparing the Measured Performance of a Material, Process, or Product to a Standard"; "Comparing the Measured Performance of Two Materials, Processes, or Products"; "Comparing the Measured Performance of Several Materials, Processes, or Products"; and "Characterizing Linear Relationships between Two Variables." The clear explanations and step-by-step JMP instructions make mastering the techniques easy.

Ramírez and Ramírez have given us a well-written, practical guide that helps us get started using JMP to solve a range of common application problems that span many areas of engineering, science, business, and industry. Readers with just a modest background in statistics and some experience with JMP will find this book a useful addition to their reference shelf.

Douglas C. Montgomery
Regents' Professor of Industrial Engineering and Statistics
Ira A. Fulton School of Engineering
Arizona State University, Tempe, AZ

Acknowledgments

Two decades ago John Sall had the vision of creating easy-to-use software where scientists and engineers could explore and model data using interactive statistical analysis and data visualization. Many years later he suggested we write a JMP book for engineers and scientists. This book would not have been possible without his vision and encouragement.

Special recognition and appreciation go to our friend, engineer, and statistician Dr. Jesús Cuéllar for his critical chapter-by-chapter review of this book, and for his valuable suggestions for how to make statistics more accessible to engineers and scientists.

SAS Press acquisitions editor, Stephenie Joyner, gave us guidance and support throughout all the phases of the production of this book. She was always ready to answer or find answers to our questions, and to provide the help we needed when it was needed.

The SAS technical reviewers assigned to this project: Annie Zangi, Duane Hayes, Paul Marovich, Tonya Mauldin, and Katrina Hauser who reviewed the document at different stages, and who asked clarifying questions that improved the readability of the text, and suggested enhancements to the JMP instructions.

Thanks goes to the SAS Press team who patiently worked with us during this process and made the final product a reality: Julie Platt (Editor-in-Chief), Candy Farrell (Technical Publishing Specialist), Jennifer Dilley and Stacy Suggs (Graphics Specialists), Brad Kellam (Technical Editor), Mary Beth Steinbach (Managing Editor), Patrice Cherry (Cover Designer), and Shelly Goodin and Stacey Hamilton for handling the promotional and marketing activities for this book.

Víctor Fiol brought his artistic and technical skills to Figures 2.6, 4.17 and 4.31, while Mark Bailey provided the Graphical ANOVA script, and collaborated with us in writing the One-Sample Equivalence Test script. Xan Gregg provided JSL help to generate the subscripts in Figure 6.1.

We feel honored that Professor Doug Montgomery wrote the foreword to our book. His contributions to industrial statistics, and in particular his books, have always been relevant and useful in our work with engineers and scientists.

Finally, we want to thank all of the engineers and scientists we have collaborated with, learned from, and taught over our many years in industry.

Chapter 1

Using This Book

1.1 Origins of This Book

In 1963 the National Bureau of Standards published the NBS Handbook 91 *Experimental Statistics*, which was "intended for the user with an engineering background who, although he has an occasional need for statistical techniques, does not have the time or inclination to become an expert on statistical theory and methodology" (Natrella 1963). Our years of experience teaching and working closely with engineers and scientists suggest that more than 45 years since these words were published they still ring true. So when the opportunity came to write a JMP book for engineers and scientists, the NBS Handbook 91 became our inspiration.

The NBS Handbook 91 *Experimental Statistics* brought together a series of five pamphlets that were commissioned by the Army Research Office. The handbook was prepared in the Statistical Engineering Laboratory (SEL) under the leadership of Mary Gibbons Natrella, and was written "as an aid to scientists and engineers in the Army Ordnance research and development programs and as a guide for military and civilian personnel who had responsibility for experimentation and tests for Army Ordnance equipment in the design and development stages of production" (Natrella 1963). Although intended for the army personnel, the NBS Handbook 91 became the National Bureau of Standards second best-selling publication because of its emphasis on solving practical problems using statistical techniques. The chapter headings are written in a language that points to a particular application or problem, rather than to a statistical technique. Within each chapter the user can find step-by-step instructions for how to carry out the analysis, including examples that have been worked out and discussions of the results.

Out of print for many years, the NBS Handbook 91 is still relevant and a highly regarded reference for many who work in industry. So much so that in the late 90s, SEMATECH (a consortium of major U.S. semiconductor manufacturers) approached the Statistical Engineering Division (SED) at the National Institute of Standards and Technology (NIST, formerly known as the National Bureau of Standards) with a "proposal for updating and recreating the book with examples directed towards the semiconductor industry" (Croarkin 2001). The result became a Web-based engineering statistics handbook (http://www.itl.nist.gov/div898/handbook/).

In writing this book we wanted to bring the spirit and usefulness of the NBS Handbook 91 to the countless engineers, scientists, and data analysts whose work requires them to transform data into useful and actionable information. In this book you will also discover how the ease and power of JMP make your statistical endeavors a lot easier than the hand calculations required when the NBS Handbook 91 was first published.

1.2 Purpose

In the spirit of NBS Handbook 91, our book is designed to serve as a ready-for-use reference for solving common problems in science and engineering using statistical techniques and JMP. Some examples of these types of problems include evaluating the performance of a new raw material supplier for an existing component, establishing a calibration curve for instrumentation, comparing the performance of several materials using a quality characteristic of interest, or troubleshooting a yield fall-out. These problems are not unique to a particular industry and can be found across industries as diverse as the semiconductor, automotive, chemical, aerospace, food, biological, or pharmaceutical industries. In addition, these problems might present themselves throughout the life cycle of a product or process, and therefore engineers and scientists focused on manufacturing, new product development, metrology, or quality assurance will benefit from using the statistical techniques described in this book.

1.3 Audience

As with the NBS Handbook 91, our main audience is *you* the engineer or scientist who needs to use or would like to use statistical techniques to help solve a particular problem. Each chapter is application driven, and is written with different objectives depending on your needs:

- For those of you who want a quick reference for how to solve common problems in engineering and science using statistical methods and JMP, each chapter includes step-by-step instructions for how to carry out the statistical techniques, interpret the results in the context of the problem statement, and draw the appropriate conclusions.
- For those of you who want a better understanding of the statistical underpinnings behind the techniques, each chapter provides a practical overview of the statistical concepts and appropriate references for further study.
- For those who want to learn how to benefit from the power of JMP in the context described previously, each chapter is loaded with general discussions, specific JMP step-by-step instructions, and tips and tricks.

1.4 Prerequisites

Although this book covers introductory topics in statistics, some familiarity with basic statistical concepts, such as average, standard deviation, and a random sample, is helpful. You should also have a basic working knowledge of JMP, including how to read and manipulate data in JMP, maneuver around the various menus and windows, and use the online help resources.

1.5 What's Unique About This Book?

We have spent many years as industrial statisticians working closely with, and developing and teaching courses for, engineers and scientists in the semiconductor, chemical, and manufacturing industries, and have focused on making statistics both accessible and effective in helping to solve common problems found in an industrial setting. Statistical techniques are introduced not as a collection of formulas to be followed, but as a catalyst to enhance and speed up the engineering and scientific problem-solving process. Each chapter uses a 7-step problem-solving framework to make sure that the right problem is being solved with an appropriate selection of tools.

In order to facilitate the learning process, numerous examples are used throughout the book that are relevant, realistic, and thoroughly explained. The step-by-step instructions show how to use JMP and the appropriate statistical techniques to solve a particular problem, putting emphasis on how to interpret and translate the output in the context of the problem being solved. You will find this book to be a useful reference when faced with similar situations in your day-to-day work.

Throughout the book we try to demystify concepts like *p*-values, confidence level, the difference between the null (H_0) and alternative hypothesis (H_1), and tests of significance. In addition, emphasis is placed on analyzing not only the average measured performance of a characteristic of interest, which is normally the case, but also the standard deviation of the characteristic of interest. This is crucial because most practitioners in an industrial setting quickly realize that the variation in a measured performance characteristic is as important, and sometimes more important, than focusing solely on the average performance characteristic in order to meet performance goals.

As in the NBS Handbook 91, each chapter heading reflects a practical situation rather than a statistical technique. This makes it easier for you, the user, to focus on the situation at hand rather than trying to figure out if a particular statistical technique is applicable to your situation. Section 1.6 describes the chapters in more detail.

1.6 Chapter Contents

The statistical concepts and techniques are divided into six chapters in this book, although some are repeated in all chapters. The chapter headings provide insight into what is being compared or studied, as opposed to using the name of the statistical technique used in the chapter. What follows is a brief description of each chapter.

Chapter 2
Overview of Statistical Concepts and Ideas

This chapter serves as the foundation of the many topics presented in Chapters 3 through 7, including the 7-step problem-solving framework. In this chapter you will learn the language of statistics, and how statistics and statistical thinking can help you solve engineering and scientific problems.

Descriptive statistics, such as the mean and standard deviation, and visualization tools, such as a histogram and box plots, are described, along with instructions on how to access them using the different platforms in JMP. Tests of significance as a signal-to-noise ratio, statistical intervals, the different measurement scales (nominal, ordinal, interval, and ratio), and how to set these appropriately in JMP are also discussed.

Chapter 3
Characterizing the Measured Performance of a Material, Process, or Product

Many situations call for the characterization of the measured performance of a material, process, or product. For example, if we have collected data on the performance of a particular product you may want to know, what is the average performance of the product, and what level of performance variation should we expect? It is important then to be able to use summary measures that efficiently represent the overall performance of the sampled population, and graphics that highlight key information, such as trends or different sources of variation. Once the data is summarized, statistical intervals, such as a confidence interval for the mean, or a tolerance interval to contain a proportion of the individual values in a population, are used to make performance claims about our data.

The industrial example presented in this chapter involves the qualification and characterization of a second source raw material that is used to make sockets in an injection molding operation. A four-cavity injection molder is used in the qualification, and the effective thickness of the sockets is the key quality characteristic. This data includes different sources of variation often found in manufacturing, and lends itself well to slicing and dicing the data in a variety of ways using different statistical techniques and graphs. Three JMP platforms are used to explore the industrial example: the **Analyze > Distribution**, **Graph > Control Chart**, and **Graph > Variability / Gage Chart** platforms.

Chapter 4
Comparing the Measured Performance of a Material, Process, or Product to a Standard

This chapter shows how one-sample tests of significance for the mean and standard deviation can be used to address the common engineering problem of comparing a quality characteristic to a standard. We show you how to use the Student's *t*-test to compare the average performance to a standard, and how to use the Chi-square test to compare the average variation to a standard. We also discuss how to set up the null and alternative hypothesis to answer the question of interest, and highlight the difference between a one- and two-sided alternative hypothesis.

We use a semiconductor industry example involving the qualification of a new three-zone, temperature-controlled vertical furnace used for thin film deposition on wafers. The goal of the qualification is to show that the thickness of the silicon dioxide layer, a key fitness-for-use parameter, meets the target value of 90 Angstrom, and to predict how much product will be made below the lower specification limit of 87 Angstrom. The 7-step problem-solving framework is applied to this situation using several JMP platforms. The **Analyze > Distribution** platform provides the required Student's *t*-test and Chi-Square test for the comparisons.

We also show how to use a test of equivalence for those situations where we want to show that the average performance is equivalent to a standard, within a given performance window. Equivalence tests combine practical significance with statistical significance in order to produce a more targeted result. A JMP script is provided for this test.

Chapter 5
Comparing the Measured Performance of Two Materials, Processes, or Products

Many simple experiments are carried out in industry to compare the performance of two different things such as two suppliers, two temperature settings on a piece of equipment, or two gauges. We show how the two-sample Student's *t*-test and the *F*-test can be used for comparing the average performance and the performance variation of two groups of samples coming from normally distributed populations, as well as how to interpret them and make sense of the JMP output.

Guidance is also provided on how to set up a two-sample study, including how to determine how many samples to include, as well as how to sample them from the population. As was the case in Chapter 4, equivalence tests are introduced as an alternative way to show that two populations are equivalent within some pre specified performance window.

The comparison of a newly purchased mass spectrometer to an existing mass spectrometer in an analytical laboratory serves as the case study for this chapter. These instruments are used to determine the isotopic ratio composition of a variety of elements with high precision. Since a key customer in the chemical industry is using silver as a catalyst in an oxidation reaction, the atomic weight of silver is used to compare the performance of the two instruments using the 7-step problem-solving framework and JMP.

Chapter 6
Comparing the Measured Performance of Several Materials, Processes, or Products

In this chapter, the simple experiments discussed in Chapter 5 that involve the comparison of two means or standard deviations are extended to include the comparison of the measured performance of three or more materials, processes, or products. Analysis of variance (ANOVA), the statistical technique at the heart of the analysis of experimental studies, is introduced, as well as the assumptions required for its use. Using the 7-step problem-solving framework, we walk through the JMP ANOVA output and review its key components, including the variation and degrees of freedom contributed by the factor (the signal), and by the experimental error (the noise).

We use an example from the home building industry in connection with a housing development of 100 semi-custom homes built in four phases. Shortly after the fourth and last phase of the development was completed, the residential developer responsible for its construction started to receive a number of complaints about premature cracking in the cement sidewalks leading from the driveways to the front doors. The compressive strength for the cement used in the different phases of construction, which is a key performance characteristic related to cement cracking, is used to compare the cements used in the four phases of construction.

Chapter 7
Characterizing Linear Relationships between Two Variables

When both the performance characteristic and the explanatory variable are measured on a continuous scale, we can fit a simple linear regression model, of the form $Y = \beta_0 + \beta_1 x + \varepsilon$, to examine the relationship between them. This simple model forms the basis of linear models, and can be easily expanded to include higher order terms, such as a quadratic term; that is, models of the form $Y = \beta_0 + \beta_1 x + \beta_2 x^2 + \varepsilon$.

Simple linear regression methods are used in the generation of calibration curves in engineering and science. We focus on an example for the development of a calibration curve for a canister-style load cell design that is used in cargo truck weighing systems. We walk through all of the necessary steps needed to produce the calibration curve, including collecting the measurements and specifying the model, with emphasis placed on checking the adequacy of the model using diagnostic statistics and plots. In addition, inverse prediction is used to demonstrate how the calibration curves are used in practice.

In addition, to show the power of regression techniques we look at the analysis presented in 1901 by a young Albert Einstein in his first published paper "Conclusions Drawn from the Phenomena of Capillarity," in which he investigated the nature of intermolecular forces.

1.7 Chapter Layout

The layout of Chapters 3 through 7 contains the same basic elements, as described in Table 1.1, to make it easier for you to find the information you need when solving a specific problem. Each chapter starts with the chapter goals followed by a description of the problem, or practical situation, that the chapter focuses on. A table with key questions, concepts, and tools is provided for easy reference.

The descriptions, examples, and discussion of the statistical underpinnings provide a "behind the scenes" look at the tools for those of you who want to deepen your understanding of statistics. Note that the chapter sections that address the statistical underpinnings of a technique might have slightly different headings to reflect the particular technique. For example, in Chapter 5 the statistical underpinnings are presented in sections titled "5.3.2 Comparing Average Performance of Two Materials, Processes, or Products," "5.3.3 What to Do When We Have Matched Pairs," "5.3.4 Comparing the Performance Variation of Two Materials, Processes, or Products," and "5.3.5 Sample Size Calculations."

Each chapter also includes a 7-step problem-solving framework to guide you through the different stages of problem solving in the context of engineering and science.

Table 1.1 Chapter Sections and Descriptions

Section	Description
Chapter Goals	These goals reflect what we hope you will be able to accomplish when you complete the chapter. In essence, you will become familiar with the technique, and will be able to follow along using JMP to solve the practical problem presented.
Problem Description	A real problem from engineering, science, or both is used in each chapter to illustrate how statistics and JMP can be used to solve the problem. The chapter starts with what we know about the problem, clarifies the main questions that need to be answered or the uncertainty that needs to be resolved, and reveals the statistical technique that will be used to solve the problem. This sets the stage for what comes next.
Key Questions, Concepts, and Tools	A table outlines the key statistical questions, concepts, and tools that are highlighted in each chapter and how they relate to the practical problem outlined in the Problem Description section. This provides you with a quick preview of the concepts that will be discussed throughout the chapter.
Description and Some Applications	An introduction to the statistical techniques used in the chapter is provided in this section, including why we would want to use it, and some of the common statistical nomenclature and terminology surrounding it. We also provide you with several examples of common questions from engineering and science that could be answered using the technique described in each chapter.
Statistical Underpinnings of Technique	These sections provide a more in-depth review of the statistical techniques and concepts that are featured in the chapter. Examples include: specifying the null and alternative hypothesis, calculating and interpreting the test statistic, obtaining and interpreting the p-value, checking model assumptions, and determining sample sizes. In addition, relevant JMP output is included in discussions of key concepts.
Step-by-Step Analysis Instructions	A 7-step problem-solving approach is used to solve the practical problem described at the beginning of the chapter using both the statistical techniques presented in the chapter, and the appropriate JMP platforms, functions, and, in some cases, scripts. The steps begin with a clear statement of the question and end with practical recommendations for next steps. These steps are shown in Section 1.8.
Summary	The techniques and concepts discussed in the chapter are summarized and key practical points are outlined.
References	A list of books and technical papers that support or further develop the ideas presented in the chapter.

Because Chapter 2 is a foundation chapter, it does not follow the layout in Table 1.1. However, it provides an overview of many statistical concepts used throughout the book, and introduces the 7-step problem-solving framework.

1.8 Step-by-Step Analysis Instructions

We realize that some of you might have no experience or limited experience with applying statistical methods, and that even those of you who use statistical methods on a more regular basis might not be fluent in all situations. For this reason, we have included a 7-step problem-solving framework to facilitate the application of statistical methods and JMP, and for solving the practical engineering and scientific problems that are presented in Chapters 3 through 7.

Table 1.2 gives an example of how the 7-step problem-solving framework is used in the context of solving the practical problem in Chapter 3 "Characterizing the Measured Performance of a Material, Process, or Product." This framework helps us to follow the scientific method in our investigations by clearly defining the questions, or uncertainties, of interest (Step 1), by making sure those questions are carefully translated into key hypotheses to be examined (Step 2), by conducting our studies using relevant data and adequate samples sizes (Steps 3 and 4), by using appropriate statistical techniques that help us answer our questions (Step 5), by not getting lost in the statistical output (Step 6), and by interpreting the results within the context of the problem and the original questions (Steps 7). In our investigations we want to prevent key ideas from being overlooked or omitted all together, and having a structured approach helps. A structured approach also gives us the ability to critically review our studies and those of others.

The third column in Table 1.2 shows the different JMP platforms that are needed to carry out the sample size calculations, the analysis, and the visualization of relevant relationships and information. As you will discover throughout the book, JMP is an integral part of the analysis, making it easy to generate the necessary output and graphs that help trigger insights, answer our questions, suggest future paths, and support our conclusions.

Table 1.2 Step-by-Step Analysis for Socket Qualification

Step	Objectives	JMP Platform
1. Clearly state the question or uncertainty.	Make sure that we are attempting to answer the right question with the right data.	Not applicable.
2. Specify the expected performance of a material, product, or process.	Cleary state the performance criteria for the socket qualification, using the existing raw material performance as a baseline.	Not applicable.
3. Determine the appropriate sampling plan and collect the data.	Identify how many sockets will be manufactured with the new raw material and what sources of variation should be included in the study. How well do we want to estimate μ and σ?	• **DOE > Sample Size and Power**
4. Prepare the data for analysis and conduct exploratory data analysis.	Make sure all stratification factors (sources of variation) are included in the JMP table, along with the measured response. Set column properties as needed. Check for any outliers or discrepancies with the data.	• **Cols > Column Info** • **Analyze > Distribution** • **Graph > Variability/Gage Chart**
5. Characterize your data and answer the questions of interest.	Calculate appropriate descriptive statistics and statistical intervals and understand sources of variation in the data. Check the analysis assumptions to make sure your results are valid.	• **Analyze > Distribution** • **Graph > Variability/Gage Chart** • **Graph > Control Chart**
6. Summarize the results with key graphs and summary statistics.	Find the best summary measures and graphical representations of the results that give us insight into our uncertainty. Select key output for reports and presentations.	• **Analyze > Distribution** • **Graph > Variability/Gage Chart**
7. Interpret the results and make recommendations.	Translate the statistical jargon into the problem context. Assess the practical significance of your findings.	Not applicable.

1.9 JMP Software

Several JMP platforms are used in the analyses presented in this book, and are highlighted in the corresponding chapters. Since we are primarily dealing with studies involving one response and up to one factor, the **Fit Y by X** platform will be our primary focus. However, we will also demonstrate other JMP platforms and functionality that are essential for studying the measured performance of a material, product, or process. The JMP platforms used in this book are shown in Table 1.3.

Table 1.3 JMP Platforms Used in This Book

JMP Platform	Purpose	Chapters
Analyze > Distribution	Study the distribution of data using a histogram, normal quantile plot, and descriptive statistics.	2 through 7
	Calculate process capability indices and percentage of samples outside of specification limits.	3
	Calculate confidence, prediction, and tolerance intervals.	2 and 3
	Conduct one-sample tests of significance for determining if the mean or standard deviation is equal to a standard value.	4
Analyze > Fit Y by X	Conduct two-sample tests of significance for comparing means and variances from two populations, including tests of equivalence.	5
	Carry out a one-way ANOVA for comparing three or more populations, conduct multiple comparison tests, and check for equal variances.	6
	Fit a simple linear regression model, or quadratic model, to study the linear correlation among two variables.	7
Analyze > Matched Pairs	Use to examine the difference between two populations when the observations are sampled in matched pairs.	4
Analyze > Fit Model	Fit a simple linear regression model or quadratic model to obtain additional row diagnostics to study residuals and influential observations, and conduct inverse prediction.	7

(*continued*)

Table 1.3 *(continued)*

JMP Platform	Purpose	Chapters
DOE > Sample Size and Power	Derive sample sizes for estimating the sample mean for one population.	3 and 4
	Derive sample sizes for estimating the sample variance for one population.	4
	Determine sample sizes for estimating the difference between two sample means.	5
	Determine sample sizes for estimating the difference among k sample means.	6
Graph > Overlay Plot	Study patterns in residuals to check for violations in model assumptions or trends in performance measures.	3 through 7
Graph > Bubble Plot	Enables information to be displayed for up to four variables. Used to examine influential observations or outliers in simple linear regression.	7
Graph > Control Chart	Used to examine process stability over time for performance measures or homogeneity assumption for residuals.	2, 3, 6, and 7
Graph > Variability/Gauge	Examine the different sources of variation and systematic patterns in the data, and estimate variance components.	3 and 4

In addition to the platforms listed in Table 1.3, there are illustrations of table functions and manipulations interspersed throughout the chapters. We show, for example, how to add columns to JMP tables, use some functionality in the Formula editor, set row states, summarize data tables, stack columns, sort tables, exclude or hide rows, and set column properties.

Although the analyses presented in this book were done using JMP 7, where appropriate we have included information about the differences in JMP 8.

1.9.1 JMP Help and Resources

If you are not familiar with the basic functionality of JMP, we encourage you to take advantage of the information available in the JMP **Help** menu, in particular the **Tutorials** and **Books**, as well as the different resources such as podcasts, JMP blog, JMP data stories, newsletters, and so on that are provided at the JMP.com Web site. These resources will definitively increase your level of expertise using JMP.

JMP Tutorials

The JMP tutorials are intended for you to learn how to do common tasks in JMP. You can access the tutorials from the **JMP Help** menu. These tutorials use a combination of textual references and demo tables to guide you through a hands-on activity. The tutorials that are directly applicable to this book include the Beginners, One Mean, Two Means, Many Means, and Paired Means tutorials.

JMP Books

You can access a list of books that are available within the **JMP Help** menu. These books provide a comprehensive overview of the features and functionalities of the software and often use different JMP tables to illustrate some of these points. However, these books are not geared towards interpretation of the output or describing the statistical underpinnings of the approaches. Some of the JMP data sets used in these books might be located under the Sample Data folder where the installation resides.

The **JMP User Guide** is an appropriate reference for learning about data manipulation. Important concepts, such as creating and opening files, editing data tables, selecting rows and columns, and assigning properties to columns are included in this user guide. You can print the **JMP Menu Card** to have easy access to the options available in the main JMP menus: **File**, **Edit**, **Tables**, **Rows**, **Cols**, **Analyze**, and **Graph**. Similarly, the **JMP Quick Reference Card** has tips on using the **Formula Editor**, **JMP Tools**, Working with Files and Editing Files, for example. Finally, selecting **Help > Contents** in JMP pulls up all of the books into one frame.

JMP.com Resources

The JMP Web site, http://www.jmp.com/about/resources.shtml, has a wealth of resources to help increase your knowledge of JMP, acquire new skills, learn about new features and events, and connect with the JMP community worldwide. These resources include the following:

- Podcasts
  Hear from who's who in the world of statistics, what they're doing, and what's on their minds.
- JMP Demos
  Demos of the new capabilities of JMP.
- JMP Blog
  A Web log devoted to all things related to data visualization, visual Six Sigma, design of experiments and other statistical topics.

- JMP Data Stories
  JMP Data Stories show you how you can use the advanced data visualization techniques in JMP software to explore your data, answer your questions, and find out what your data is trying to tell you.
- *JMP Foreword* Magazine
  The statistical discovery magazine from SAS, which is available in PDF.
- *JMPer Cable* Newsletter
  A technical publication for JMP users.
- JMP Scripting Library
  The JMP Scripting Language (JSL) enables JMP users to program JMP to repeat analyses, write custom programs to manipulate data in complex ways (including full matrix algebra support), and even extend the graphical and analytical capabilities of JMP.
- JMP File Exchange
  The JMP Community File Exchange contains files contributed by users and includes sample data files, JMP Scripting Language (JSL) files, and SAS stored processes.
- Seminars
  Because there's no substitute for hands-on learning, JMP offers seminars featuring world-renowned leaders in statistical thought.
- White Papers

1.10 Scope

The statistical analyses presented in this book are appropriate when the performance characteristic of interest is measured on a continuous scale. The variables used to explain the behavior of the performance characteristic can be either categorical or continuous. This, we believe, covers a large number of applications that you might encounter in industry and science. A companion book dealing with categorical performance characteristics will be published at a future date.

The following statistical techniques and concepts are covered in this book:

- descriptive statistics such as the mean, median, mode, standard deviation, range, and process capability indices
- graphics including histograms, trend plots, control charts, normal quantile plot, box plots, scatter plots, and residual plots
- tests of significance
    - one-sample tests for the mean and standard deviation
    - two-sample tests for means and standard deviations
    - paired *t*-test
    - equivalence tests for the mean
    - analysis of variance (ANOVA)
    - multiple comparison tests
    - Brown-Forsythe test for comparing variances
    - simple linear regression
- measurement scales: nominal, ordinal, interval, and ratio
- experimental and observational units
- response vs. factor
- signal-to-noise ratio
- statistical intervals: confidence, prediction, and tolerance
- Type I and Type II errors
- *p*-value and confidence level
- degrees of freedom
- randomization
- stratification factors
- statistical inference
- sampling plans, including random and stratified random samples
- sample size calculations
- probability distributions, for example, normal, Student's *t*, *F*, and Chi-square
- variance components

1.11 Typographical Conventions

Several typographical conventions are used in this book to point you to important information related to both statistical content and JMP software usage.

Statistics Typographical Conventions

Two different types of typographical conventions are used throughout this book to highlight important statistical content. This was done to draw your attention to key concepts that are important when applying statistics to engineering and science problems. The first statistical convention is the use of statistics notes. These are numbered according to the chapter they are in, and their order of appearance. Statistics notes appear throughout a chapter to emphasize main points and things worth remembering. An example of a statistics note (Statistics Note 4.2) from Chapter 4, "Comparing the Measured Performance of a Material, Process, or Product to a Standard," summarizes the three main assumptions that are required in many statistical analyses.

Statistics Note 4.2: The validity of this procedure (for determining if the average performance is equal to a standard) depends upon the assumptions that were discussed in Chapter 2 and are reiterated here. The underlying population must be normally distributed, or close to normally distributed, the data are homogeneous, and the experimental units must be independent from each other.

The second type of statistics typographical convention that the reader will encounter is a callout box, like the one below. The information in the callout box is a snippet of a key point that is presented in the main text of the chapter. They are intended to be short and memorable.

> *A p-value is an area under a probability density curve that quantifies the likelihood of observing a test statistic as large as, or larger than, the one obtained from the data.*

JMP Typographical Conventions

In each chapter, JMP software is discussed in a general sense in order to highlight the various elements that support the statistical techniques and analyses used to solve a practical problem. In addition, specific instructions are provided so the reader can reproduce the JMP analysis using the data showcased in each chapter. The following typographical conventions are used in reference to JMP software in this book.

- *JMP platforms* will be displayed using a bold Arial font and platform paths will be shown using an '>' to distinguish the individual selections needed. For example, the Distribution platform is labeled as **Analyze > Distribution**.
- *JMP dialog boxes* that get launched after selecting a JMP platform and require user input are also shown in bold Arial font. For example, to create a histogram the user must highlight a response variable in the **Select Columns** window, and then click **Y, Response**.
- *JMP column names,* which are unique to a JMP data table and are used to populate JMP dialog boxes, are shown in unbolded Arial font. For example, we must highlight Effective Thickness, which is the name of the response of interest in our JMP table.

When these elements come together to provide instructions for creating, for example, a histogram for a response in JMP, it will look like the following: "We use the **Analyze > Distribution** platform to generate the histogram and the descriptive statistics for the data. In the Distribution dialog box, select Effective Thickness, click **Y, Columns**, and then click **OK**."

Similar to statistics notes, JMP notes are scattered throughout the chapters to highlight information that is relevant to using JMP but that might not be obvious. The reader will find that they are numbered according to the chapter number they are in and the order in which they appear. Shown below is a JMP note from Chapter 7 "Characterizing Linear Relationships between Two Variables."

> **JMP Note 7.2**: In JMP raw residuals can be saved to the data table from the **Analyze > Fit Y by X** platform, but the studentized residuals must be saved from the **Analyze > Fit Model** platform.

We also include notes pertaining to JMP 8 usage when there is a significant difference between how something is handled in JMP 8 as compared with JMP 7. Below is an example from Chapter 7.

> **JMP 8 Note:** To get the same output in JMP 8 from within the **Distribution** platform, we select **Continuous Fit > Normal**.

JMP Figure Annotations

Each chapter has numerous figures that capture the JMP instructions for the analyses together with the corresponding output. At times, these figures are annotated with additional information for ease of use or interpretation. Most of the time these annotations are presented to the left- and right-hand side of the output. Figure 7.12 below shows an example of an annotation outside of the output.

Figure 7.12 Example of Annotation Outside of JMP Output

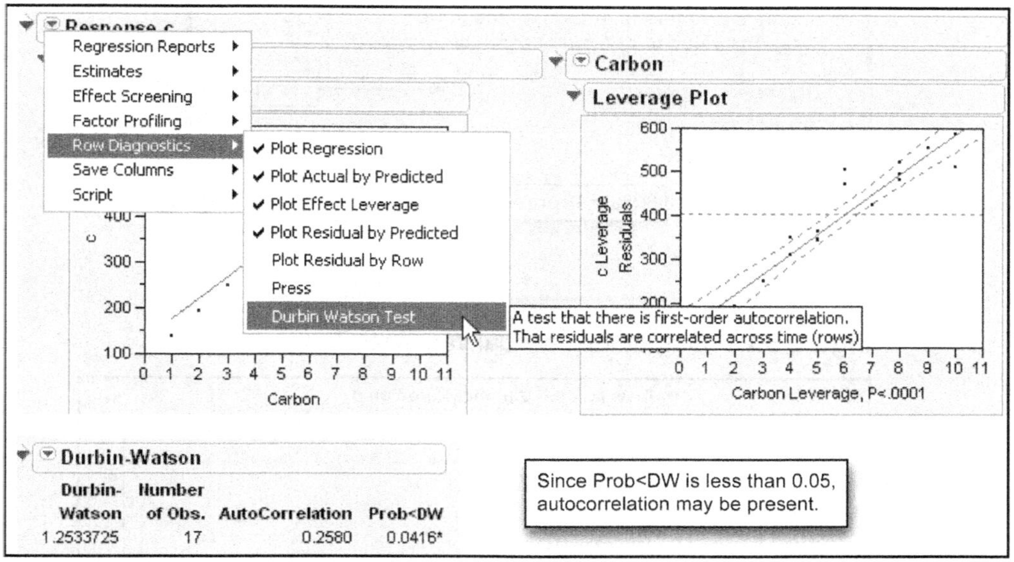

There are a few instances, however, where the annotations are included within the JMP input window or the JMP output window. These annotations are not generated by JMP and cannot be reproduced by the user, but they have been added to facilitate the understanding of the inputs needed in a particular analysis or the interpretation of the output. Figure 4.8 shows an example identifying the inputs required in the **Sample Size and Power** calculator, while Figure 5.23 shows the annotated output for the **Unequal Variances** test.

Figure 4.8 Sample Size Calculation for One Sample Mean

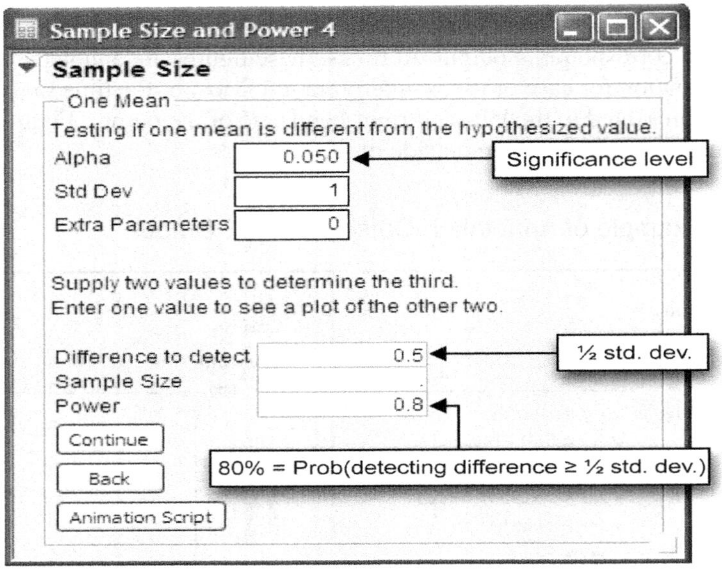

Figure 5.23 Output for Unequal Variances Test

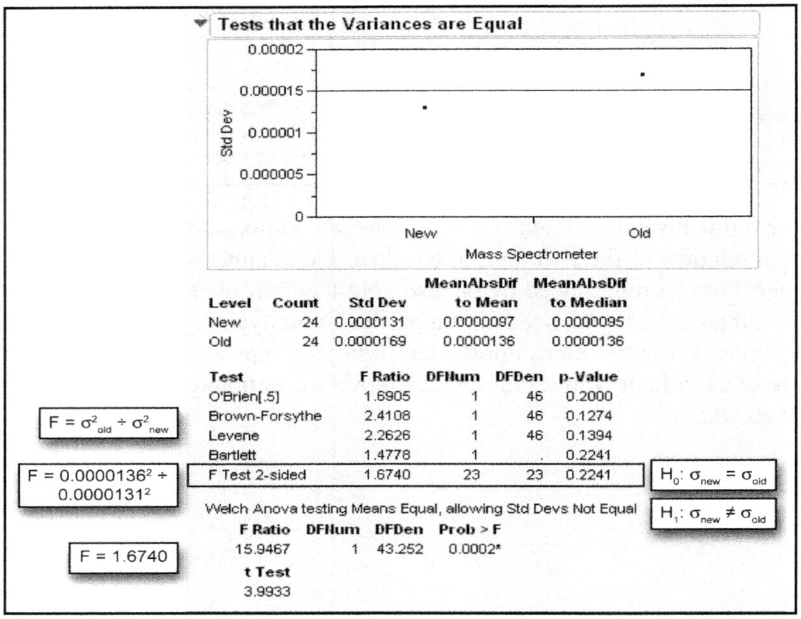

1.12 References

Croarkin, M.C. 2001. "Experimental Statistics" NIST SP 958, *A Century of Excellence in Measurements, Standards, and Technology: A Chronicle of Selected Publications of NBS/NIST, 1901-2000*, 132-134.

Natrella, M. G. 1963. *Experimental Statistics, NBS Handbook 91*. Washington, D.C.: U.S. Department of Commerce, National Bureau of Standards. Reprinted 1966.

NIST/SEMATECH e-Handbook of Statistical Methods, 2006. Available http://www.itl.nist.gov/div898/handbook/.

Chapter 2

Overview of Statistical Concepts and Ideas

Chapter Goals

- Learn how statistical thinking and statistical methods can help support scientific hypotheses and solve common problems in engineering and science
- Become familiar with the language of statistics
- Learn to summarize data down to a few useful measures and incorporate a margin of error to account for uncertainty
- Provide a framework for answering key uncertainties with statistical rigor

2.1 Why Statistics?

The twentieth century witnessed a revolution in science and engineering due to the widespread use of "statistical models of reality" (D. Salsburg, 2003) to predict and quantify the uncertainty in random events, and due to the recognition of *statistical thinking* as a philosophy of learning and action. At the core of statistical thinking is the fact that variation exists in all processes, and that understanding and reducing this variation is crucial, not only in science and engineering, but in other disciplines as well.

As engineers and scientists we are familiar with the ever-present variation in our measurements and the phenomena we deal with. We put a lot of effort in trying to understand, eliminate, or compensate for variation, for example, when using a thermostat to regulate the temperature variation in a car or home, or when using a proportional-integral-derivative (PID) controller to regulate the output in an extrusion line. We conduct studies, monitor processes, or design experiments with the goal of discerning the signals from the noise, but variation tends to mask the signals that we are trying to identify. In some cases we can use sound engineering judgment based on past experiences, or use interpretations of data in order to (typically) make good decisions. This is usually the case if the signals in our systems are large, with respect to the noise in the system, or if we can explain them with mechanistic models. That being the case, we still need to ask: how do we understand noise, how can we quantify it, study its effects, and maximize our signal-to-noise ratio? This is the why of statistics: statistics is a discipline that enables us to discern signals from the noise, and that serves as a catalyst for scientific discovery.

Another element of statistical thinking is that all work occurs in a system of interconnected processes. Each process, and each component within each process, contributes to the overall variation in the system, i.e., components of variation are *always* additive, making it even harder to discern the signals from the noise. As an example, consider the system for laminating two pieces of materials together that is represented in Figure 2.1.

Figure 2.1 Detailed Lamination System

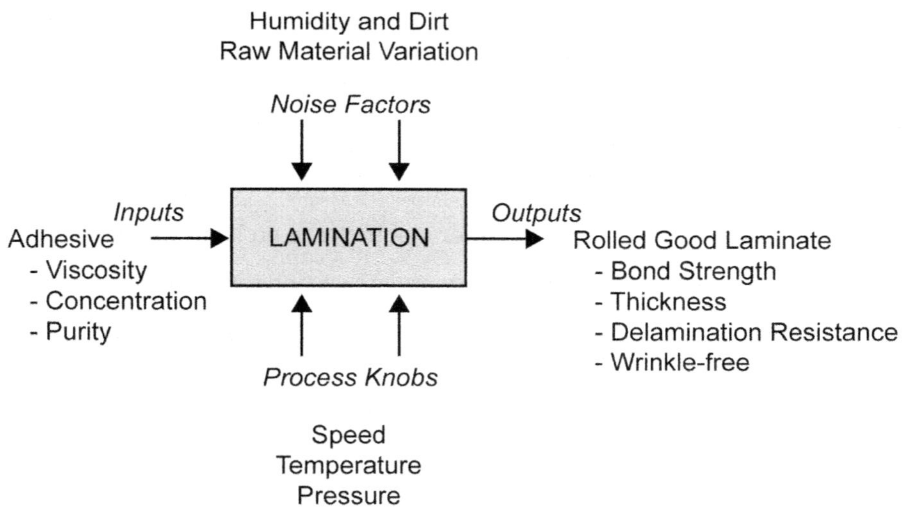

There are four major dimensions that should be considered when describing a process:

1. Inputs
2. Outputs
3. Process knobs
4. Noise factors

For the lamination system in Figure 2.1 we should consider the following:

- the inputs, or raw materials, that are consumed in the lamination step, including their important attributes or features
- the output form and its key characteristics
- the process knobs or settings that we have control over and that change the inputs in a particular way
- the noise in the system or factors that we cannot control but that may affect the quality of the lamination

Statistical thinking informs us that every time we use this laminator, using the same process settings, the bond strength or thickness of our rolled good laminates will vary. In order to produce high quality and consistent laminates we need to minimize this variation to a level that does not affect fitness-for-use. Statistics can help us to further our understanding of how to consistently make good laminates.

Throughout this book we will highlight some good habits that statistical thinkers routinely follow in their scientific and engineering explorations, which include:

- using the scientific method for inquiries and problem solving
- obtaining relevant data to answer the questions at hand
- studying both the average performance (signal) and the variation (noise) of the system
- interpreting the results in the original context of the problem
- recommending specific actions based on the results

It is important to be methodical when solving problems, answering questions, or driving improvements, and the scientific method provides a framework for the effective use of statistics in this context. Figure 2.2, based on Figure 1 of McLaren and Carrasco-Aquino (1998), is a representation of this process. The scientific method begins with a problem or uncertainty that we are trying to answer and, through our knowledge of the subject matter and analytical reasoning, we can derive a scientific hypothesis that is our best guess at providing clues for answering the problem (hypothesis generation). The use of statistics within this framework can help us to gather and interpret evidence by means of determining ways to sample our data, put forth a statistical hypothesis, conduct the appropriate test of significance, and interpret the results in the context of our original hypothesis (knowledge generation).

Figure 2.2 Statistics within the Context of the Scientific Method

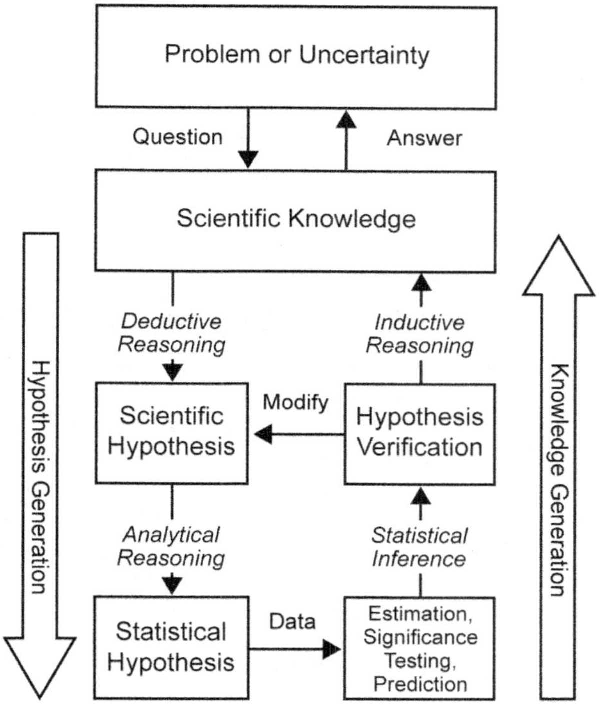

Our knowledge then increases through a never-ending cycle of deductive and inductive reasoning, going from the theory to the experiment (data), and from the particular to the general. (See, for example, Box, Hunter and Hunter 2005, p. 2.) Statistics acts as a catalyst in this process of learning in the presence of variation and uncertainty.

As Wheeler (2005) points out there are four major ways in which we can use statistics:

1. descriptive statistics to summarize the information contained in a set of data (Section 2.4).

2. probability modeling to quantify the uncertainty and to model the likelihood of different possible outcomes based on samples drawn from a known universe. Here we use deductive and analytical reasoning to go from the general to the particular (Section 2.5).

3. statistical inference based on a sample drawn from an unknown universe, and the statistics calculated from the sample. Here we use inductive reasoning, going from the particular to the general, to make claims about this unknown universe (Section 2.3).
4. checking for homogeneity to make sure that we can reasonably assume that our data came from a single universe rather than multiple ones. This is important because the other three ways to use statistics depend on this assumption to be of any practical value.

2.2 Measurement Scales, Modeling Types, and Roles

Although when Lord Kelvin lectured at the Institution of Civil Engineers in 1883 he was referring to numerical quantities, his statement about being able to quantify phenomena is still relevant: "In physical science the first essential step in the direction of learning any subject is to find principles of numerical reckoning and practicable methods for measuring some quality connected with it. I often say that when you can measure what you are speaking about, and express it in numbers, you know something about it; but when you cannot measure it, when you cannot express it in numbers, your knowledge is of a meager and unsatisfactory kind; it may be the beginning of knowledge, but you have scarcely in your thoughts advanced to the state of *Science* whatever the matter may be."

In order to answer the engineering or scientific questions of interest we need relevant data, which in turn can be obtained if we have ways to measure, either by using an instrument or by classification, the characteristics linked to our objectives. For example, in order to know whether the lamination system shown in Figure 2.1 is capable of producing a good roll, we need to be able to measure the quality characteristics that make a roll "good." Among these characteristics we have bond strength, thickness, delamination resistance, and the lack of certain types of defects. We need to think ahead and decide which characteristics are directly related to the objectives of our study, so we can achieve results that are valid, useful, and lead to actions.

When measuring a given attribute or quality characteristic, there are four types of measurement scales that can be used to classify the data at hand: nominal, ordinal, interval, and ratio scales. These measurement scales help us identify the amount of information contained in a set of measurements, and serve as a guide for the types of summary statistics and analyses that can be performed with a given data set. Table 2.1 shows these measurement scales along with definitions, appropriate mathematical operations, examples, the relevant summary statistics that can be computed for that scale, and the corresponding modeling type assigned by JMP. The level of complexity, in terms of the mathematical operations that can be performed and the summary statistics that can be calculated, increases as we go from the simple nominal scale up to the ratio scale.

2.2.1a Nominal Scale

A nominal scale does not provide a lot of information since it is just a classification into predefined categories. These categories can be defined using names or numbers, but the numbers just act as labels without an intrinsic order. We can compare the categories to determine equality or inequality, and we can count the number in each category.

2.2.1b Ordinal Scale

In an ordinal scale the categories have an implicit order enabling us to compare them in terms of equality, less than (before), or greater than (after), as well as count the number in each category. For example, in a car race we can classify the cars as finishing first, second, third, fourth, and so on. We know that the first car was faster than all the other cars, and that the third car was faster than the fifth, but we cannot say that the fourth car was twice as slow as the second car. In the ordinal scale we can answer questions of ranking (bigger, faster, longer, and so on).

2.2.1c Interval Scale

The interval scale is such that the distance between two adjacent values, or interval, is the same across all the scale. We can add and subtract values in the scale but multiplication and division are not meaningful. In the Celsius temperature scale, 10ºC + 45ºC = 55ºC, and 87ºC – 13ºC = 74ºC. However, the operation 100ºC/10ºC = 10 does not mean that 100ºC is 10 times hotter than 10ºC because in the Fahrenheit scale the ratio 100ºC/10ºC is equivalent to 212ºF/50ºF = 4.24, which gives a different ratio. This is a consequence of not having a fixed, or meaningful, zero point. On the other hand, one can calculate the ratio of differences; e.g., (60ºC–40ºC) / (20ºC–10ºC) =2 means that the difference between 60ºC and 40ºC is twice the difference between 20ºC and 10ºC. In the interval scale we can answer questions of ranking as well as of the difference between two quantities.

2.2.1d Ratio Scale

All the mathematical operations (addition, subtraction, multiplication, and division) are permissible in the ratio scale because the scale has an absolute 0. A person weighing 250 pounds is twice as heavy as a person weighing 125 pounds; i.e. 250 pounds/125 pounds =2. Money is another example of a ratio scale.

Table 2.1 Measurement Scales for Factors and Responses

Measurement Scale	Definition	Quantitative Information	Meaningful Mathematical Operations	Examples	Relevant Summary Statistics	JMP Modeling Type
Nominal	Classification into different categories	Equality/inequality	Counting	Good/bad, scratch/wrinkle/bump, pass/fail, female/male	Counts mode	Nominal
Ordinal	Classification into different categories that have an order	Equality/inequality, ranking	Counting order (> <) linear transformations, monotone transformations	Low/medium/high, satisfaction scores (e.g., 1, 2, 3, 4, 5), finishing place in a race (1st, 2nd, 3rd, 4th, etc.)	Counts mode, median, minimum, maximum	Ordinal
Interval	Equal distance between intervals, arbitrary 0	Equality/inequality, ranking, equal intervals, relative differences, ratio of differences	Addition, subtraction, linear transformations	Temperature (°C/ °F)	Mode, median, mean, std. dev.	Continuous
Ratio	Equal distance between intervals, fixed 0 point	Equality/inequality, ranking, equal intervals, differences, ratios of differences, ratio	Addition, subtraction, multiplication, division, and linear transformations	Temperature (Kelvin), age, height (feet), weight (lbs), proportions	Mode, median, mean, geometric mean, harmonic mean, std. dev., C.V.	Continuous

2.2.2 Which Scale?

If you deal with data on a regular basis you quickly realize that this classification is not exhaustive. For example, percentages (yield), even though they are bounded between 0 and 100, have a natural 0, and therefore possess all the attributes of a ratio scale. We can claim that 40% is twice as large as 20%. However, proportions are derived from count data (nominal or ordinal) and most analyses, by default, treat them as such. It is very important, then, to keep in mind that this "typology" of data types is just a guideline for the types of data that we are dealing with, and that the context of the data and the goals of the study will, and should, often determine how we are going to use the data to answer the questions of interest. In other words, real life data is not always as clear-cut as a classification may imply. In the same analysis the scale of one variable can change from nominal, if we are just interested in comparing the levels they represent, to interval or even ratio. Something as simple as a lot number in a production line, which is "obviously" nominal, is normally assigned sequentially and can, implicitly or explicitly depending on what is recorded, contain a time sequence for analysis not deemed appropriate for a nominal scale. Or consider the common practice of using the average of numbers in an ordinal scale; although this is "not permitted" by the classification it nevertheless can lead to meaningful results, and useful decisions.

> *Scale type is not an attribute of the data, but rather depends upon the questions we intend to ask of the data, and any additional information we may have.*
>
> *Velleman and Wilkinson (1993)*

Table 2.1 shows how JMP uses this classification for the modeling roles of variables into nominal, ordinal, and continuous (interval and ratio) scales, and in turn, for recommending an analysis (as shown in Table 2.2). The power of JMP is that we are not constrained by its automatic analysis recommendations because it enables us to quickly change the modeling type of our variables. After all, good engineering and science depend on our ability to look and analyze our data in different ways, and to discover and explore the hidden information contained in our data.

2.2.3 Responses and Factors

In the previous section we saw how the measurement scales can suggest certain analyses that we can perform on our data. Another consideration is the role that a measurement will play. Do the measurements relate to the characteristics of interest, or do they help explain the behavior of other variables? JMP denotes these as follows:

Response: Response is also known as the dependent variable, or "what we want." A response is a variable that we want to characterize or optimize in some way in order to keep it on target, maximize it, or minimize it. When we talk about cause-and-effect relationships, a response is thought of as the "effect." Responses can take any of the four measurement scales described in Section 2.2.

Factor: Factor is also known as the independent or explanatory variable, or "how to achieve it." A factor is a variable that we want to characterize or optimize in some way because it has an impact on a response that we are interested in, and it can be integral to helping us achieve our goals for the response. When we talk about cause-and-effect relationships, a factor is thought of as the "cause," or what can explain the responses behavior. Factors can take any of the four measurement scales described in Section 2.2.

How do we decide which role is appropriate for our measurements? If we look back to the lamination system, the purpose of this conversion step is to stick two materials together using an adhesive, and we have been successful if our outputs, such as bond strength and thickness, are within specification limits. These are the quality, or fitness-for-use, characteristics that we are trying to achieve with this system, and therefore can be considered as responses in our analyses. Some of the factors in our lamination system that will help us achieve our goals may include the process knobs, such as temperature and speed, or the raw materials and their attributes, such as the viscosity of the adhesive, that have an effect on the quality of the bond strength and thickness. Factors can be further categorized as *control factors*, those that we can control or change to affect our responses, or *noise factors*, those that we normally do not control but that can affect our responses. When designing experiments it is sometimes possible to "control" some of these noise factors so we can understand their impact, and minimize their effect, on our responses.

JMP enables us to either pre-select the role of a variable (response denoted as **Y**, or factor denoted as **X**, as shown in Figure 2.3a), or to select the modeling type (measurement scale) as shown in Figure 2.3b.

Figure 2.3a Variable Role Selection

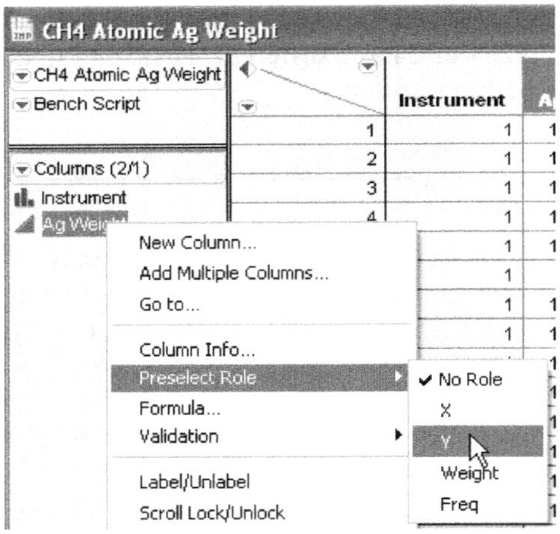

Figure 2.3b Modeling Type Selection

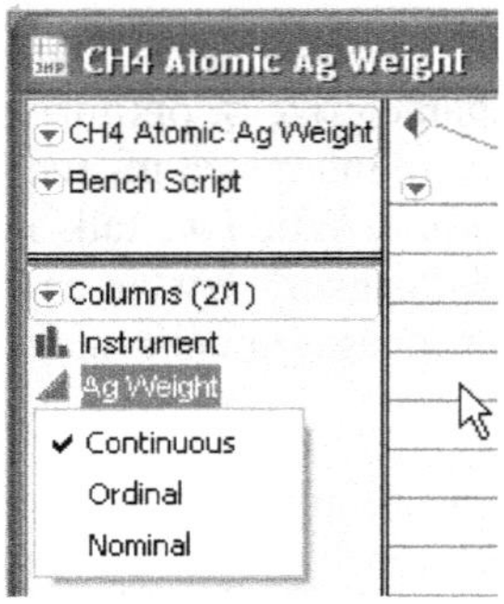

When we launch the **Fit Y by X platform** JMP uses the role and modeling type information to suggest an analysis, and we need to specify which variable is the **Y, Response** and which variable is the **X, Factor**. If these Y and X roles were pre-selected in the data table the variables will automatically appear in these windows; if not we can select the response and the factor before we proceed.

Figure 2.3c Fit Y by X Variable Selection Window

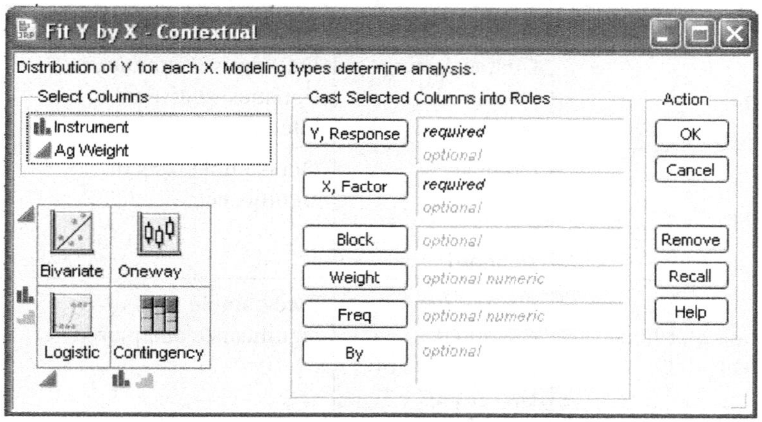

The lower left corner of Figure 2.3d shows how the modeling type of the factor (columns) and response (rows) determine the analysis that JMP will perform. A right-angle triangle (blue) is used to denote a continuous variable, while bar icons are used to denote nominal (red), or ordinal (green) variables. Note that there are two choices for the response, and two choices for the factor.

Figure 2.3d Fit Y by X Modeling Types

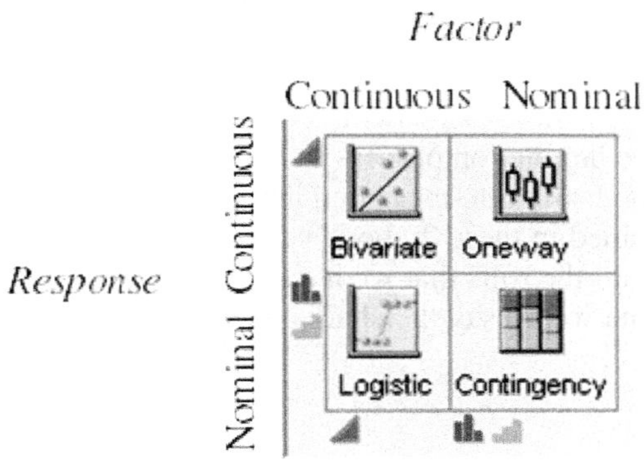

These four combinations generate four types of analyses: **logistic, bivariate, one-way, and contingency**. In this book we cover the **bivariate** and **one-way** analysis types. Table 2.2 shows these combinations plus those cases where we have only responses, the suggested analysis technique, and the chapter covering that particular topic.

Table 2.2 Chapter Reference for Analysis Techniques

Modeling Type		Purpose	Analysis Technique	Chapter
Response (Y)	**Factor (X)**			
Continuous	None	Characterizing measured performance	Descriptive/summary statistics, statistical intervals	3
Continuous	None	Comparing measured performance to a standard	One-sample tests of significance	4
Continuous	Nominal/ordinal, 2 levels only	Comparing measured performance of two groups	Two-sample tests of significance and paired *t*-tests	5
Continuous	Nominal/ordinal, > 2 levels	Comparing measured performance of several groups	One-way analysis of variance (ANOVA)	6
Continuous	Continuous	Studying linear relationships	Simple linear regression	7

2.3 Statistical Inference: From a Sample to a Population

As the engineer responsible for the lamination process shown in Figure 2.1, you have collected measurements on the bond strength, measured in kg/cm^2, of 20 rolls. How can you use the information contained in these 20 bond strength measurements to make claims about the bond strength of all the rolls that we have made or we are going to make in the future? This is the goal, some would say the challenge, of *statistical inference*: to be able to make a claim, or claims, (inference) about some characteristics of a complete, and possibly unknown, collection of objects (population), based on the characteristics of just a portion of this collection (sample).

> ***Statistical inference:*** *to be able to make some claims (inference) about some characteristics of a complete, and possibly unknown, collection of objects (population) based on the characteristics of just a portion of this collection (sample).*

Statistical inference, then, deals with populations and samples. We think of a population as a very large, perhaps infinite, collection of objects whose membership in the population is determined by a characteristic, or by characteristics, that describes them. A sample is just a portion, usually small, of this population. Measurements are taken on this sample to be able to make generalizations about the population. Why? Because in most fields, particularly in engineering and science where we need to make inferences about populations that may not yet exist, it is usually not practical or feasible to obtain measurements for the entire population. There are situations, however, where we do sample the entire population, such as when we use 100% inspection to sort out a defective product. In this case we are not making inferences about the product population but describing it based on some summary statistics.

The objectives of our study play a critical role in defining the population of interest. Once this population is defined we need a random and representative sample that is large enough to give us confidence in our results. The measurements collected on this random sample are then used to test those hypotheses that relate to the objectives of the study, and to calculate statistics of interest. The results from the analyses help us make inferences about the large population from which the sample was drawn. This is a never-ending cycle in which the inferences from our sample generate new hypotheses about the population, that lead us to select new random samples, that in turn generate new inferences, and so on (Figure 2.4). It is important to remember then that samples, and the measurements taken on them, do not exist in isolation but are determined by the questions we are trying to answer in our investigations. The same is true of the statistical analysis performed using the data. The quantities derived from the data (statistics) are meaningful *only* in the context of the data. In other words, for any meaningful analysis, we need to understand how the data is going to be collected, what they represent, and how many we need.

Figure 2.4 Statistical Inference

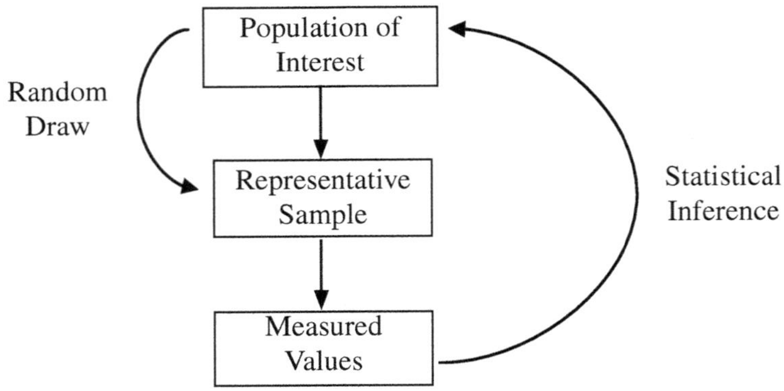

We take a random and representative sample from the population in order to make some statistical inferences. What constitutes a representative sample? As the following examples show, a representative sample depends on the objective of the study and the sample should accurately represent the population. Please read the descriptions carefully. Which ones provide examples of samples that appear to be representative of the population? Which ones do not? Are the objectives clearly stated?

1. **Objective**: Develop a new marketing campaign for a popular soft drink.
   Population: All U.S. consumers of a popular soft drink.
   Sample: 1,000 consumers, both males and females, of three different soft drinks from 29 states in New England, South, East, West, Mid-Atlantic, and Mid-West regions, ages 20 through 50.

2. **Objective**: Verify the efficacy of a new drug to reduce high cholesterol in obese adults between the ages of 25 and 55.
   Population: Obese males and females between 25 and 55, diagnosed with high cholesterol.
   Sample: 500 males, ages 35 and 50, diagnosed with high cholesterol, 10 pounds overweight, on a diet and exercise program.

3. **Objective**: Understand the largest sources of variation in bond strength that occur in the lamination step.
   Population: All relevant inputs, process knobs, and noise factors in the lamination system.
   Sample: Include variations in adhesive properties, temperature, speed, pressure, and different lots of raw materials according to a statistically designed experiment.

4. **Objective**: Ship customer orders with product that is fit-for-use and scrap non-conforming product.
   Population: All products manufactured and sold to customers.
   Sample: Two pieces randomly selected from each lot of 1,000 pieces in a production batch and inspected for defects.

In the first example, the objective is to develop a new marketing campaign for a popular soft drink and the population is broadly defined as "all U.S. consumers of a popular soft drink." The sample definition begins to provide more insight into the consumer base with details about sex, age, purchasing information, and geographic location. However, it is difficult to know whether this sample is representative of the population because the population is too broadly defined. Even if the sample is well defined, when we are finished collecting, analyzing, and interpreting the data it is impossible to draw inferences back to the population of "all U.S. consumers of a popular soft drink." We need to define our population with the same level of detail that we use for defining our sample so that the inferences we make from the data are relevant.

In the second example, our sample may not be totally representative of the population of interest. The objective of this study is to verify the efficacy of a cholesterol-reducing drug for obese adults, and our sample includes only adult males within a narrower age range than the population. Another other potential problem is that the sample includes males that are only 10 pounds over their ideal weight, while the population includes obese adults whose weight may be greater than 10 pounds over their ideal weight. Yet a third problem is that the sample asks for males that are currently on a diet and exercise program, but the population description makes no claims about lifestyle factors that may improve the efficacy of this drug. Can we draw inferences back to the broader population of obese males and females using only the efficacy findings of adult males, who are 10 pounds overweight, are on a diet, and are exercising?

The third example shows good agreement between the objective of the study, the description of the population, and the sample that will be used to draw inferences. The objective of the study is to determine sources of variation in the bond strength produced by the lamination step. In order to accurately define our population, we must consider the impact of variations in raw materials, process knobs, and noise factors (factors that we normally do not control during lamination) that contribute to the bond strength produced by this conversion step shown in Figure 2.1.

In the fourth example, we select only 2 samples out of a lot of 1,000 pieces. How representative is a sample size of 2? If the percent of non-conforming pieces is small (5% or less, for example) a sample size of 2 will very rarely contain a non-conforming item, but the remaining 998 pieces in the lot may contain up to 50 non-conforming items. In other words, there is a very high probability of shipping the customer non-conforming product.

2.3.1 Random Sampling

Our *sampling plan* will describe exactly how we sample our data from the population under investigation. This includes how many samples to take, how frequently, and how to take them from the population of interest (the sampling scheme). In other words, the sample size, *n*, and the sampling scheme for how to select the *n* samples defines the sampling plan for our study. There are many ways of selecting random samples from populations. Below, we discuss three popular sampling schemes that are commonly used in engineering and scientific studies: the simple random sample, the stratified simple random sample, and the cluster random sample.

- **Simple Random Sample:** For this sampling scheme, every unit in the population has an equal chance of being selected for the study. Selecting a random sample can be as simple as selecting at random *n* samples from the population of interest, like selecting things out of a hat. However, you probably want to be more rigorous than that. This can be easily done in JMP by clicking on the **Rows** menu and selecting **Row Selection > Select Randomly**. This brings up the Select Randomly window, where we can either type a sampling rate as a percentage of the population size, or as the number of random samples we want to select. Let's say we need to select a random sample of 20 items out of a population of 500. We type 20 in the given space (Figure 2.5a) and click **OK**.

Figure 2.5a Random Row Selection

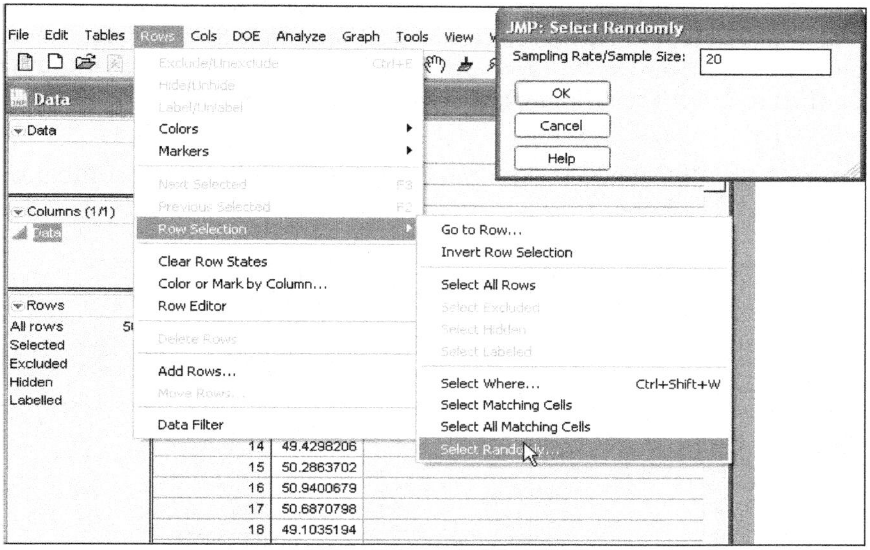

JMP selects 20 random rows from the given table as shown in Figure 2.5b.

Figure 2.5b Table Showing the Randomly Selected (Highlighted) Rows

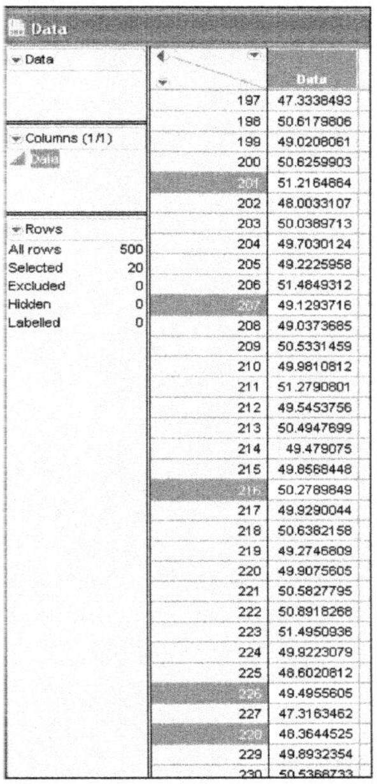

Now that the rows have been selected we can create a table with just the randomly selected rows by clicking on the **Tables** menu and selecting **Subset**. The Subset window has **Selected Rows** selected, and allows you to give the output table a new name. We type `Twenty Random Samples from Data` in the **Output table name** box (Figure 2.5c). The resulting table is shown in Figure 2.5d.

Figure 2.5c Selecting 20 Random Samples from the Original Table Using Subset

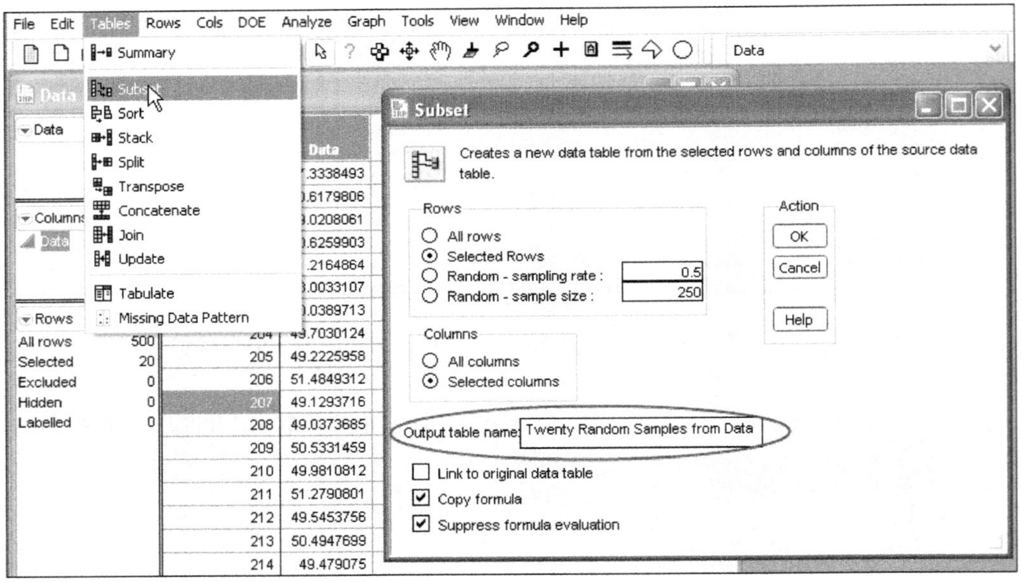

Figure 2.5d JMP Table Containing Twenty Random Samples from Data Table

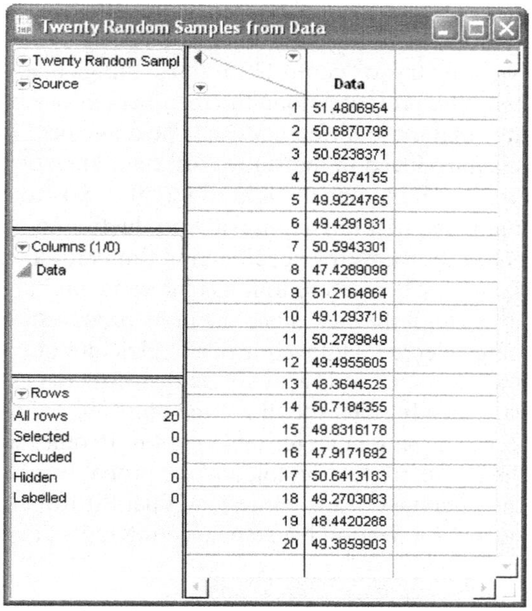

	Data
1	51.4806954
2	50.6870798
3	50.6238371
4	50.4874155
5	49.9224765
6	49.4291831
7	50.5943301
8	47.4289098
9	51.2164864
10	49.1293716
11	50.2789849
12	49.4955605
13	48.3644525
14	50.7184355
15	49.8316178
16	47.9171692
17	50.6413183
18	49.2703083
19	48.4420288
20	49.3859903

- **Stratified Simple Random Sample:** What happens if there is a natural division of the population into non-overlapping groups or different categories (*strata*), or subpopulations, like Female/Male; Supplier A/Supplier B/ Supplier C; Shift 1/Shift 2/Shift 3, and so on. For example, if we are conducting a study to determine the impact of a cholesterol medication on reducing "bad" cholesterol levels in humans, then we might want to first stratify our population into male and female and randomly select people from each subpopulation. A simple random sample will not guarantee that we are going to have the same number of samples from each of the categories, or that we will have a proportional number from each of the subpopulations. A *stratified random sample* selects units at random *from each* subpopulation or strata to form the sample.

- **Cluster Random Sample:** A cluster sample also takes advantage of the natural hierarchy or clustering of the population, but samples the clusters rather than the members within a cluster. All the members of the randomly selected cluster are included in the sample. This can be advantageous if, for example, we have information only about the clusters and not the individual members of the cluster, or if a single random sample will produce a sample that is costly or difficult to sample.

2.3.2 Randomization

How do we obtain an *unbiased* representative sample of our population? The concept of randomization was introduced by Sir Ronald Fisher as the physical basis for the validity of statistical tests. Even though we may try to "control" for those things that we are not studying, differences will always be present among the members of our population. Selecting a random sample ensures that these differences will "average out" in the long run, and that no perceptible bias is introduced in the study. Yes, randomization sounds simple and seems like common sense, but no one thought of it before Sir Ronald Fisher! For example, what if we only sample the first three units of a production lot, and use them to determine whether the entire lot meets the specification limits for a thickness measurement. Is this a random sample, where every unit has the same probability of being selected? As you can see, this is not the case since the units produced fourth, fifth, and sixth have zero chance of being selected. Did we introduce bias into our sample? Although the issue of bias is context-specific, it is easy to see that if there is a start-up effect, i.e., it takes 10 units to reach steady state, the first three units are going to be different from the rest since we are sampling in an unstable region. In other words, the sample is not representative of the entire production lot, and we probably will not be able to make good inferences about the thickness of the entire lot. What Sir Ronald Fisher called the "essential safeguard" of randomization leads to selecting units at random from the entire production lot.

Randomization forms the physical basis for the validity of statistical tests.

Any study can be made more sensitive by increasing the number of units sampled from the population. But how many samples are enough? This is perhaps the most frequently asked question by experimenters. How many samples should we include in our study in order to answer our question with a high degree of confidence? In the following chapters we will see how the **JMP Sample Size and Power** calculator makes it easier for us to determine the appropriate sample size for our study. The specific details will depend upon the type of study that we are conducting.

Apart from having a sample size of n, for example, we also take a measurement or multiple measurements of each unit in our sample. It is important to make a distinction between the sample units and the measurements taken of them. For this we must understand the difference between an *experimental* and *observational* unit.

- **Experimental Unit (EU):** This is the smallest unit that is affected by the "treatment," or factors, we want to study. The output from the JMP sample size calculations will dictate how many experimental units we will need in our study. Note, this is not necessarily the same as the number of measurements that we will have at the end of our study. If we decide, for example, to measure each experimental unit more than once, then we will end up with more readings than we have experimental units.
- **Observational Unit (OU):** This is the smallest unit on which we take a measurement.

In the lamination example our experimental unit (EU) is a roll of material, which gets processed under a given temperature, speed and pressure, while our observational units (OU) are the different places within the roll where we take thickness measurements. To further illustrate the difference between an experimental and observation unit, let us consider a study involving the deposition of an oxide layer on wafers, which requires a sample size of 20 experimental units. Figure 2.6 shows the five thickness measurements taken on one wafer in the upper left (UL), upper right (UR), Center (C), lower left (LL), and lower right (LR) side of the wafer. Do these five measurements count toward our required sample size of 20 experimental units?

Figure 2.6 Thickness Measurements Taken on One Wafer

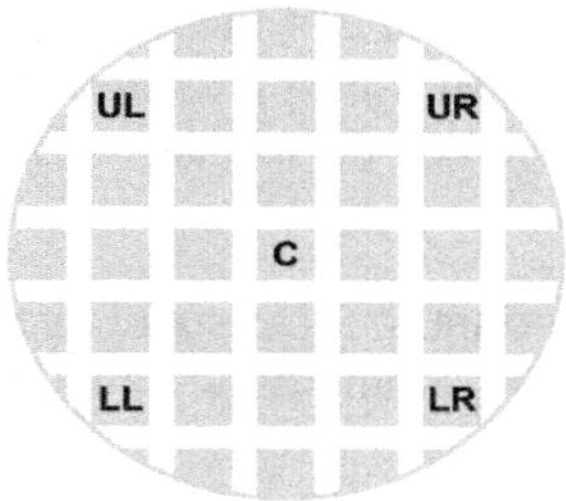

Well, perhaps. If the deposition factors that we are studying (for example, the temperature of the deposition furnace) get applied to the entire wafer at once, then the experimental unit is the wafer and the five thickness readings are five multiple readings on the same experimental unit. The five individual sites (UL, UR, C, LL, and UR) become the five observational units (OU) where the thickness measurements are taken. In this case we need 20 wafers (EU) for a total of 20×5 = 100 (OU) measurements. On the other hand, if we can subject the five individual sites (UL, UR, C, LL, and UR) to different temperatures of the deposition furnace that are run at different times, then the experimental unit and the observational unit are one and the same: one site within a wafer. Four wafers will provide the required 20 experimental units needed for the study. Of course, any deposition engineer will tell you that this is not valid because a deposition process is an example where multiple units, wafers and wafer sites, are processed simultaneously, as opposed to piece-part processes where units are processed one at a time. As a general rule, if units are processed simultaneously there is a very good chance that the experimental units are different from the observational units, while in a piece-part process the experimental units are the same as the observational units.

2.4 Descriptive Statistics and Graphical Displays

What can we say about a group of data, i.e., how can we describe a collection of measurements? Even before we make a claim about the performance of a certain material with respect to a standard (Chapter 4), what can we say about the performance of the material? How can we summarize its performance using a few quantities? These questions are particularly important if the number of measurements is large. Fortunately for us there are several quantities that can be used to characterize and describe the properties of a set of measurements, as well as different graphical displays to bring to light interesting facts about our data. Statistical visualization software like JMP makes this very easy. Because calculations and graphs are easy to obtain, we just need to understand what is behind these summary statistics and graphs, and how they can help us gain insight from our data. In this section we describe some of the most common and useful summary statistics, as well as some of the most useful graphical displays. In Chapter 3 we cover the characterization of the measured performance of a material, process, or product in more detail.

When we summarize a set of measurements it is important to understand two key measures that can help us to describe our data (Section 2.5) and draw a few conclusions from it. These two measures are:

- the central tendency of the data—where the *center of gravity* is
- how far the data is from its center of gravity—its *spread* or *radius of rotation*

An example could include opinion polls that report the percentage of respondents for a given issue (the central tendency), plus or minus the margin of error (a function of the spread). If the margin of error is large we are less confident about the claims that are made about the results. Table 2.3 shows some common descriptive statistics that are used to describe the central tendency, or average performance, of the data; the spread, or performance variation, of the data; and some statistics that combine both center and spread. Some of these summary statistics are affected by observations that are different or far away from the mass of the data. We call those unusual observations *outliers*.

The good news is that most of the summary statistics presented in Table 2.3 can be easily calculated in JMP by selecting **Tables > Summary** (Figure 2.7).

Table 2.3 Descriptive Statistics for Center, Spread, and Center and Spread

Property	Summary Statistic	Description	How to Calculate It
Center	Mean	The average of a set of data—its center of gravity. This estimate is usually affected by outliers (unusual values).	Add all the measurements and divide by the number of samples: $\bar{X} = \frac{\sum_{i=1}^{n} X_i}{n}$
	Median	The value that divides the ranked data (from smallest to largest) in half. It is also referred to as the 50th percentile of the data. The median is robust to outliers.	Order the data from smallest (1) to largest (n), $x_{(1)}$, $x_{(2)}$, $x_{(3)}$, . . , $x_{(n)}$. For n even, average the ordered values at positions n/2 + (n/2+1). For n odd, it is the ordered value at (n–1)/2 + 1.
	Mode	The most frequently occurring value in the data. It is most useful for nominal and ordinal data. There may be more than one mode.	Count the occurrences for all unique data values. The mode has the largest number of occurrences.

(continued)

Table 2.3 *(continued)*

Property	Summary Statistic	Description	How to Calculate It
Spread	Range	The difference between the maximum and minimum values in the data. This estimate is usually affected by outliers.	$X_{max} - X_{min}$
	Standard deviation	Square root of the average squared distance from the data points to the mean—its radius of rotation around the mean. Most data falls in the interval of mean ± 3s.	$s = \sqrt{\frac{\sum_{i=1}^{n}(X_i - \overline{X})^2}{n-1}}$
	Interquartile range (IQR)	When an ordered data set is divided into four equal parts, the four division points are called quartiles. The IQR is the difference between the 3rd and 1st quartiles.	Rank order data and subtract the ordered values located at $q_3 = x_{(3n/4)}$ and $q_1 = x_{(n/4)}$.
Center and Spread	Coefficient of variation	The ratio, in percent, of the standard deviation to the mean. This is useful for comparing measures with different units.	$100\frac{s}{\overline{X}}$
	Percentiles	The kth percentile is a data value such that approximately 100k% of the ordered data is below it and 100(1–k)% of the ordered data is above it.	Rank order data from smallest to largest. To get the p_k percentile, multiply n by k% and round to nearest unit and find corresponding rank ordered value.
	C_{pk}	Process capability index. It normalizes the distance of the closest specification limit (lower specification limit or upper specification limit) to the mean by 3 standard deviations.	$\min\{\frac{\overline{X} - LSL}{3s}, \frac{USL - \overline{X}}{3s}\}$

This in turn launches a window to select the data (measurement) and the summary statistics that we want to calculate (Figure 2.7).

Figure 2.7 Summary Platform

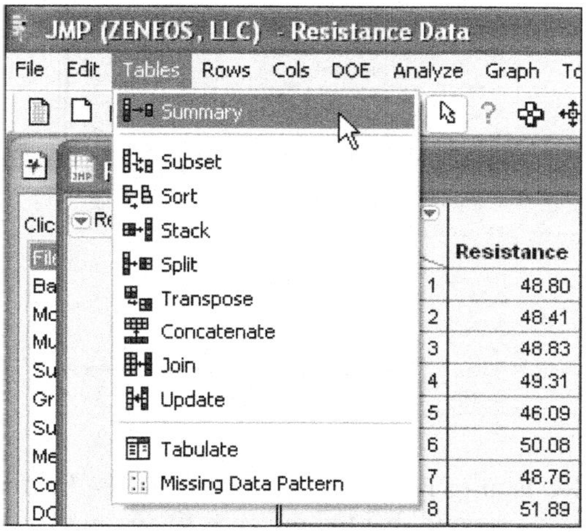

First we must select the column in our JMP table that we want to summarize under the Select Columns window and then click on the **Statistics** button to select the summary statistics of interest. Each time one of these statistics is selected it is displayed in the window to the right of this menu (see Figure 2.8.) When we are finished selecting our summary statistics we give a name, `Summary of Resistance`, to the new data table that will be created when we click **OK**.

Figure 2.8 Summary Statistics Window

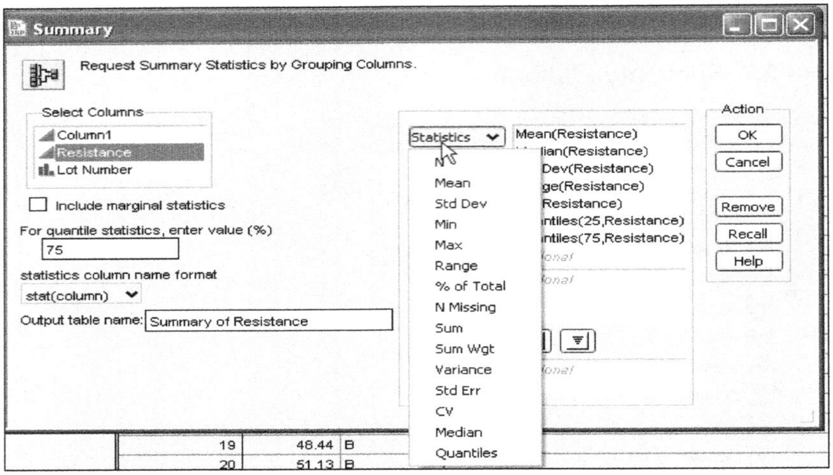

The summary statistics table for Resistance is shown in Figure 2.9. The first column tells us that we have 40 measurements, the second that the mean is 49.86, and the third that the median is 49.63. An estimate for the standard deviation of the data is 1.9612 and the coefficient of variation is 3.93%. These summary statistics are quite useful in reducing our 40 resistance measurements down to a few useful numbers, giving us a good feel for its central tendency and spread; 49.86 Ohms ± 1.96 Ohms. If you would like to see how we can use these summary statistics to characterize the measured performance of a material, process, or product feel free to jump (pun intended) to the next chapter.

Figure 2.9 Summary Statistics Output for Resistance Data

Summary of CH2 Resistance Data

	N Rows	Mean (Resistance)	Median (Resistance)	Std Dev (Resistance)	Range (Resistance)	CV (Resistance)
1	40	49.86	49.63	1.9612	7.40	3.93

Although summary statistics are helpful in describing our data, graphical displays can reveal unexpected patterns and key information that is not easy to see with just these measures. It is one thing to say that our resistance measurements are 49.86 Ohms ± 1.96 Ohms, and quite another to visualize it. JMP provides us with many different types of graphical displays that we can use to gain insight from our data. A selection of these graphical displays is listed in Table 2.4, Figure 2.10, and in the JMP online Help via the **Help** menu. The type of graph we use depends on the type of measurements we have collected, i.e., continuous or attribute.

Table 2.4 Common Graphical Displays for Continuous Data

Graph Name	Graph Description	JMP Menu
Histogram	A frequency distribution of data that shows the shape, center, spread, and outliers in a data sample. When the histogram is overlaid with specification limits, we can see how much room we have to move within the specifications. The number of bars in the histogram should be approximately 1 + Log_2 (sample size).	**Analyze > Distribution**
Trend Plot	A plot that is easily constructed by plotting our measurements in a time order, with the response on the y-axis and time on the x-axis, and connecting the points with a line. You might find it useful to add a horizontal reference line that represents the mean, or lines representing boundary limits when looking for trends, outliers, or unusual patterns in the data.	• **Graph > Overlay Plot** • **Graph > Control Chart > Run Chart**
Process Behavior Chart	An individuals and moving range chart is a plot that helps characterize the underlying system that generates our data. If the data falls within the three-sigma control limits, then we can assume that the data is homogeneous, which is a key assumption in many statistical techniques.	• **Graph > Control Chart > IR**
Normal Quantile Plot	A plot that has the normal quantile on the y-axis and the observed measurement on the x-axis. The normal quantile is determined by using the ranked order values of the data and using the cumulative probabilities from a normal distribution to get the corresponding normal quantiles. This plot is useful to decide whether the data can be approximated using a normal distribution.	**Analyze > Distribution**
Box Plot	A plot that shows the frequency distribution of the data using the *5-number summary* (minimum, 1st quartile, median, 3rd quartile, maximum) of the data to show its center (median), spread (IQR), and outliers. This plot is very useful for comparing more than one population of data or stratifying one population by different factors.	• **Graph > Variability/ Gauge Chart** • **Analyze > Fit Y by X**
Scatter Plot	A plot that shows the relationship between two measurements. A scatter plot is typically used to look for correlations, where, for each pair of points, the dependent variable is plotted on the y-axis and the independent variable is plotted on the x-axis. Some relationships that we look for with a scatter plot include: none, linear, or quadratic.	• **Analyze > Fit Y by X** • **Graph > Overlay Plot**

Figure 2.10 shows you where some of these graphs can be found in the **Graph** and **Analyze** platforms in JMP.

Figure 2.10 Graphing Platforms in JMP

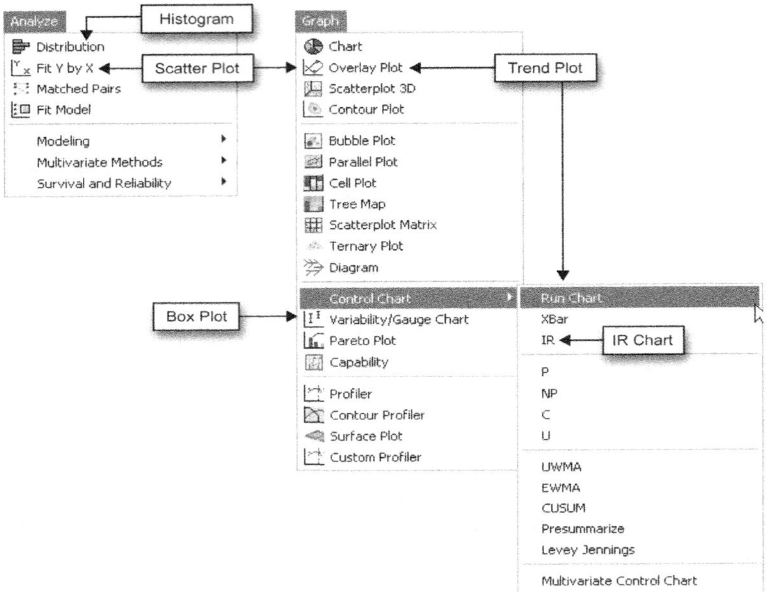

As an example let us create a histogram for the resistance data that we summarized as 49.86 Ohms ± 1.96 Ohms. Select **Analyze > Distribution** from the primary menu bar in JMP (Figure 2.11).

Figure 2.11 Analyze Distribution Platform

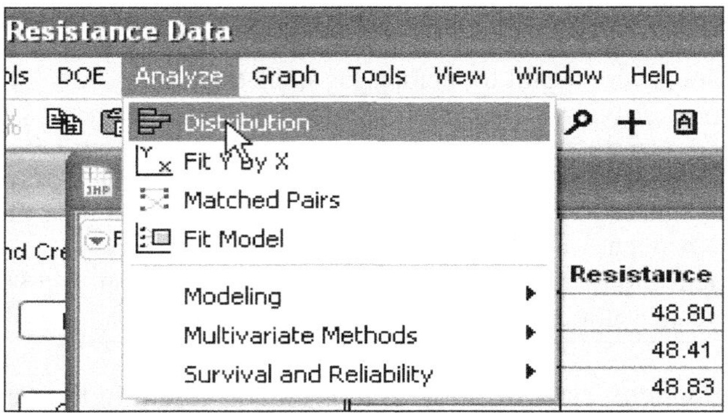

This launches a window to select which measurement to use for the histogram (Figure 2.12). We select the variable **Resistance** and click on **Y, Columns** to place Resistance in the Y window.

Figure 2.12 Distribution Window

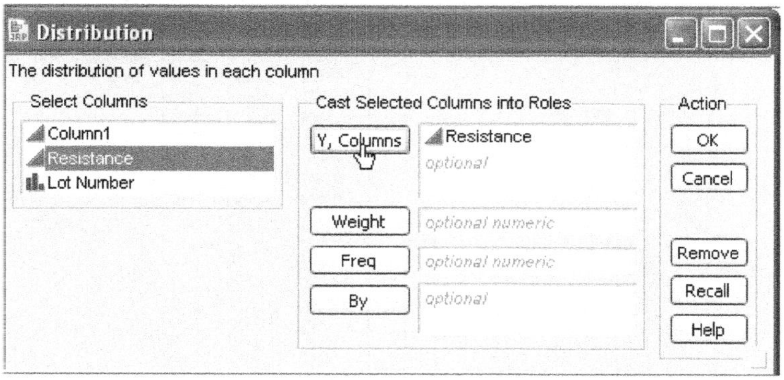

After we click **OK** the histogram is shown in a new window along with quantiles and summary statistics to left of the histogram (Figure 2.13). On top of the histogram there is an **Outlier Box Plot** and the **Shortest Half Bracket** representing the densest 50% of the observations, i.e., the shortest half of the interquartile range (IQR), which is from about 48 Ohms to 50 Ohms in this case. The histogram shows the central tendency of the data to be around 49 Ohms, although it is not that obvious.

JMP Note 2.1: By default JMP displays the histograms in a vertical way. To display the histogram in a horizontal layout, as in Figure 2.13, we need to click the red arrow next to **Distributions** and select **Stack**. If you like to display your histograms horizontally but don't want to do this every time, you can select **Preferences (CTRL+K)** from the **File** menu, and then select **Platforms > Distribution** where you can select the **Horizontal Layout** option.

Figure 2.13 Histogram of Resistance Data

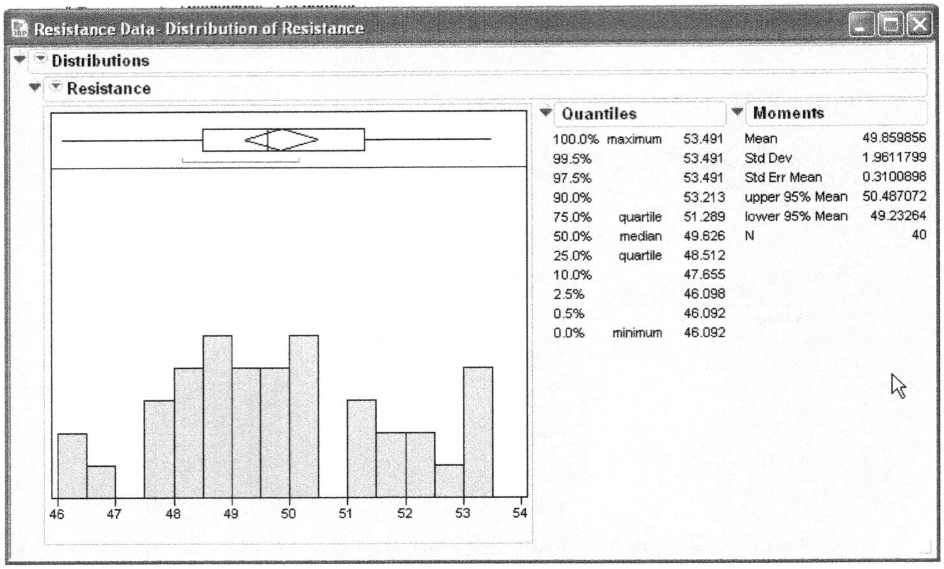

We can enhance this output by adding more summary statistics. This is done by clicking the red triangle next to the **Resistance** label and selecting **Display Options** and **More Moments** as shown in Figure 2.14. The output is updated to include, for example, the sum of all the observations, the variance (square of the standard deviation), and coefficient of variation (CV) as shown in Figure 2.15.

Figure 2.14 Adding More Summary Statistics

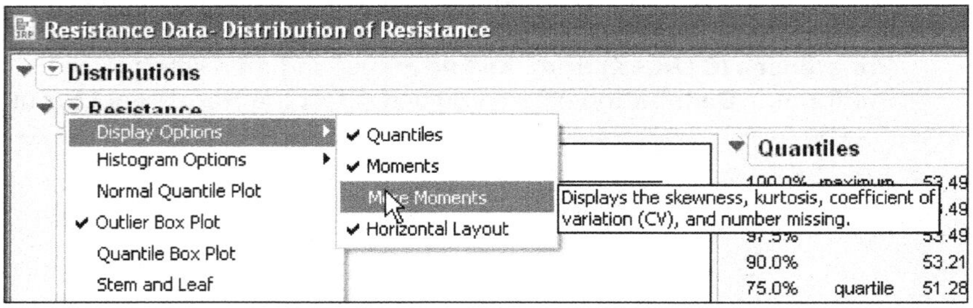

Figure 2.15 Histogram and Summary Statistics for Resistance Data

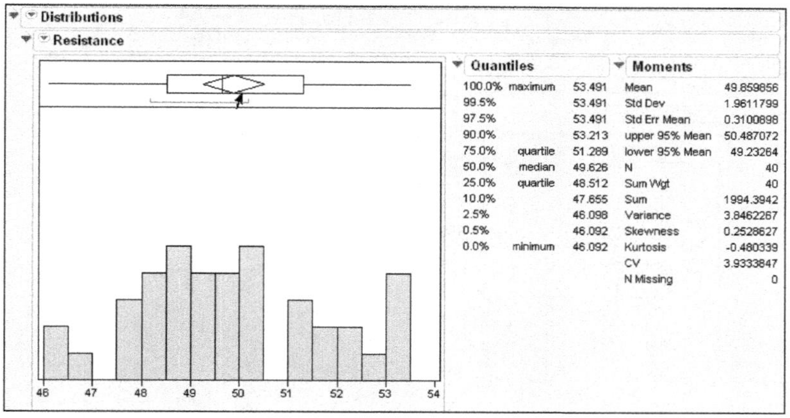

Notice that the histogram in Figure 2.15 has a large number of bars (JMP defaults to between 2 and 20 bars), which makes it difficult to visualize the center and shape of the data. The rule of thumb suggests that, based on the number of data points in our sample, we should get approximately $1+\text{Log}_2(40)$ or, rounding down, six bars in our histogram. This can be done easily by changing the increment on the x-axis. To do this, double-click on the x-axis below the histogram. This brings the **X Axis Specification** window where we can change the increment to 1.3 (see Figure 2.16). The resulting histogram is shown in Figure 2.17a now clearly showing the center of the distribution around 49 Ohms, and spread around 2 Ohms. The bars at each side of the central bar are at 48 Ohms and 51 Ohms, while the IQR = 51.29 Ohms – 48.51 Ohms = 2.8 Ohms.

Figure 2.16 X Axis Specification Window

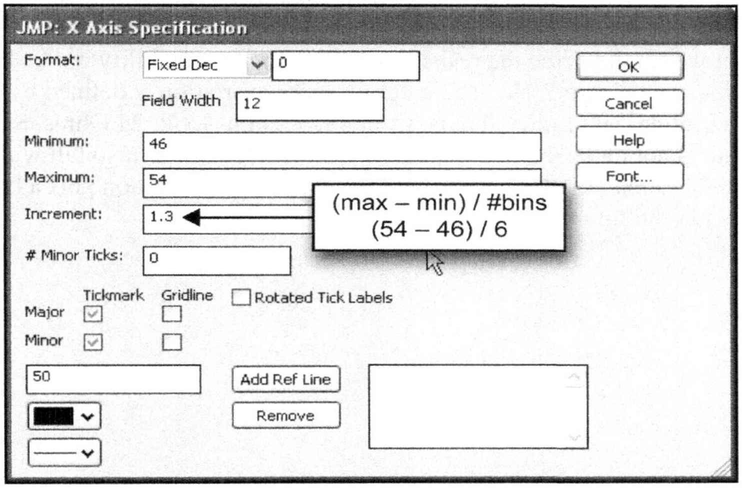

Figure 2.17a Histogram for Resistance Data with 6 Bins

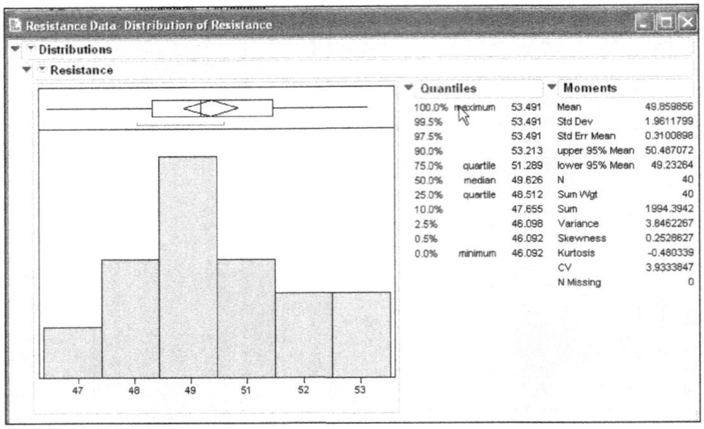

Are the data homogeneous?

Figure 2.17b shows an individuals and moving range chart (**Graph > Control Chart > IR**) for the resistance data. This is one of the simplest, but most useful, of the different process behavior charts. This chart is a running record of the data values against natural process limits defined as Average ± 3 (Local Measure of Variation), where the local measure of variation is based on the moving ranges (see Chapter 3). If the data is homogeneous virtually all the data should fall within this local interval (Wheeler 2005). Homogeneous means that the process generating the data can be characterized by a single probability distribution. In the next section we discuss how probability distributions can be used to quantify the uncertainty in our data, and to make predictions about future observations.

All the 40 resistance data points appear to vary randomly inside the upper control limit (UCL) and lower control limit (LCL), indicating homogeneity of the data. We feel confident that we can describe the resistance data using a probability model like the normal distribution (Figure 2.19). The natural process variation is defined by the lower and upper natural process limits, that is, from 43.48 Ohms to 56.24 Ohms. Since the resistance data is homogeneous we expect future observations to also fall within 43.48 Ohms and 56.24 Ohms. In other words, the natural process variation puts a bound on what to expect in the future.

Figure 2.17b Individual Measurements Chart for Resistance Data

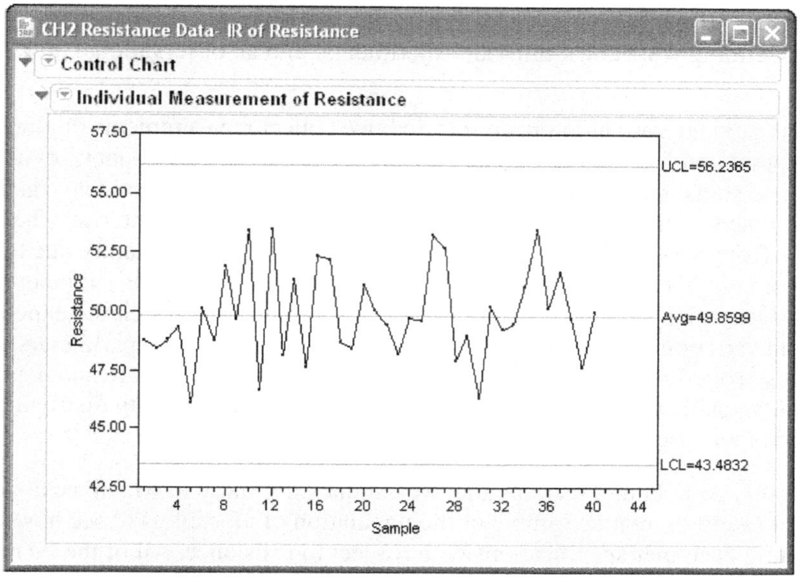

In this section we wanted to provide a quick introduction to some useful graphs and summary statistics. In Chapter 3 we will show how to use them to characterize the measured performance of a material, process, or product, and in subsequent chapters we will continue to use them, as appropriate. Every data analysis and formal tests of significance should include a thorough *exploratory data analysis* (EDA) using the summary statistics and graphical displays discussed in this section. Apart from helping us identify unusual or outlying observations, EDA can help answer questions of interest, and it is a great springboard for suggesting hypotheses that can be explored with further data analysis techniques.

2.5 Quantifying Uncertainty: Common Probability Distributions

In Section 2.3 we discussed the need for considering the different aspects that make for a good study. These include the following:

- clearly defining the reasons for our study
- being detailed about our population of interest from which a sample will be selected

- determining the size and how a random and representative sample will be selected
- defining what constitutes an experimental and an observational unit.

Once this random sample has been selected, we collect measurements on the experimental units and observational units of this random sample, analyze the data using appropriate statistical techniques, draw conclusions based on the evidence gained from the data analysis, and make inferences back to the population of interest. The inferences we make from our conclusions will involve some degree of uncertainty due to a sample, perhaps small, that does not include all the members of the population; inherent errors in our measurement techniques and instruments; variations between the experimental units; and variation between the observational units (within the experimental units). If the study was properly designed and randomization was used, the nature and degree of this uncertainty can be described and quantified by means of probability distributions, and their corresponding probability density functions.

To drive this point home, let's say that we conducted a study in which we collected 20 cables (a representative sample of the population of all cables) to see how well we are meeting customer specifications with respect to resistance. All of the 20 resistance readings were inside of our specification limits. Based on this information can we conclude that the resistance of our population of cables has a 100% yield, i.e., that all previously produced cables had, and all future cables will have, a resistance reading within specification limits? This certainly would be a very risky claim if we could not describe and quantify our uncertainty, especially if our yield loss and sample size are small. For example, if our true yield loss was around 2% then we would expect to observe about 0.4 (20 x 0.02) defective cables in our sample of 20 cables—not even one cable. However, if we are making claims about 100,000 cables then we expect 2,000 defective cables, which is a far cry from 0! We will see below that using a probability distribution, along with some of the key summary statistics, enables us, in this case, to make better yield predictions about future samples.

Probability density functions (pdf) are mathematical models, i.e., equations, that mimic the frequency distribution of the measurements of interest, and which are functions of a few key parameters that define their behavior. Most probability density functions have parameters that control the central value around which the measurements are distributed, the spread of the measurements around the central value, and the shape of the curve. Probability density functions take on a variety of different forms based upon the nature of our responses (see Section 2.2 if you need a refresher), and based upon the physical mechanism behind our data. Some examples of different probability density functions are provided in Table 2.5, along with the measurement scale they correspond to, a description of the distribution, and some typical applications in science and engineering. We use Greek symbols to denote the key parameters of each density function. Note that each of the probability density functions in Table 2.5 can be represented by a curve, and that the area under the curve corresponds to the total probability, which is 1.

Table 2.5 Common Probability Distributions

Distribution	Measurement Scale	Probability Density Function Description	Typical Applications
Normal	Ratio and interval	Defined by two parameters μ (central tendency) and σ (spread). Symmetrical around μ. μ ± kσ brackets different proportions of the population. $\frac{1}{\sigma\sqrt{2\pi}}e^{\frac{-(x-\mu)^2}{2\sigma^2}}$	Heights, dimensions, bond strength, and so on.
Student's *t*	Ratio and interval	Defined by ν, degrees-of-freedom. Symmetrical around 0. As ν increases the distribution rapidly approximates the normal distribution. $\frac{\Gamma\left(\frac{\nu+1}{2}\right)}{\Gamma\left(\frac{\nu}{2}\right)\sqrt{\nu\pi}}\left(1+\frac{t^2}{\nu}\right)^{-\frac{\nu+1}{2}}$	Heights, dimensions, bond strength, and so on, when sample sizes are small (≤30).
Log normal	Ratio and interval	Log(data) has a normal distribution. Skewed to the left or right. Also defined by μ and σ. $\frac{1}{x\sigma\sqrt{2\pi}}e^{\frac{-(\log(x)-\mu)^2}{2\sigma^2}}$	Distribution of particle size, dosages in drug applications, and lifetime data.
Exponential	Ratio and interval	Defined by a rate parameter λ. Asymptotic lower bound. $\lambda e^{-\lambda x}$	Random recurrent events in time, decay phenomena, and lifetime data.
Weibull	Ratio and interval	Defined by a scale parameter α and a shape parameter β. $\alpha\beta x^{\beta-1}e^{-\alpha x^{\beta}}$	Time to failure of different physical properties like the breaking strength, or the time to failure of integrated circuits.

(*continued*)

Table 2.5 *(continued)*

Distribution	Measurement Scale	Probability Density Function Description	Typical Applications
Binomial	Nominal	Defined by the probability, p, of observing one of the two possible outcomes of an event, and the number of trials, N. Gives the probability of observing k occurrences of the event in a sample of size N. $\binom{N}{k} p^k (1-p)^{N-k}$	Pass/fail data like wafer yield.
Multinomial	Ordinal	Extension of the binomial distribution. When there are more than two possible outcomes for an event.	Multiple defect types or customer satisfaction scores.
Poisson	Ordinal	Defined by the average number of counts in the interval, λ. $\frac{\lambda^x e^{-\lambda}}{x!}$	Number of counts occurring at random in an interval. Particle, defect, or blood counts, arrival rates, and queue lengths.

Any distribution that is used to approximate the behavior of our measurements provides two meaningful things about our data:

- a way to describe, and visualize, all of the possible values that our measurements can take
- a way of quantifying the likelihood of occurrence of a value or of a set of values that our measurements can take

For example, if we assume that the resistance measurements (Figure 2.17a) can be adequately described with a normal distribution, then we can make statements about the resistance of the population of cables. Such statements can include the range of resistance measurements that we can expect to observe in our population, and what percentage of them should fall between an upper and lower specification limit. However, before we can provide exact details for these statements, we must know the mathematical form of the distribution, its equation, and provide estimates for the parameters that described the distribution.

Note that the last three distributions, Binomial, Multinomial, and Poisson are used with discrete or categorical types of data. These distributions will not be covered in this book.

2.5.1 Normal Distribution

The normal distribution, also known as the Gaussian distribution, is one of the most used probability models in statistics, the famous bell-shaped curve. As noted in Table 2.5, the normal distribution is defined by its central tendency, μ, and its standard deviation (spread), σ.

Figure 2.18a Normal Distribution pdf

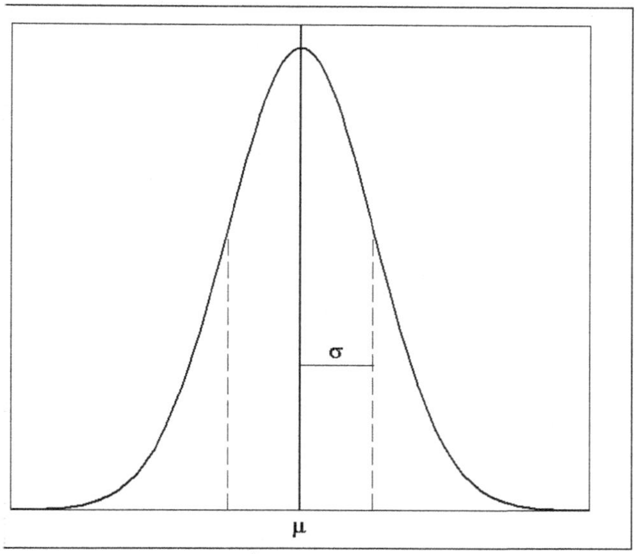

In Figure 2.18a we can see that μ falls right on the axis of symmetry of the curve, and therefore represents the center of gravity, or centroid, of the distribution. The standard deviation, σ, is the radius of rotation from the centroid, such that the quantity σ^2 (which is proportional to $\int(y - \mu)^2dy$) represents the moment of inertia, or second moment of the area under the curve. As you can see, the parameters that describe the behavior of the normal distribution have quite specific physical meanings. In statistical jargon we call σ the *standard deviation* and σ^2 the *variance*. For a process, product, or material performance measure, Y, the notation $Y \sim N(\mu, \sigma)$ implies that our performance measure follows a normal distribution with parameters μ and σ. In order to use this mathematical equation, we must specify the values for μ and σ. These parameters can be estimated from the data by using the mean and standard deviation formulas in Table 2.3. For the resistance data Resistance ~ N(50, 2) indicates that our resistance measurements are centered at 50 Ohms, and have a spread, or standard deviation, of 2 Ohms.

Standard Normal Distribution $\mu = 0$ and $\sigma = 1$

A special case of the normal distribution, so special that it's called the *standard normal distribution*, has $\mu = 0$ and $\sigma = 1$. This special case, shown in Figure 2.18b, serves as a reference for probability calculations using the normal model, and has the following important properties:

- The center of gravity (centroid) of the curve is 0 (where the axis of symmetry is).
- The radius of rotation of the curve, its spread, is 1 (the distance from the centroid).
- Since the curve is symmetrical about 0, and the area under the curve is 1, it follows that the area under the curve to the right of 0 is 0.5 and the area under the curve to the left of 0 is also 0.5.
- With a very high probability (99.9937%) most values will occur between ± 4.
- Approximately 68% of the values will occur between ± 1.
  Conversely, 32% of the values will equally occur outside of ± 1.
  In other words, 16% of the values will occur above +1 and 16% of the values will occur below −1.
- Approximately 95% of the values will occur between ± 2.
  Conversely, 5% of the values will equally occur outside of ± 2.
  In other words, 2.5% of the values will occur above +2 and 2.5% of the values will occur below −2.
- Approximately 99.73% of the values will occur between ± 3.
  Conversely, 0.27% of the values will equally occur outside of ± 3.
  In other words, 0.135% of the values will occur above +3 and 0.135% of the values will occur below −3.
- The six sigma ($\pm 6\sigma$) interval contains 99.9999998% of the distribution.

Figure 2.18b Standard Normal Distribution with Mean = 0 and Standard Deviation = 1

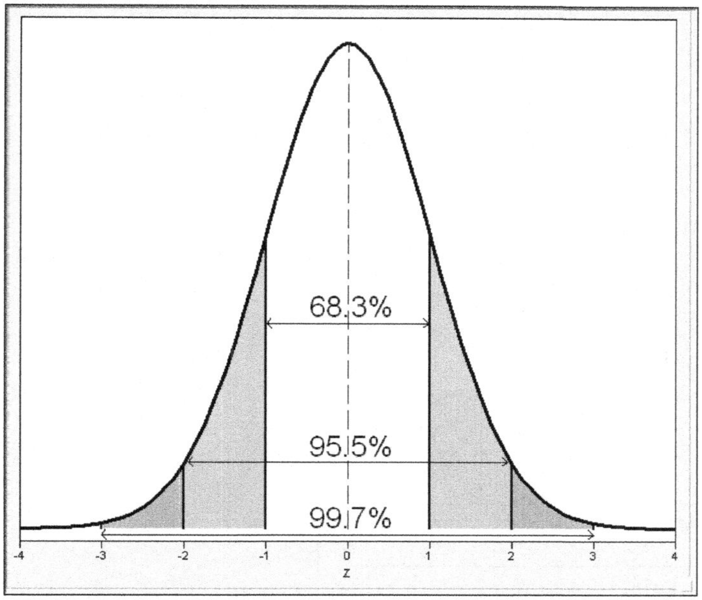

To see why it is helpful to use the standard normal in probability calculations let us look at the resistance data shown in the histogram Figure 2.19. Forty resistance measurements gave an estimated mean of 49.86 Ohms and standard deviation of 1.9612 Ohms. The histogram looks symmetric around the mean suggesting that a normal distribution may be a good approximation for the behavior of the resistance measurements population (in Figure 2.20 we introduce the normal quantile plot, which is a better graphical tool for assessing normality). We can now make a probability statement and say that 99.7% of the resistance measurements will between 49.86 Ohms ± 3(1.9612) Ohms, or between (43.98 Ohms, 55.74 Ohms). Based on this information, what yield loss can we expect if our specification limits are LSL = 45 Ohms and USL = 55 Ohms?

Figure 2.19 Resistance Data Histogram with Specification Limits

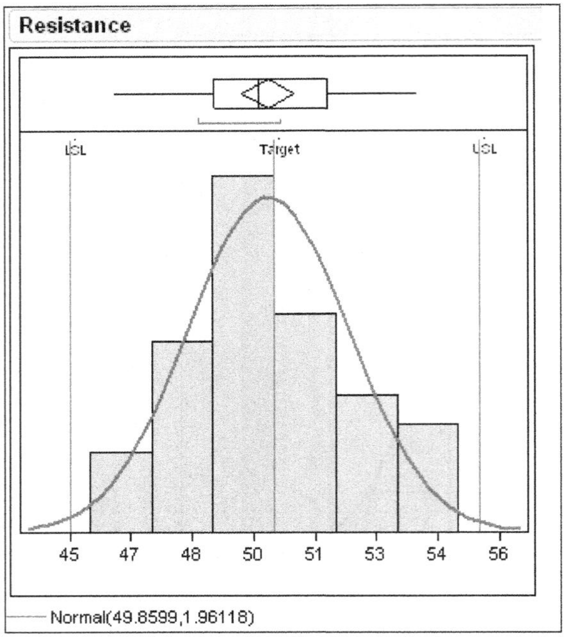

We can already see from Figure 2.19 that the tails of the superimposed normal curve fall outside the specification limits. How much yield loss can we expect? We can use the normal distribution to predict the yield loss at each side of our specification limits, and add them up to get the overall predicted yield loss. The steps are provided below.

Resistance data yield loss calculations:

1. We need to transform the lower specification limit to a *standard normal score* (Z-score) using the mean and standard deviation estimated from the data.

   $$Z_{lower} = (LSL - \overline{X}) / s = (45 - 49.86) / 1.961 = -2.48.$$

   Remember that the areas under probability density functions are probabilities. We need to find the area under the standard normal to the left of -2.48. But how? We can use the **Formula Editor** in JMP to find the exact probability.

2. Start by creating a new table with just one row. Then, right-click on the **Column 1** name and select **Formula**, then select **Normal Distribution** from the **Probability** menu (Figure 2.19a).

Figure 2.19a Selecting the Formula Editor and the Normal Distribution

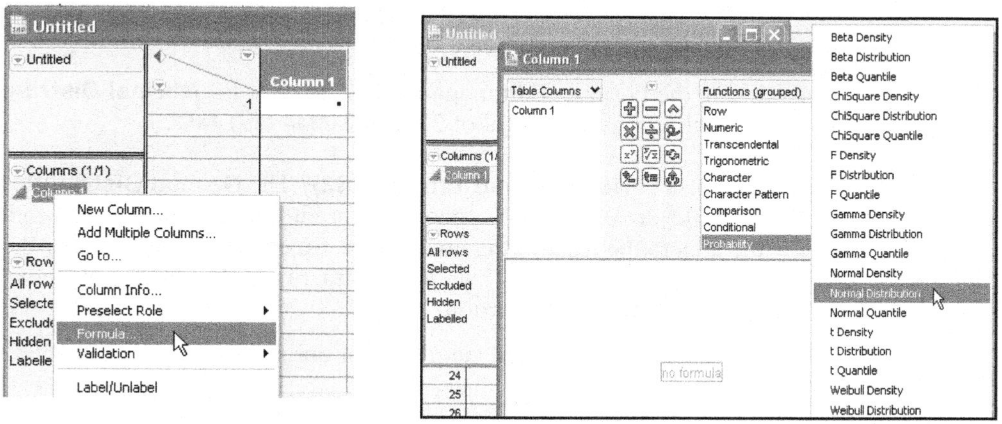

Once the **Normal Distribution()** is in the **Formula Editor** just click in the red box and type `-2.48`. Once you click **OK** the answer appears as 0.00656912 or 0.66% (Figure 2.19b). Of course, you can just type the expression `(45 - 49.86) / 1.961` inside the red box and JMP will calculate the Z-score before computing the area.

Figure 2.19b Normal Distribution for -2.48 and Probability Result

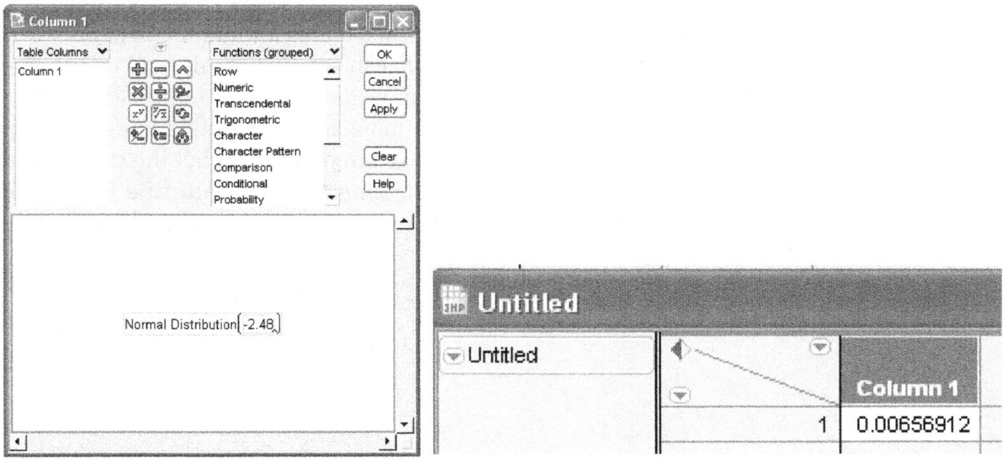

3. A similar calculation is done to transform the upper specification limit to a Z-score.

$$Z_{upper} = (USL - \overline{X}) / s = (55 - 49.86) / 1.961 = 2.62.$$

4. Using the Formula Editor again this time with 1 – **Normal Distribution(2.62)** gives the area to the right of 2.62 as 0.0044 or 0.44%.

   Why use 1–**Normal Distribution(2.62)**? The **Normal Distribution()** function gives the area to the left of the given point. For the upper specification we need the area to the right of the given point. Since the total area under the curve, or the total probability, is 1, we just subtract from 1 the value from the **Normal Distribution(2.62)** function.

5. The combined predicted yield loss for our resistance data is 0.66% + 0.44% or 1.1%.

Goodness of Fit

How do we know that the normal distribution is a good approximation for our data? Although there are statistical tests to test for normality, in our experience it is best to check for normality using graphical tools. This is because the tests for normality tend to be too sensitive to slight departures from normality, especially when the sample size is large. Remember that no data is going to follow a normal distribution exactly, and what we can hope for is that the normal distribution is a "good" approximation to the behavior of the data. To check for normality graphically we plot our measurements in a normal quantile plot and see if the resulting plot is close to a straight line. How close? JMP makes this easy by providing confidence bands around the data line. If most points fall inside of the confidence bands then we can say that the normal distribution is a good approximation for our data. Figure 2.20 shows a normal quantile plot for the resistance data with all the points inside the two confidence bands (dotted lines). The normal quantile plot was generated from the histogram output by clicking the red triangle to the left of the **Resistance** label, and then selecting **Normal Quantile Plot**. JMP also plots a solid line within the confidence bands to aid the eye in deciding if the line defined by the data points is straight.

Figure 2.20 Normal Quantile Plot for Resistance Data

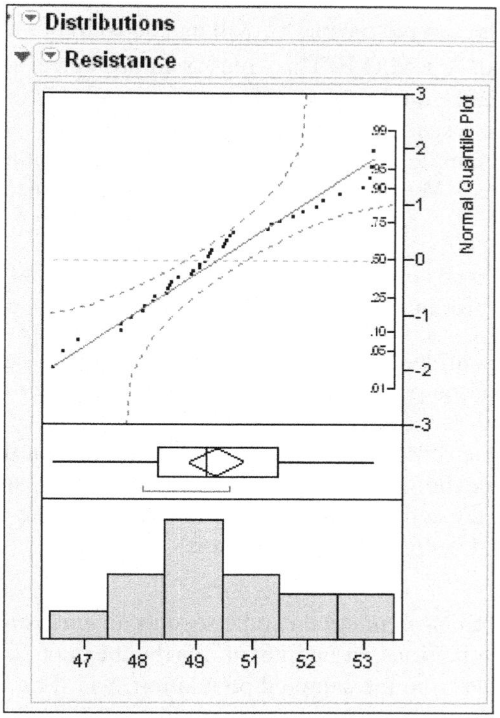

2.6 Useful Statistical Intervals

In Section 2.3 we discussed how a well-chosen sample (representative, randomized, and so on) enables us to make inferences about the population from which the sample was drawn. Statistics calculated from this sample (Section 2.4) help us describe characteristics of the population, like center and spread, while probability distributions (Section 2.5) provide models that help us quantify the degree of uncertainty that is always present when we argue from the particular (the sample) to the general (the population). Since there are inherent errors in any sample we take, what confidence do we have about the statistics computed from the sample? How "noisy" are these estimates? In other words, can we quantify the degree of uncertainty in an estimate of location like $\overline{X}$, or in an estimate of spread like the sample standard deviation? In this section we introduce three types of *statistical intervals* that will enable you to put uncertainty bounds on your estimates, as well as on the value of future observations.

Statistical intervals enable us to quantify the degree of uncertainty in our sample estimates, and get an understanding of how much confidence we can place in them. For example, a statistical interval for the sample mean, $\overline{X}$, will enable us to put a *margin of error* (M.O.E.) around our estimate, $\overline{X} \pm$ M.O.E. This "plus or minus" gives us a sense of the "goodness" of the estimate since the length of this interval is a reflection of the uncertainty in our estimate. For the resistance data a confidence interval (we'll get to the definitions shortly) around the sample mean takes the form of 49.86 Ohms ±0.63 Ohms (or 49.23 Ohms, 50.49 Ohms). We are able to estimate the mean resistance of our population to about ±0.6 Ohms.

Although there are many different types of statistical intervals we have found the following three types to be very useful in engineering and science.

1. **Confidence Interval:** A confidence interval is ideal for quantifying the degree of uncertainty around common parameters of interest like the center of a sampled population, μ, or its spread, σ. Confidence intervals for the mean (μ) add a margin of error to where the center of the population is, and the confidence intervals for the standard deviation (σ) add a margin of error to the spread of the measurements around the center of the population, Yes, even the estimate of noise, σ, has noise in it and we have to account for it!

2. **Prediction Interval:** A prediction interval can be seen as an *enclosure* interval for one or more future observations, or an interval for the mean or standard deviation of a future sample from the sampled population. You'll be surprised to know that this type of interval is not commonly used, except in regression analysis. Since in engineering and science we are interested or concerned, depending on the kind of industry, in the future performance of the products that we design and make, this type of interval has important applications.

3. **Tolerance Interval:** A tolerance interval is an *enclosure* interval for a specified proportion of the sampled population. For example, we may want to determine lower and upper bounds such that 99% of the population is contained within them. This is a very useful interval for yield predictions since it puts bounds on the "yield," as given by a percentage, of the sampled population. These intervals can also be used to set up specification limits based on our sampled data. Remember this is an interval for a specified proportion of the sampled population *not* its mean or standard deviation.

While the equations for each type of interval depend on the type of interval and the quantity the interval is for, they all have similar ingredients. A statistical interval requires a sample estimate of the parameter of interest or a specified quantity of interest; an estimate of the noise in the sample; a confidence level; and a quantile from an appropriate distribution. The choice of distribution depends on the statistic, or quantity of interest, and the nature of the data.

2.6.1 Confidence Interval for the Mean

A *confidence interval* (we explain the meaning of confidence in Section 2.6.4) for the center of gravity, μ, is essentially $\overline{X}$ ± Margin-of-Error (M.O.E.). When the spread of the data, σ, is unknown, as is usually the case, the confidence interval is given by this formula:

$$\overline{X} \pm t_{1-\frac{\alpha}{2},n-1} \, s\sqrt{\frac{1}{n}}$$

In this case the sample estimate for the mean is $\overline{X}$, the estimate of noise (spread) in the sample is given by *s*, the sample size is given by *n*, the degree of confidence is given by the quantity $(1-\alpha)$, and $t_{1-\alpha/2,\, n-1}$ is a quantile from a Student's t distribution with $(n - 1)$ degrees of freedom. The Student's *t* distribution, Table 2.5, is a symmetric distribution that behaves like the normal distribution except that it depends on the numbers of observations in the sample through the parameter $(n - 1)$. The M.O.E. is given by $t_{1-\alpha/2,\, n-1} \times (s/\sqrt{n})$. For the resistance data, a 95% confidence interval for the center of the sampled population is $49.86 \pm 2.02\times(1.961 / \sqrt{40}) = 49.86 \pm 0.63$ = (49.23 Ohms, 50.49 Ohms).

2.6.2 Prediction Interval for One Future Observation

A *prediction interval* for a single future observation from the same sampled population has this form:

$$\overline{X} \pm t_{1-\frac{\alpha}{2},n-1} \, s\sqrt{1+\frac{1}{n}}$$

Note that this interval is almost identical to the confidence interval for the mean except that we are allowing for the uncertainty in the future value by the additional 1 inside the square root. As a consequence, for a given confidence level, a prediction interval for a single future observation is always wider than a confidence interval for the mean.

2.6.3 Tolerance Interval to Contain a Given Proportion *p* of the Sampled Population

A *tolerance interval* to contain a given proportion *p* of the sampled population, with a given confidence (1-α) is given by this formula:

$$\overline{X} \pm g_{(1-\alpha/2,p,n)}\, s$$

This is similar to the confidence interval for the mean, but the t-quantile is replaced by the function g, which is a function of the degree of confidence, (1−α), the proportion, *p*, that the interval is going to contain, and the sample size *n*. In other words, the margin-of-error M.O.E. = $g_{(1-\alpha/2,\, p,\, n)} \times s$. The function $g_{(1-\alpha/2,\, p,\, n)}$ is somewhat complicated for hand calculations (although tables are available), so it is better to let JMP evaluate it for us.

Prediction and tolerance intervals provide ways to characterize the output performance of a process by giving us a window where future measured performance values will fall, or a window where a large proportion (e.g., 95%, 99%) of the measured performance values will fall.

Luckily for us the calculations for these statistical intervals are readily available in JMP through the Distribution Platform. In Chapter 3, we will show you how to calculate these intervals in the context of a problem. However, to give you a feel for them Table 2.6a shows four statistical intervals for the Resistance data.

Table 2.6a Statistical Intervals for Resistance Data

Interval Type	Confidence Level	Lower Limit (Ohms)	Upper Limit (Ohms)
Confidence interval for μ, center of the sampled population	95%	49.23	50.49
Confidence interval for σ, spread around center of the population	95%	1.61	2.52
Prediction interval for one future observation	95%	45.84	53.88
Tolerance interval containing 99% of population	95%	43.55	56.16

We already mentioned that with a 95% degree of confidence the uncertainty window for the average resistance goes from 49.23 Ohms to 50.49 Ohms. On the other hand, we can say with a 95% degree of confidence that the spread in the resistance data can be as low as 1.61 Ohms, and as large as 2.52 Ohms. If we want to predict where the resistance of one future cable is going to fall, based on our sample of 40 cables, we are 95% confident that it will be between 45.84 Ohms and 53.88 Ohms. Note how this interval is much wider that the confidence interval for the mean (8 Ohms vs. 1.25 Ohms). Finally, based on our sample of 40 cables, we expected that 99% of the resistance values are going to fall between 43.56 Ohms and 56.16 Ohms.

Statistics Note 2.1: Note that the natural process variation of 43.48 Ohms to 56.24 Ohms, defined by the individuals and moving range chart in Figure 2.17b, is a very good approximation for the 99% tolerance interval 43.56 Ohms to 56.16 Ohms. The natural process variation can then be used to predict where future observations generated by the process should fall. In other words, if the process remains stable we are willing to bet money that successive resistance measurements will fall between 43.48 Ohms to 56.24 Ohms.

JMP 8 Note: JMP 8 gives us the ability to compute a prediction interval not only for one future observation, but also for *k* future observations. Table 2.6b shows the 95% prediction interval for a batch of 10 future cables. This is a simultaneous interval in the sense that we expect the resistance of *ALL* 10 future cables to fall within 43.95 Ohms and 55.77 Ohms. This interval is useful when we are making claims about small batches.

Table 2.6b Prediction Interval for 10 Future Resistance Values

Interval Type	Confidence Level	Lower Limit (Ohms)	Upper Limit (Ohms)
Prediction interval for 10 future observations	95%	43.95	55.77

The length of a given interval influences how much credence we want to place in our results, i.e., wider intervals reflect a higher degree of uncertainty. Is it possible to control this length? The length of these intervals depends on several factors, including the size of our sample, the spread (noise) in our data, and the degree of confidence ($1 - \alpha$) we choose. Let's look at the impact the sample size and the degree of confidence have on a confidence interval for the mean. It is not difficult to see that the more samples we have, the less the uncertainty, and the closer we will get to the true values of the distribution. For given estimates of the sample mean, $\overline{X}$, and spread, s, the M.O.E. = $t_{1-\alpha/2,\, n-1} \times (s/\sqrt{n})$, is a function of the degree of confidence, through (1−α/2), and the spread, *s*, in the data. From this formula it is easy to see that the larger the spread in the data, the wider the interval. This is intuitive because the more noise there is in the data, the less we can trust the results. It is easier to visualize the impact of the degree of confidence on the t-distribution quantile, which gets larger the more "confident" we want to be, by means of a graph.

In Figure 2.21 we can clearly see that as we increase our degree of confidence, the M.O.E. increases, which results in a wider interval. For example, holding the sample size constant at 10, as we increase the degree of confidence from 80% to 99% our M.O.E goes from 0.86 Ohms to 2.02 Ohms, more than a 200% increase (the price of confidence). On the other hand, if we hold our degree of confidence fixed at 95%, and increase our sample size from 10 to 100, our M.O.E. decreases from 1.40 down to 0.39 (the gain from a larger sample size). It is worth noting from Figure 2.21 that there is a point of diminishing returns with respect to an increase in sample size. The biggest reduction in the M.O.E. occurs for sample size increases between 10 and 50. However, increasing our sample sizes to 70 and beyond has a minimal reduction on our margin of error.

Figure 2.21 Impact of Sample Size and Degree of Confidence on M.O.E.

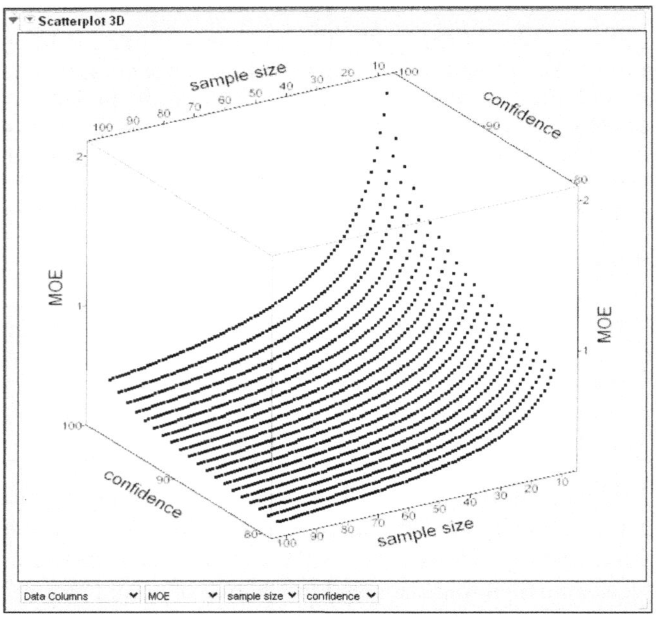

2.6.4 What Does Confidence Level Mean?

Unfortunately for practitioners and statisticians, the concept of confidence level is one of the most (if not the most) misunderstood concept in statistics. Since we are quantifying uncertainty in our estimates and we are casting a "net" around them, it is very, very, very (did we write very?) tempting to say that the confidence level is the same as the probability that the calculated interval contains the population parameter of interest. This is intuitive but, we are sorry to say, wrong. After the interval has been calculated the probability that it contains the population parameter of interest is 1 or 0. That is, it either contains the population parameter or it doesn't. Surprised? You're not alone.

So, what does it really mean to have a confidence level of 95%, say? The *degree of confidence* refers to the confidence we have in the procedure of making these "nets." It refers to the yield of this "net" making process. Imagine, if you will, that you are a process engineer in a manufacturing line that makes statistical "nets" (intervals) for capturing, or enclosing, the population mean being estimated from a set of data. A work order comes down for a 95%, the biggest seller, confidence interval. The standard operating procedure (or "recipe") for making the 95% confidence interval calls for the following:

- the sample mean, $\overline{X}$
- the sample standard deviation, s
- the sample size
- a 95% confidence level
- a quantile from the t-distribution for the given sample size and confidence level

What is particular (very particular) about your process is that the confidence level determines the *yield* that your process is going to have on average. So when you are making a 95% confidence interval the yield of your process is, on average, 95%. What does this mean and how does it relate to the confidence interval? What it means is that, on average, 95% of all the statistical nets produced by your process work. That is, these statistical nets are going to capture the true population mean, and 5% of them, on average, are known to fail, meaning that they do *not* capture the true population mean. But once you make this statistical "net" it either works or it doesn't, just like any product you may buy. In other words, *before* you make the 95% confidence interval there is a 95% chance that your statistical "net" is going to work, capture the true population mean, but *after* it was made it either captured the true population mean or it did not.

> *It's just like a game of poker: before the hand is dealt you have a 42.3% chance of getting one pair, but after you get the hand you either have the pair or you don't!*

Some of you may be thinking that since we don't know the true mean (or any other parameter) of the population then, unless we sample the whole population, there is no way for us to know whether our statistical "net" works; i.e., there is no way to know whether the interval contains the true mean or not. And yes, that is true, but we can use simulations to generate multiple samples from a population with known parameters, compute a 95% confidence interval for the mean for each sample, and see how many of these intervals contain the mean. Figure 2.22 shows a simulation of 100 samples, each of size 50, taken from a population of resistance measurements with a true average of 50 Ohms, and a true standard deviation of 2 Ohms. You can see that six out of the 100 intervals, 6%, fail to capture the true 50 Ohms mean. The other 94 did. In other words, this simulation shows that the 95% confidence-interval-making process had a yield of 94%, which is close to the stated 95% confidence level. When you run the JMP script, these intervals will be the ones shown in red in the JMP output.

Figure 2.22 95% Confidence Intervals for Random Samples of Size 50

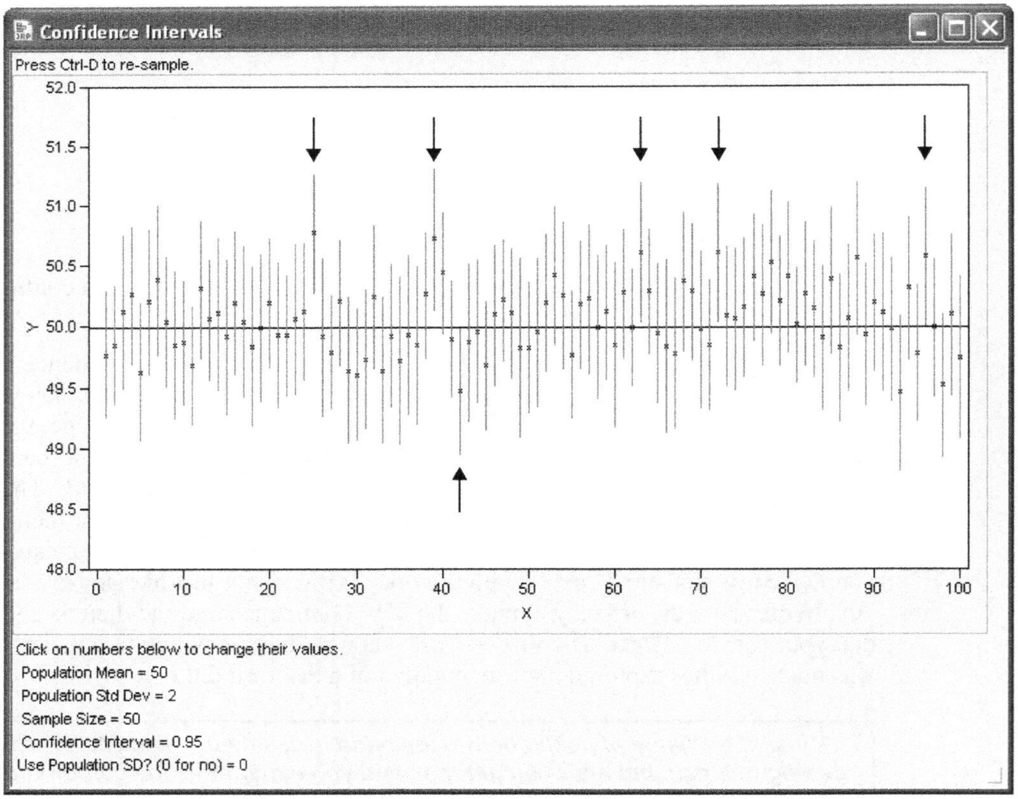

The graph in Figure 2.22 was generated using the Confidence.jsl script that is located in C:\Program Files\SAS\JMP7\Support Files English\Sample Scripts. You can open this script file from the **File > Open** menu. Make sure you select the **JSL file type** as shown in Figure 2.23. Once the script opens you can run it by pressing **CTRL+R**.

Figure 2.23 Opening the Confidence.jsl Script

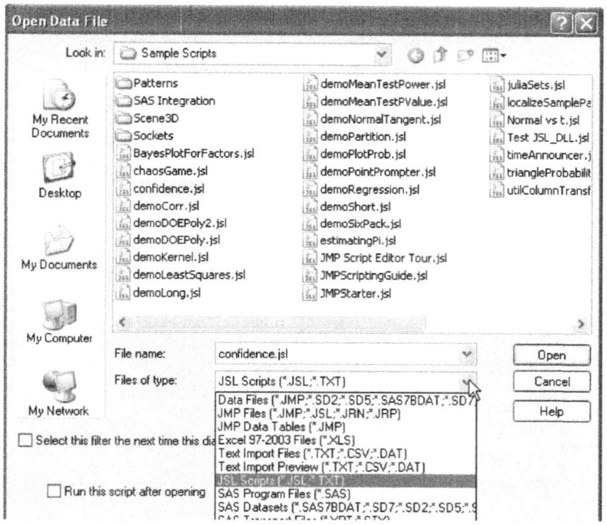

Once this is done the simulation window will come up (Figure 2.24) where you can specify the population mean (50 Ohms), the population standard deviation (2 Ohms), the sample size (50 components), and the confidence level (95%), in other words, the first four ingredients in the "confidence interval for the mean" recipe (the fifth ingredient, t-quantile, is calculated by JMP). You can change the values by clicking on the numbers. By pressing **CRTL+D** repeatedly new samples are generated enabling you to see that the "yield" of this process is, on average, 95%.

Figure 2.24 Simulation Window for Confidence Intervals for the Mean

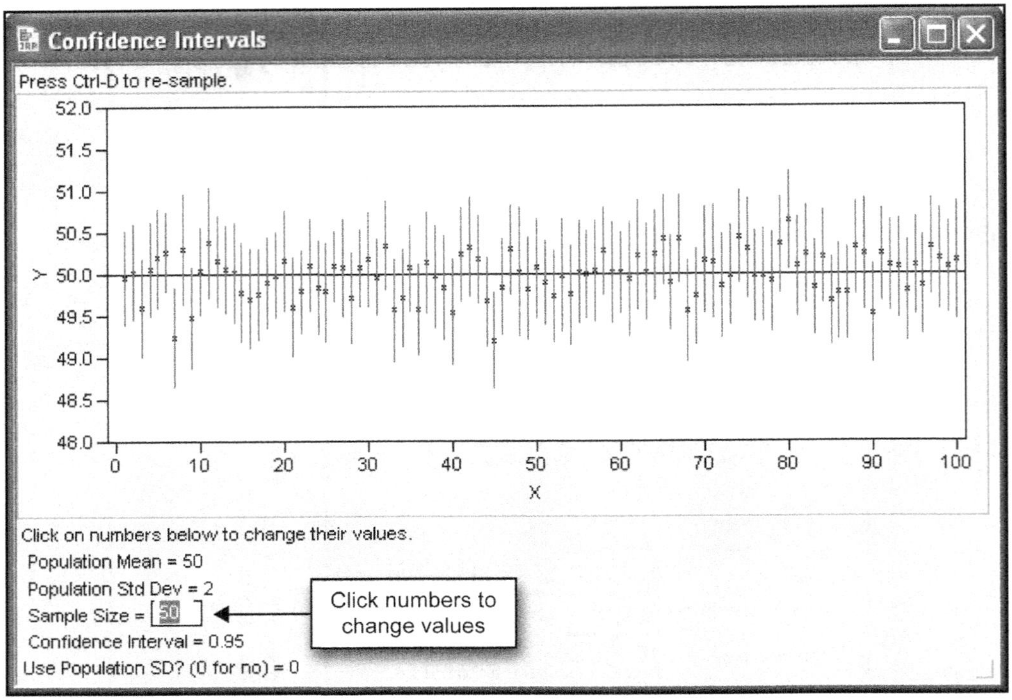

2.7 Overview of Tests of Significance

Along with statistical models, statistical tests of significance have played a crucial role in helping answer the endless questions that arise in science and engineering. Tests of significance provide a sound methodology for testing hypothesis using data collected from designed or observational studies. As we shall see in Chapters 4, 5, 6, and 7, a statistical test of significance provides a way of demonstrating (with data) a significant discrepancy from a given hypothesis. In other words, if we believe that our hypothesis is true the test of significance demonstrates how uncommon or exceptional our data is. If there is a discrepancy between the data and our hypothesis, we reject the hypothesis in favor of an alternative that is likely to be true.

There are different types of statistical tests of significance, some of which we will cover in this book. Table 2.7 describes the type of hypothesis the statistical test is designed for, a description of the statistical test, and the chapter in which it is covered.

Table 2.7 Different Types of Statistical Tests of Significance

Test of Significance	Description	Chapter
One sample t-test and Chi-square test	Used to test the performance of a single process, product, or material against a specified standard value. We can use a one-sample test to qualify a new piece of equipment to ensure that its output is hitting a specified target value.	4
Two Sample t-test and F-test	Used to determine whether the performances of two processes, products, or materials are different with respect to their means or variances. We can use a two-sample test to select a supplier of a raw material for which a key attribute has less performance variation.	5
Paired t-test	Used for making "before" and "after" comparisons by sampling the *same* experimental unit under each condition. A paired t-test can be used to determine whether two different non-destructive test instruments are providing similar results by measuring the same parts on both instruments.	5
ANOVA (analysis of variance) F-test	Used to determine whether the performances of three or more processes, products, or materials are different in their mean or variance. For example, we can use ANOVA to determine which of several die types maximizes yield for a die cutting operation.	6
Linear Regression t-test / F-test	Used to study a linear relationship between a factor and a response when the factor is continuous in nature. For example, we can use linear regression to determine how increasing pressure on a laminator affects the bond strength of the materials being laminated together.	7

2.7.1 Critical Components of a Test of Significance

While each test of significance has its own formula and characteristics, which we will cover in subsequent chapters, there are several components that are common to all of them and deserve some discussion.

1. ***The Null Hypothesis, H_0*** — The null hypothesis is a statement about the population under study. It should be explicitly formulated in the initial stages of a study, and should relate to the questions that we are trying to answer. The null hypothesis establishes the frame of reference for the test and can never be proven. The study will provide data that may disprove the null hypothesis in favor of an alternative hypothesis.
2. ***The Alternative Hypothesis, H_1*** —This is what we prove to be true when we reject the null hypothesis (the straw man, argument) and give evidence in favor of the alternative hypothesis. Proving the alternative hypothesis means that we have enough evidence (proof) to demonstrate, or establish, the validity of claims in the alternative hypothesis.
3. ***Level of Significance (α risk)*** — Tests of significance are based on calculating the probability, based on the null hypothesis, of how rare the observed outcomes of the study are. The level of significance defines the probability threshold under which we are willing to admit that the alternative hypothesis is true. The level of significance is stated in advance before the data is collected and analyzed. A level of significance of 5% (0.05) is one of the most common values used in science and engineering. However, other values are sometimes chosen depending on the objective of the study.
4. ***Decision Making*** — If the calculated probability of observing the results, *p-value*, is smaller than the level of significance, then we have enough evidence provided by the data to conclude the following:

   a. The null hypothesis is false.

   b. The alternative hypothesis is true.

   If the calculated probability, *p-value*, is greater than the level of significance, then we do not have enough evidence to reject the null hypothesis. Therefore, we assume that the null hypothesis is true.

We must be careful, however, not to think that investigations in engineering and science are this clear cut, i.e., a null vs. an alternative hypothesis. Although tests of significance help us to, as Sir Ronald Fisher put it, "...ignore all results which fail to reach this standard (p-value), and, by this means, to eliminate from further discussion the greater part of fluctuations which chance causes have introduced into their experimental results..." (1937), the power of statistics also lies in its ability to serve as a catalyst for discovering unexpected results, and in helping us to generate new hypothesis.

It is also important to note that the lack of rejection of the null hypothesis does not prove the null hypothesis. To quote Sir Ronald Fisher again:

> For the logical fallacy of believing that a hypothesis has been proved to be true, merely because it is not contradicted by the available facts, has no more right to insinuate itself in statistical than in other types of scientific reasoning . . . it would therefore, add greatly to the clarity with which the tests of significance are regarded if it were generally understood that tests of significance, when used accurately, are capable of rejecting or invalidating hypotheses, in so far as they are contradicted by the data: but that they are *never capable of* establishing them as certainly true. (1937; emphasis added)

Statistics Note 2.2: The null hypothesis (H_0) is set up in a way that a significant result provides a way of demonstrating a significant discrepancy from it. There are situations, however, where the goal of the study is to gain some assurance that the performance of two drugs, for example, do not differ by much—in other words, that they are equivalent. In Chapters 4, 5, and 6 we will give examples of such tests.

2.7.2 A 7-Step Framework for Statistical Studies

As we mentioned in Section 2.1 we believe it is important to be methodical when solving problems, answering questions, or driving improvements. And, even though a common or standard methodology does not exist for all scientific and engineering investigations, we believe it is helpful to have a framework (Table 2.8), or step-by-step sequence, to aid our efforts. We suggest the following 7-step framework to focus our attention to the key things that we need to accomplish for a successful implementation of the statistical techniques presented in this book. In each of the following chapters we will introduce real life examples from science and engineering, and we will follow most (if not all depending on the application) of these steps to solve the given problem. We regularly follow these steps in our collaborations with scientists and engineers, so much so that they have become second nature. Note that each step is somewhat dependent on the previous one. The scientific and statistical hypothesis that will guide our investigations need to be derived from the uncertainties that we are trying to solve.

Table 2.8 7-Step Framework for Statistical Studies

Step	Objectives
1. Clearly state the question or uncertainty.	Make sure that we are attempting to answer the right question with the right data.
2. Specify the hypotheses of interest.	What claims do we want to make? Translate scientific hypothesis into statistical hypothesis.
3. Determine the appropriate sampling plan, and collect the data.	Identify how many samples, experimental units, and observational units are needed. State the significance level (α), i.e., the level of risk we are willing to take.
4. Prepare the data for analysis, and conduct exploratory data analysis.	Plot the data in different ways. Look for outliers and possible data entry mistakes. Do a preliminary check of assumptions.
5. Perform the analysis to verify your hypotheses and to answer the questions of interest.	Are the observed differences (discrepancies between the data and our hypothesis) real or due to chance? We want to be able to detect these differences and estimate their size. Check the analysis assumptions to make sure your results are valid.
6. Summarize the results with key graphs and summary statistics.	Find the best graphical representations of the results that give us insight into our uncertainty. Select key output for reports and presentations.
7. Interpret the results and make recommendations.	Translate the statistical jargon into the problem context. Assess the practical significance of your findings.

These steps are related to the five-stage model Problem, Plan, Data, Analysis, and Conclusions (PPDAC) of MacKay and Oldford (2000). Steps 1 and 2 correspond to the problem definition, step 3 to the plan, step 4 to the data, steps 5 and 6 to the analysis, and step 7 to the conclusion. As MacKay and Oldford point out, "A structure for statistical method is useful in two ways: first to provide a template for actively using empirical investigation and, second, to critically review completed studies" (2000). The second point is the cornerstone of good science and engineering: the ability to critically review our studies and those of others.

2.8 Summary

In this chapter we discussed how the ideas, concepts, and methods of statistics are a catalyst for solving common problems encountered by engineers and scientists. The four major areas in the use of statistics: descriptive statistics, probability models, statistical inference, and the assumption of homogeneity make it possible to learn in the presence of variation and uncertainty and, when coupled with the scientific method, lead to the generation of new knowledge.

Statistical thinking focuses our attention and helps us to recognize that all work occurs in a system of interconnected processes, that variation exists in everything we do, and that understanding and reducing this variation is a key for success. At the system level we pay attention to the inputs to our processes, the process knobs that we can control, the noise factors that affect our processes but we cannot control, and the outputs and their characteristics produced by the process.

We also introduced a 7-step framework to help focus our attention to the key things that we need to accomplish for a successful implementation of the statistical techniques presented in this book.

2.9 References

Box, G.E.P., J.S. Hunter, and W.G. Hunter. 2005. Statistics for Experimenters: Design, Innovation, and Discovery. 2d ed. New York, NY: John Wiley & Sons.

Fisher, Sir R. A. 1929. "The statistical method in psychical research." *Proceedings of the Society for Psychical Research* 39: 189–192.

Fisher, Sir R.A. 1937. *The Design of Experiments*. 2d ed., Edinburgh: Oliver & Boyd.

MacKay R.J., and R.W Oldford. 2000. "Scientific Method, Statistical Method and the Speed of Light." *Statistical Science* 15:254–278.

McLaren C.G., and M. Carrasco-Aquino. 1998. "The Role of Statistics in Scientific Endeavor." *Stats* 21: 3–7.

Salsburg, David. 2001. *The Lady Tasting Tea: How Statistics Revolutionized Science in the Twentieth Century.* New York, NY: W.H. Freeman/Owl Book.

Thomson, Sir William. 1889. "Electrical Units of Measurement." *Popular Lectures and Addresses*, Vol. I, 2d ed. London: Macmillan and Co.

Velleman, P.F., and L. Wilkinson. 1993. "Nominal, Ordinal, Interval, and Ratio Typologies Are Misleading." *The American Statistician* 41:65–72.

Wheeler, D.J. 2005. *The Six Sigma Practitioner's Guide to Data Analysis*. Knoxville, TN: SPC Press.

Chapter 3

Characterizing the Measured Performance of a Material, Process, or Product

Chapter Goals

- Become familiar with the terminology and concepts needed to characterize the measured performance of a material, product, or process in terms of location and spread.
- Carry out an exploratory data analysis (EDA) using the **Analyze** and **Graph** platforms in JMP to generate descriptive statistics, statistical intervals, and graphs.
- Learn how to identify sources of variation in the data through stratification factors.
- Translate JMP statistical output in the context of a problem statement.
- Present the findings with the appropriate graphs and summary statistics.

3.1 Problem Description

What do we know about it?
You are a quality engineer for an injection molding operation. You recently learned that one of your suppliers is no longer supplying you with a key raw material that is used to make sockets. You were able to identify a second source for this raw material, and in looking over some preliminary performance data supplied by them, it appears that their raw materials should meet the incoming specification requirements. However, because of requirements set up with your customers, you still need to characterize and qualify this raw material substitution in your manufacturing processes to make sure that the final product performance is not compromised.

What is the question or uncertainty to be answered?
The injection molding operation is the manufacturing step that uses this particular raw material and is the focus of the qualification. The injection molder has four cavities and is able to produce four sockets at a time. The business and the customers have no issue with this new raw material as long as the sockets perform the same, or better, as the current supplier against the fitness-for-use criteria.

How do we get started?
A characterization and qualification of this new raw material in the injection molding process must be carried out. In order to get started, we need to do the following:

1. Use statistics and graphs to characterize the performance of the thickness measurements.
2. Be able to translate the statistical findings in a way that clearly conveys the results to our business partners and customers without getting lost in the statistical jargon.

The most important attribute of the socket is its thickness, which should be around 12 millimeters. The measured quantity is the excess thickness from 12 mm, which has a lower specification limit (LSL) equal to 0 mm, and an upper specification limit (USL) equal to 15 hundredths millimeters. The business wants to make sure that there is no additional yield loss associated with this new raw material. The process capability using the current raw materials is $C_{pk} = 1$ (see Table 3.2 for the C_{pk} formula).

What is an appropriate statistical technique to use?
In Section 3.3 we discuss the importance of exploratory data analysis (EDA) as a way of gaining an understanding of the data before statistical inferences can be made, and we review the key concepts and tools that are relevant to EDA, such as the importance of stratification, descriptive statistics, and graphs. In Section 3.4 we introduce a step-by-step approach for characterizing the measured performance of a material, process, or product using JMP in order to qualify the second source supplier for the sockets.

3.2 Key Questions, Concepts, and Tools

Many of the statistical concepts and tools introduced in Chapter 2 are relevant when characterizing the measured performance of a material, process, or product. Table 3.1 outlines these concepts and tools, and the Problem Applicability column helps to relate these to the characterization of the new supplier for the socket manufacturer.

Table 3.1 Key Questions, Concepts, and Tools

Key Questions	Key Concepts	Problem Applicability	Key Tools
Is our sample meaningful?	Inference from a representative sample to a population	One of the most important concepts in statistics includes selecting a representative sample from our population. When characterizing the raw material substitution used to make sockets, we need to make sure that we identify the most common sources of variation in the injection molding process and include them in the sampling plan. This will help us make a better inference into the future performance of this process under the new conditions.	Sampling schemes, experimental and observational units, degrees of freedom, and sources of variation.
Are we using the right techniques?	Probability model using the normal distribution	All of the techniques used in this book rely upon the assumption that the data can be approximated using a normal distribution. In particular, the confidence interval for the mean and the tolerance interval rely upon this assumption. Graphical methods are useful to check this assumption for the socket characterization.	Normal quantile plots to check distributional assumptions.
What risk can we live with?	Decision theory	Whenever we calculate a statistical interval, we must specify the level of risk that we want to use in the calculations. Equivalently, the flip side of risk is the confidence level. For the qualification, the most popular choice for α (= 0.05) should suffice.	Type I and Type II errors, statistical power and confidence.
Are we answering the right question?	Exploratory data analysis	The socket qualification will characterize the performance of the new supplier under normal operating conditions, including typical sources of variation and will most likely yield a large data set. Many of the tools described in Chapter 2 are helpful to estimate how well the sockets perform against the specification limits and understand our largest sources of variation in the process.	Descriptive statistics, graphs with stratification, variance components, and statistical intervals.

(continued)

Table 3.1 (*continued*)

Key Questions	Key Concepts	Problem Applicability	Key Tools
How do we best communicate the results?	Visualizing the results	Exploratory data analysis can result in lots of graphs and statistics. For this study, we will pick out the critical few that best tell the story of why we will or will not qualify the injection molding process with a new raw material supplier.	Variabity charts, histograms overlaid with specifications, and tolerance intervals.

3.3 Overview of Exploratory Data Analysis

3.3.1 Description and Some Applications

As the engineer and physicist Theodore von Kármán aptly put it, "Scientists discover the world that exists; engineers create the world that never was." As engineers and scientists we observe and try to understand the different phenomena that surround us, and the different systems and manufacturing process that we create. We collect representative samples and, with statistics as a catalyst for discovery and knowledge generation, we try to make inferences about either populations in the past, or, more importantly, populations in the future that have not been observed or created. When we use inductive reasoning and statistics to extrapolate from the known (sample) to the unknown (population), and make claims about the population from which the sample came, many assumptions are required for these claims to be meaningful, applicable, and valid. For example, if we want to find prediction limits with a certain confidence level within which the thickness of five future parts will fall, we need to make sure that our sample is taken from a homogeneous population.

In this journey of discovery, questions, not answers, make science the ultimate adventure (Greene 2009). It seems that we are always trying to test different hypotheses about the phenomena we are observing. Where do these hypotheses come from? Can statistics help us with their generation? Professor John Tukey "expounded a practical philosophy of data analysis that minimizes prior assumptions and thus allows the data to guide the choice of appropriate models" (Velleman and Hoaglin 1981). Exploratory data analysis (EDA) lets the data reveal its underlying structure helping us to do the following:

1. Suggest plausible hypothesis about the phenomena at hand, which can later be tested with statistical tests of significance.
2. Guide the selection of appropriate models to describe the phenomena and relationships.
3. Evaluate the assumptions that might be needed for valid statistical inferences.

The most insight into the engineering and science behind our study is gained by carrying out an exploratory data analysis (EDA) on the sampled data from a population or populations. The EDA philosophy is predicated on our ability to carry out good detective work by iterating through many cycles of questions and answers as we uncover interesting patterns in the data. This is done most effectively using a combination of the most meaningful descriptive statistics and insightful graphs and visualization tools that enable us to stratify and slice-and-dice our data in order to uncover interesting or unexpected features in the data, such as relationships and sources of variation.

Exploratory data analysis is the springboard for three of the four different ways of using statistics that we introduced in Section 2.1, namely descriptive statistics, probability models, and statistical inference. Note that the fourth way, checking for homogeneity, is part of the EDA phase.

The ideas presented in this chapter are the foundation of the remaining chapters in this book. Even when more formal tests of significance are being carried out in a study, every analysis should include a thorough characterization of the data and good detective work using EDA. In the remainder of this chapter, we will briefly review some of the concepts and tools that we are using to characterize the new raw material on the injection molding process.

3.3.2 Descriptive Statistics

Some popular descriptive statistics were introduced in Table 2.3 in Chapter 2. The ones that we are using in the socket example are reproduced here in Table 3.2.

Table 3.2 Descriptive Statistics for Center, Spread, and Center and Spread

Summary Statistic	Description	How to Calculate It	JMP Menu
Mean	The average of a set of data; its center of gravity. This estimate is usually affected by outliers (unusual values).	Add all the measurements and divide by the number of samples: $\bar{X} = \frac{\sum_{i=1}^{n} X_i}{n}$	**Analyze > Distribution**
Median	The value that divides the ranked data (from smallest to largest) in half. It is also referred to as the 50th percentile of the data. The median is robust to outliers.	Order the data from smallest (1) to largest (n), $x_{(1)}, x_{(2)}, x_{(3)}, \ldots, x_{(n)}$. For n even, average the ordered values at positions n/2 + (n/2+1). For n odd, it is the ordered value at (n–1)/2 + 1.	**Analyze > Distribution**
5-Number Summary	The extreme values (min and max), the median, and the 1st (25%) and 3rd (75%) quartiles.	See Table 2.3 in Chapter 2.	**Analyze > Distribution**
Standard Deviation	Square root of the average squared distances from the data points to the mean. It represents the radius of rotation around the mean. The interval mean ±3s usually covers more than 99% of the data.	$s = \sqrt{\frac{\sum_{i=1}^{n}(X_i - \bar{X})^2}{n-1}}$	**Analyze > Distribution**
DNS	Distance to nearest specification in standard deviation units. Describes how the process is centered relative to the specifications.	$DNS = \min\{\frac{\bar{X} - LSL}{s}, \frac{USL - \bar{X}}{s}\}$	**Formula Editor**
C_{pk}	Process capability index. It normalizes the distance of the nearest specification limit (Lower Specification Limit or Upper Specification Limit) to the mean by three standard deviations.	$\frac{DNS}{3} = \min\{\frac{\bar{X} - LSL}{3s}, \frac{USL - \bar{X}}{3s}\}$	**Analyze > Distribution > Capability**

The mean and median are measures of location, that is, where the center of the data is. If the distribution is symmetric, then estimates for the median and the mean should be very close in value. However, the median provides a more robust estimate of the center of the data when outliers are present. The *5-Number summary* provides a quick summary of the data and is the basis for the box plot (Section 3.3.3). The standard deviation, in the units of the data, represents the average deviation of the data from the mean or center of the distribution. In addition to quantifying the noise in the measured performance, we can also use the standard deviation to provide rough limits that contain a certain percentage of the total population: for a Normal distribution, the limits are as follows:

- Approximately 50.0% of population will fall within $\overline{X} \pm 0.67s$
- Approximately 68.3% of population will fall within $\overline{X} \pm 1s$
- Approximately 95.5% of population will fall within $\overline{X} \pm 2s$
- Approximately 99.7% of population will fall within $\overline{X} \pm 3s$

Although these approximations work well for normally distributed data, a more general empirical rule (for an example, see Wheeler and Chambers 1992, p. 61) is given by the following:

- Approximately 60% to 75% of the data will fall within $\overline{X} \pm 1s$
- Approximately 90% to 98% of the data will fall within $\overline{X} \pm 2s$
- Approximately 99% to 100% of the data will fall within $\overline{X} \pm 3s$

Figure 3.1a shows the mean, standard deviation, median, and the 5-Number summary for 60 observations taken from a Normal population. The histogram reveals a unimodal distribution for which the mean = –0.9894 is slightly lower than the median = –0.869, and the estimated standard deviation is 1.4565. The 5-Number summary is displayed in the **Quantiles** section of the output as follows:

- Maximum = 1.752
- 75% quartile = 0.158
- Median = –0.869
- 25% quartile = –2.164
- Minimum = –4.092

The 5-Number summary quickly gives us a sense of where the extremes are, where the center of the data lies, and where 50% of the population occurs. In this example 50% of the data falls within the 25% and 75% quartiles or between [–2.164; 0.158]. We can compare this interval to the theoretical interval based on the normal distribution; that is, the one given by

$$\overline{X} \pm 0.67s = [-0.9894 - 0.67 \times 1.4565401; -0.9894 + 0.67 \times 1.4565401] = [-1.972; 0.007]$$

to see that it is close to the 5-Number summary estimate.

Figure 3.1a Basic Descriptive Statistics

Distributions

Response

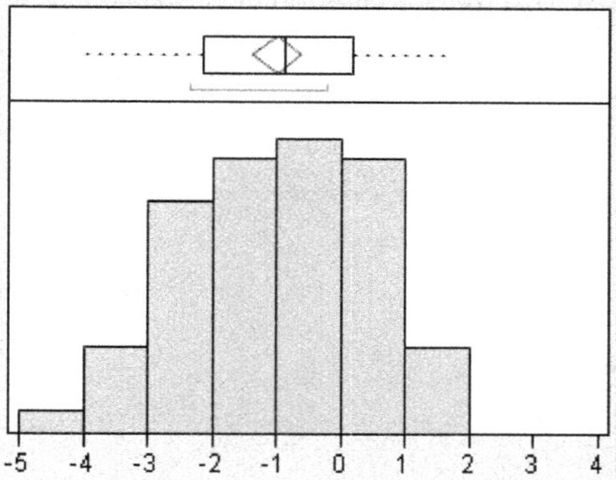

Quantiles

100.0%	maximum	1.752
99.5%		1.752
97.5%		1.713
90.0%		0.972
75.0%	quartile	0.158
50.0%	median	-0.869
25.0%	quartile	-2.164
10.0%		-2.871
2.5%		-3.916
0.5%		-4.092
0.0%	minimum	-4.092

Moments

Mean	-0.989447
Std Dev	1.4565401
Std Err Mean	0.1880385
Upper 95% Mean	-0.613183
Lower 95% Mean	-1.365711
N	60

The process capability index, C_{pk}, is a measure of process capability that standardizes the distance from the mean of the process to the closest specification limit, by three times an estimate of the standard deviation (variation). In other words, the C_{pk} is one-third the distance to the nearest specification (DNS). The farther the process average is from the specifications limits in standard deviation units, the larger the DNS and the C_{pk}, and the more "elbow room" the process has to cope with upsets. For a given set of specifications, think of the specification window (LSL – USL) as the width of a garage door and the process output (its distribution) as a car we are trying to park. A $C_{pk} = 1$ (DNS=3) indicates that the width of the car is almost the width of the garage; any deviation from the center will scratch the sides of the car! It is desirable then, to have large values of C_{pk}. A $C_{pk} \geq 1.5$ is considered good, in the sense that the car has enough room on each side and it will not get scratched even if we deviate a little from the center of the garage. In other words, a $C_{pk} \geq 1.5$ enables the process enough room to shift before making lots of out-of-specification material.

For the data in Figure 3.1a, assume that the LSL = –3.5 and the USL = 3.5. The distance to the nearest specification (the LSL in this case) is DNS = [(–0.9894) – (–3.5)]/(1.4565401) = 1.7237, indicating that the distance from the average to the LSL is 1.7237 standard deviations units. This gives a C_{pk} = 1.7237/3 = 0.575. Figure 3.1b shows the estimated C_{pk} = 0.575, along with actual and estimated percentage outside the specification limits in the **Capability Analysis** section of the output.

Another measure of process performance is its yield, or the percent of the population that falls within the specification limits. This can be calculated by counting the number of data points that fall within the specification limits, and then dividing this number by the total number of observations. However, when our sample size is small this might not provide a reasonable estimate. Even though the C_{pk} = 0.575, the actual number of measured responses that fall above the USL = 3.5 is 0.

Figure 3.1b C_{pk} and Estimate of Percent Out-of-Spec

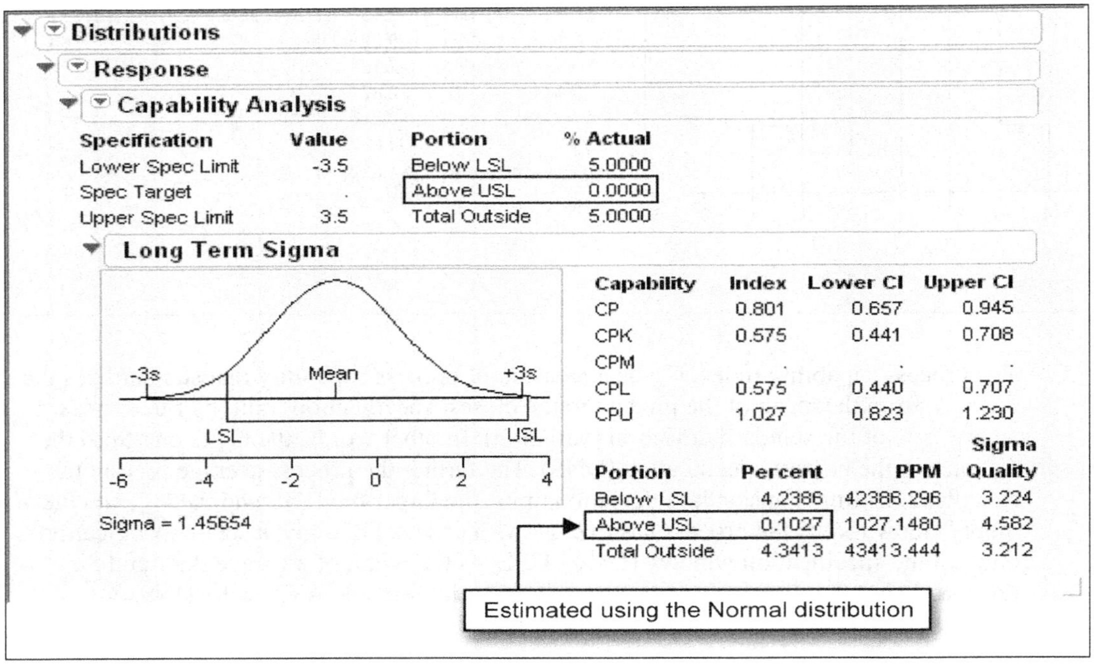

As we show in Chapter 2, we can use the sample mean and standard deviation, along with the normal distribution, to estimate the process yield by

- calculating the Z-scores $Z_{LSL} = (\overline{X} - LSL) / s$ and $Z_{USL} = (USL - \overline{X}) / s$
- using the normal distribution to calculate $Prob(Y \leq Z_{LSL})$ and $Prob(Y \geq Z_{USL})$
- adding $Prob(Y \leq Z_{LSL}) + Prob(Y \geq Z_{USL})$ to give the estimated proportion outside the specifications
- subtracting this sum from $1-[Prob(Y \leq Z_{LSL}) + Prob(Y \geq Z_{USL})]$ = process yield

The above equations are used in the **Capability Analysis** section of the output in Figure 3.1b. We can see that the predicted percent of values that are below the LSL is 4.2386% and the predicted percentage above the USL is 0.1027%. This estimate can be used to determine how many measurements in the random sample of 60 we should expect to be above the USL as follows: 0.001027*60 = 0.0616. Even though the difference between the actual and the estimate is small, it is still helpful to use a probability model to describe the underlying population from which the sample was taken, and to make predictions about future samples.

The capability analysis for the entire sample in Figure 3.1b has a $C_{pk} = 0.575$, which indicates that the process does not have enough "elbow room." Are things really as bad as they seem? As we mentioned before, within the EDA framework it is important to look at the data and calculate descriptive statistics in different ways. Sometimes we might not be aware that more than one population exists in our sample. Stratification factors (variables that enable us to separate the data into meaningful categories) are key in identifying different populations.

Consider the two distributions and capability analysis shown in Figure 3.2. The distribution representing the data from Stratum 1 has $C_{pk} = 1.041$ with 0.0861% expected to be below the LSL = −3.5. The distribution for Stratum 2 has $C_{pk} = 0.441$ with 9.2744% expected to below the LSL = −3.5, which has considerably worse performance than Stratum 1. Right away we see that the performance we saw in Figure 3.1b is mainly due to Stratum 2. Since the two means are almost a whole unit apart, we might conclude that the two strata reflect data generated from two different populations. The analysis in Figure 3.2 suggests that we should concentrate on understanding the differences between the two strata and improve the process represented by Stratum 2. If we do not stratify the data, as we did in Figure 3.1b, we might miss opportunities to improve our products and processes.

Figure 3.2 Capability Analysis by Stratum

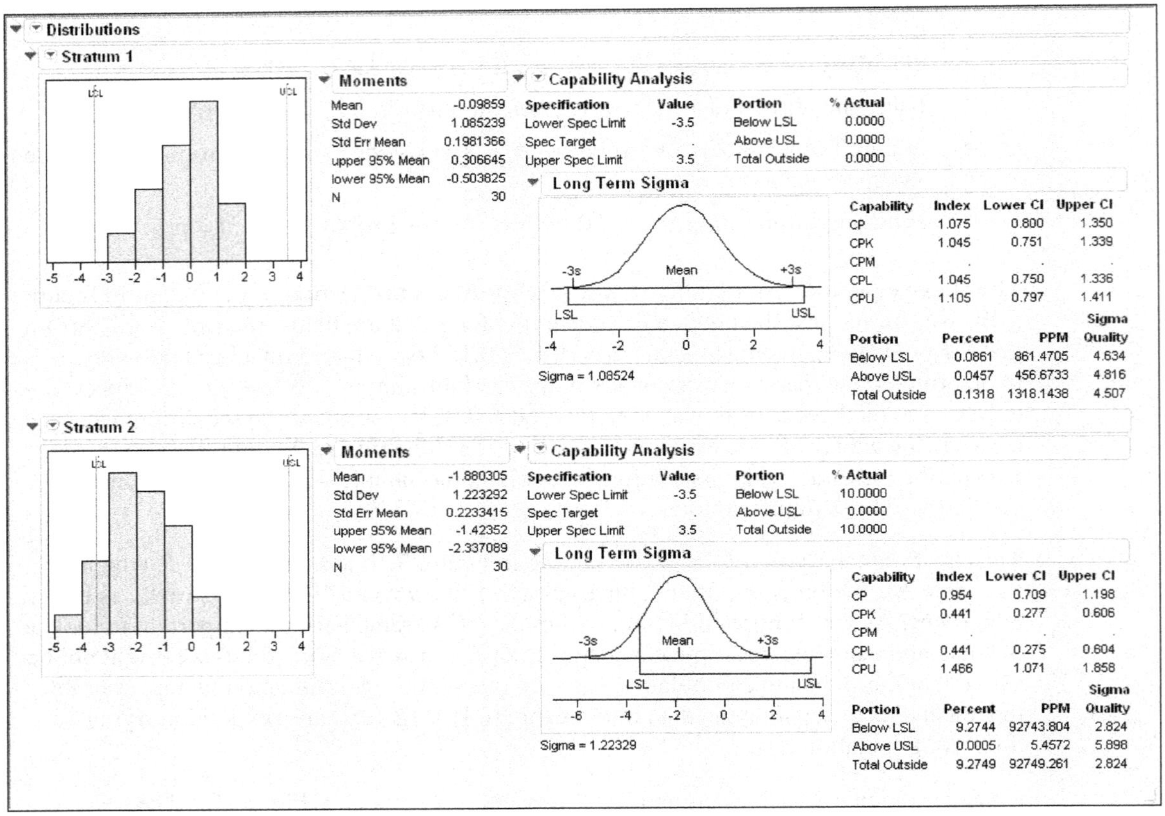

Statistics Note 3.1: In the literature you will find the terms C_{pk}, process capability ratio, and P_{pk}, process performance ratio. The difference between the two resides in the estimate of the standard deviation, or sigma, used for the calculation. The C_{pk} uses an estimate of sigma based on the short-term variation coming from a control chart, usually $\overline{R}/d_2$, where $\overline{R}$ is the average range and d_2 is a correction factor depending on the subgroup size. The ratio P_{pu} uses an estimate of sigma based on the long-term variation using the standard deviation formula in Table 3.1. For a stable or predictable process the $C_{pk} \cong P_{pk}$, but if the process is not stable the P_{pk} is just an attempt to characterize the *past performance* of an out-of-control process. Care must be taken to understand that a P_{pk} is not a "capability value" but a "(hypothetical) performance value" that is descriptive of the past but not predictive of the future (Wheeler and Chambers 1992). A process should be improved before making claims about its capability with respect to specifications. JMP calculates P_{pk}, but labels it C_{pk} by default. This can be changed in the preferences.

3.3.3 Graphs and Visualization Tools

Descriptive statistics are an important part in EDA to quantify the performance characteristics of interest, but they seldom give us a complete picture of our data. Appropriate graphs and visualization tools must be used to complement the summary statistics in order to fully characterize the material, product, or process. In Table 2.4 of Chapter 2 we introduced a number of useful graphs. The ones that are used in this chapter are provided again for reference in Table 3.3.

Table 3.3 Common Graphical Displays for Continuous Data

Graph Name	Graph Description	JMP Menu
Histogram	A frequency distribution of data that shows the shape, center, spread, and outliers in a data sample. When the histogram is overlaid with specification limits, we can see how much room the process has to move. The number of bars in the histogram should be approximately $1 + \text{Log}_2(\text{sample size})$.	• **Analyze > Distribution**
Normal Quantile Plot	A plot that has the normal quantile on the y-axis and the observed measurement on the x-axis. The normal quantile is determined by using the ranked order values of the data and the cumulative probabilities from a normal distribution to get the corresponding normal quantiles. This plot is useful to decide whether the data can be approximated using a normal distribution.	• **Analyze > Distribution**
Box Plot	A plot that shows the frequency distribution of the data using the *5-Number summary* (minimum, 1st quartile, median, 3rd quartile, maximum) of the data to show its center (median), spread (IQR), and outliers. Very useful for comparing more than one population of data or stratifying one population by different factors.	• **Graph > Variability/ Gauge Chart** • **Analyze > Fit Y by X**
Process Behavior Chart	An individuals and moving range chart is a plot that helps characterize the underlying system that generates our data, and that helps us evaluate if the data is homogeneous, which is a key assumption in many statistical techniques.	• **Graph > Control Chart > IR**

Histograms and Normal Quantile Plots

The primary performance goal, as was stated in Section 3.1 Problem Description, is to maintain the same, or better, process capability, as is measured by the performance capability ratio C_{pk}. A histogram of the sampled sockets thickness measurements, with overlaid specification limits, is a great way to visualize how well the distribution of values performs relative to the specifications. From this graph, we can see whether the thickness measurements are centered directly between the specification limits or are off-center, as well as any measurements that fall outside of these limits. Histograms can be created in JMP using the **Analyze > Distribution** platform, which also generates descriptive statistics, and enables us to calculate statistical intervals and capability measures.

A normal quantile plot can be used to verify the normality assumption, since statistical intervals and the tests of significance that we are covering in later chapters assume that the underlying population can be approximated with a normal distribution. This plot is also accessed from the **Analyze > Distribution** platform from within JMP by clicking the red triangle next to the histogram title. This plot is preferred over a histogram to assess the normality of the distribution. Figure 3.3a shows a histogram and normal quantile plot for a random sample of size 20 observations drawn from a normal distribution. As long as the points follow a straight line, and are inside of the outer bands in the normal quantile plot, we can assume that the normal distribution is a good approximation for describing the behavior of our data. Although the histogram in Figure 3.3a might seem bimodal, the normal quantile plot shows all the points following a straight line within the bands.

Figure 3.3a Normal Quantile Plot to Check for Normality

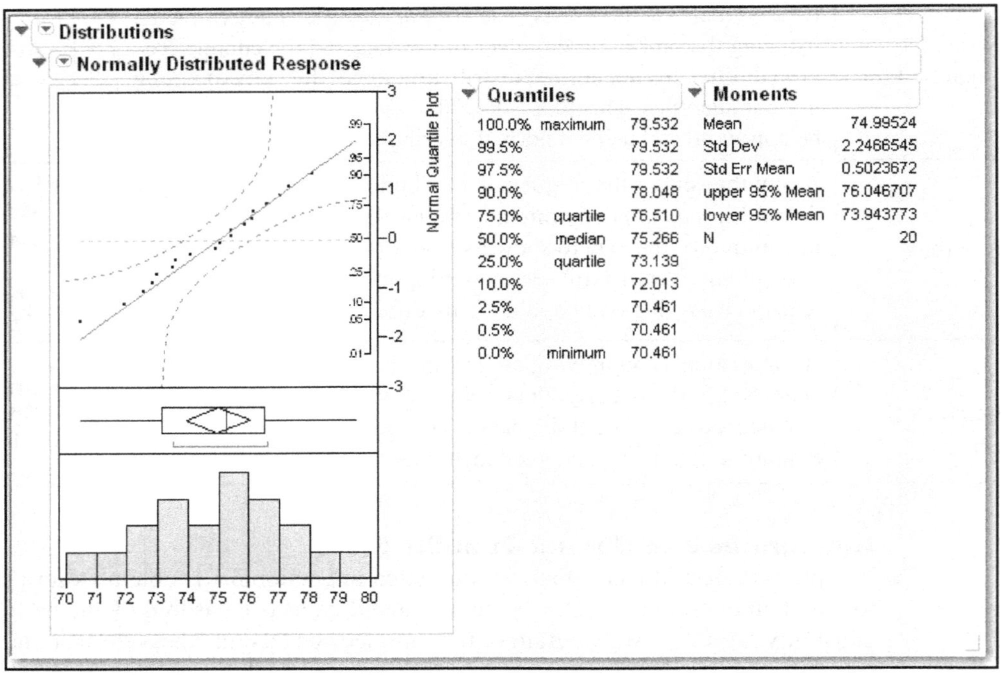

Box Plots

Box plots are a great visualization tool that shows the key features of the distribution of the data based on the 5-Number summary. The elements of generic box plot are shown in Figure 3.4 and further described below.

Figure 3.4 Generic Box Plot Elements

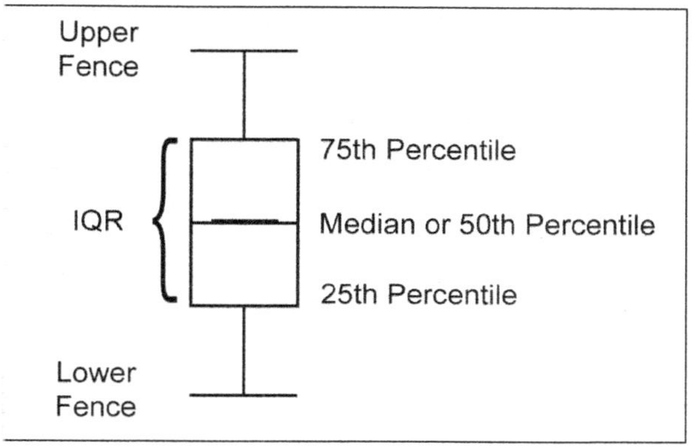

- **Median or 50th Percentile:** this is the value that cuts the data in half; that is, 50% of the ordered values are above and 50% are below the median. The box surrounds the middle half of the data, and the line that goes across the entire box represents the median.
- **25th Percentile:** this is the value that cuts the lower 50% of the data in half; that is, 25% of the ordered measurements in the sample are below it. The 25% percentile is the bottom line of the box.
- **75th Percentile:** this is the value that cuts the upper 50% of the data in half; that is, 25% of the ordered measurements in the sample are above it. The 75% percentile is the top line of the box.
- **IQR (Interquartile Range):** this is the difference between the 75th and 25th percentile and serves as a quick estimate of the variation in our sample.
- **Upper Fence:** this is located at a distance of 1.5×IQR from the median (Median + 1.5×IQR), or at the maximum (100% percentile) value, and serves as an upper bound for identifying potential outliers in our sample.
- **Lower Fence:** this is located at a distance of −1.5×IQR from the median (Median − 1.5×IQR)) or at the minimum (0% percentile) value, and serves as a lower bound for identifying potential outliers in our sample.

JMP Note 3.1: The box plot at the top of the histogram in Figure 3.3a is called Outlier Box plot in JMP, and it is the default. Note that this box plot does not have lines at the fence values. JMP also has a Quantile Box plot that displays the 5-Number summary and additional quantiles, namely the 0% (minimum), 10%, 25% 50% (median), 75%, 90%, 100% (maximum) quantiles, with lines at the lower and upper fences. This box plot can be generated from the **Analyze > Distribution** platform by clicking the red triangle to the left of **Normally Distributed Response** as shown in Figure 3.3b below. The Outlier and Quantile Box plots are shown in Figure 3.3b.

Figure 3.3b Outlier and Quantile Box Plots

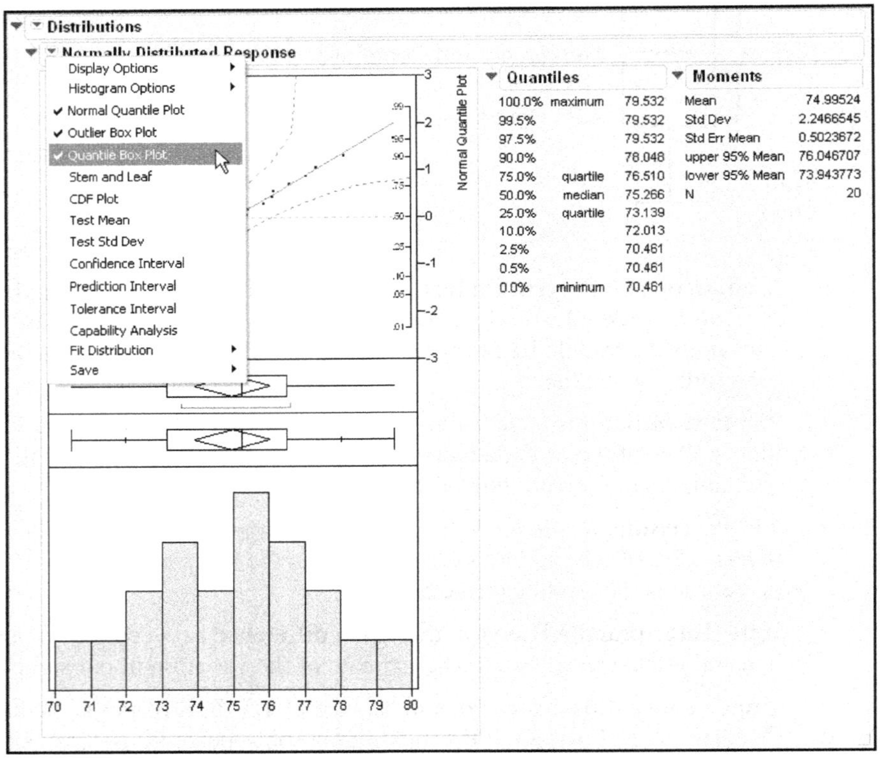

Box plots also enable us to visualize and compare the different sources of variation that might be present in our data. If stratification variables are present we can use them to generate a box plot for each level, or combination of levels, of the stratification variables. For example, what if we randomly sampled electronic devices and measured their resistance in the months of August through November, operating five days per week, running two shifts per day, using two dedicated manufacturing lines, and taking five devices every hour.

In the **Graph > Variability/Gauge Chart** platform, we can easily visualize the resistance values by month or by shift and month, for example. The **Variability/Gauge Chart** dialog box is shown in Figure 3.5 for the resistance data described above. The measured response, Resistance, should be highlighted and then selected for **Y, Response** and Month is our first **X, Grouping** factor. This produces a box plot with one factor, shown in Figure 3.6. Note that a number of options were turned on to alter the appearance of the plot, and they are also shown in Figure 3.6.

Figure 3.5 Dialog Box for Variability/Gauge Chart

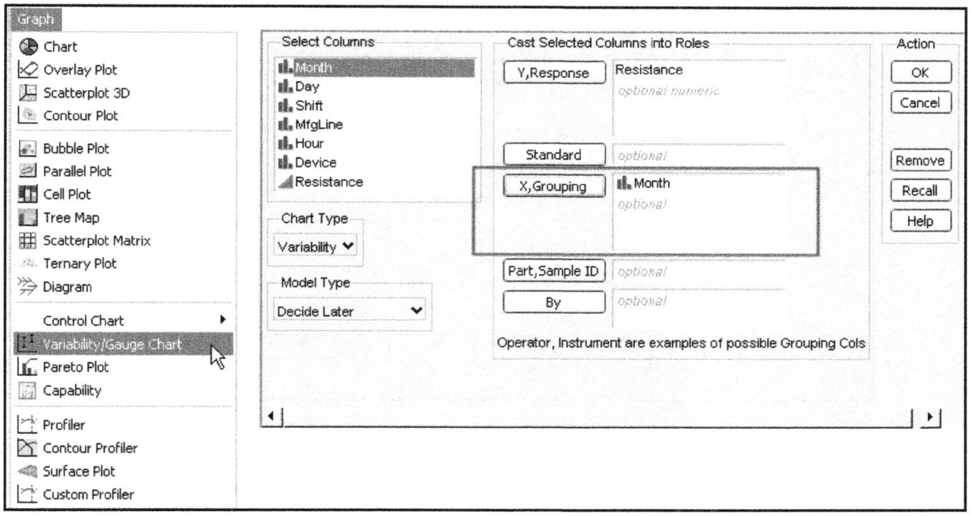

Since we selected Month, which has four levels, as our grouping factor, there are four box plots in Figure 3.6, one for each month. This plot provides us with the ability to compare the resistance measurements for each month in two distinct ways. First, we can compare the average monthly resistance values by comparing the shorter lines located in the center of each box, which are connected by selecting the option **Connect Cell Means**. We can see that August and October have lower resistance readings, on average, compared with September and November. Second, by visually comparing the size of the boxes we can assess differences in variation between the groups; that is, the variation in November seems lower than in the previous three months.

The output also shows the **Variability Summary for Resistance**. The differences in the monthly averages give us a sense of the amount of month-to-month variation that is present in the data set, while the differences between the month-to-month standard deviations give us a sense of the within-month variation, or how consistent the resistance readings are within a given month.

Figure 3.6 Variability/Gauge Chart Output

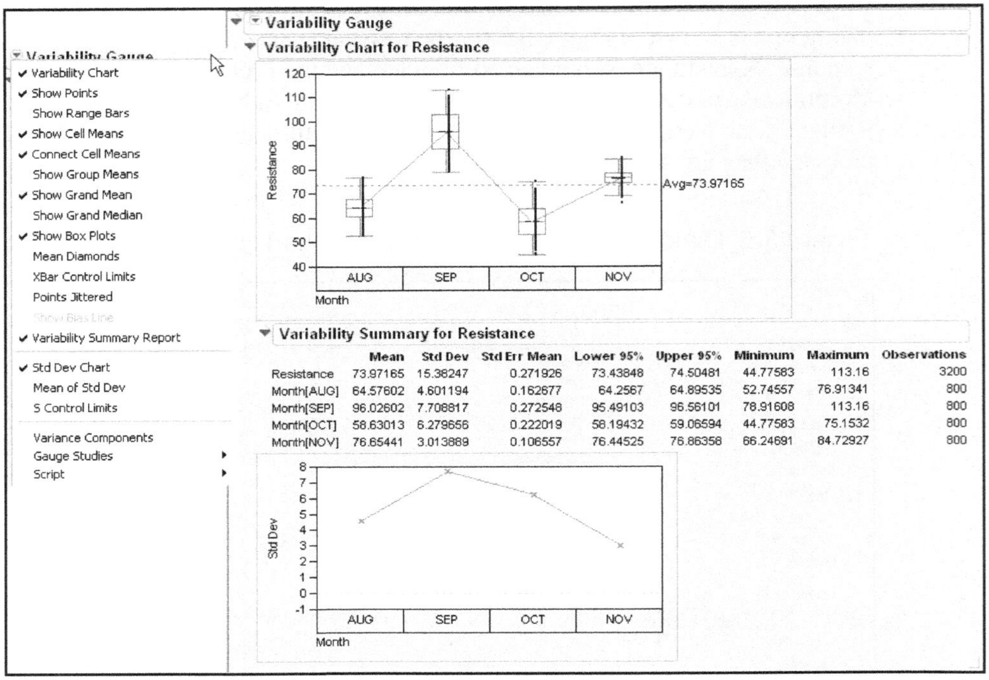

In addition to providing a graphical representation of the sources of variation in a data set, we can also obtain a decomposition of the total variation using the **Graph > Variability/Gauge Chart** platform. A thorough discussion of a variance components analysis (VCA) is beyond the scope of this book. However, because this JMP platform makes it relatively easy to obtain a breakdown into variance components, we illustrate its use in this chapter. If you are interested in learning more about variance components, see Section 9.3 of Box, Hunter, and Hunter (2005).

Statistics Note 3.2: In general, the actual components and how they are specified will depend upon the details of the study. The total variation $\sigma^2_{total} = \sigma^2_{component\ 1} + \sigma^2_{compoent\ 2} + \ldots + \sigma^2_{component\ k}$. With this decomposition we can attribute each component's contribution to the total variation by looking at the ratio of each component to the total. For example, $100 \times \sigma^2_{component\ 1} / \sigma^2_{total}$ provides us with the percentage contribution to the total for component 1.

Two sources of variability are depicted in the variability chart shown in Figure 3.6; therefore, we can break down the total resistance variation, σ^2_{total}, into two variance components, σ^2_{month} and $\sigma^2_{within\ month}$. From within the **Graph > Variability/Gauge Chart** platform, at the top of the output window, select **Variance Components** from the drop-down menu that appears when clicking the red triangle. This option is shown in Figure 3.6 and, when selected, two additional tables are added to the JMP output. Figure 3.7 shows the **Variance Components** table, which contains the columns Component; the estimate of the variation, s^2, attributed to each component (Var Component); the contribution of each component to the total variation (% of Total); a bar chart depicting the variance components; and the square root of each variance component, s, (Sqrt (Var Comp)), which shows the components in the units of the data.

Figure 3.7 Variance Components Output

Variance Components

Component	Var Component	% of Total	20 40 60 80	Sqrt(Var Comp)
Month	272.37050	89.4		16.504
Within	32.27861	10.6		5.681
Total	304.64911	100.0		17.454

The total variation in the resistance data, $s^2_{total} = 304.65$, is the sum of the variation due to month, $s^2_{month} = 272.37$, and variation within each month, $s^2_{within} = 32.28$. We can see that most of the variation (89.4%) in the resistance data can be attributed to month-to-month differences, while 10.6% of the variation is due to within-month differences. Note that the **Within** source of variation includes five other sources of variation: day-to-day, shift-to-shift, manufacturing line-to-manufacturing line, hour-to-hour, and device-to-device. A further analysis can be conducted by breaking down the total variation by these components, but the month-to-month variation being so large suggests that this is where we should start.

Process Behavior Charts and Homogeneity

Process behavior charts were discussed in Chapter 2 and, although they have not traditionally been included in the EDA tool set, we believe they are essential in assessing the basic assumption that the data is a random sample that comes from a single and stable universe or population. This assumption is critical for making performance claims about the future. For the socket qualification data, we want to make sure that the samples are selected at random, that they are representative of the population of sockets created using the new material, and that there are no unusual trends or patterns over time. Trends or patterns can be systematic or random, depending upon the root cause of the instability and, with suitable rational subgroups (see Statistics Note 3.3 below), the appropriate process behavior chart detects both types.

For the resistance data we can use the **Graph > Control Chart > IR** platform to create an individual values and moving range (IR) chart for each month and manufacturing line. An IR chart is a quick way to check for the homogeneity (the lack of patterns or trends) of a set of data. Figure 3.8a shows the process behavior chart for the resistance readings for manufacturing line A in the month of August. For a stable, predictable process the points should scatter randomly around the centerline of the chart and should be within the control limits. This is clearly not the case for the August resistance readings coming from manufacturing line A because there appears to be some unusual up-and-down patterns in the chart, and several points outside the lower and upper control limits. This indicates that the August resistance readings coming from manufacturing line A are not homogeneous.

Figure 3.8a Process Behavior Chart for August Resistance Readings from Manufacturing Line A

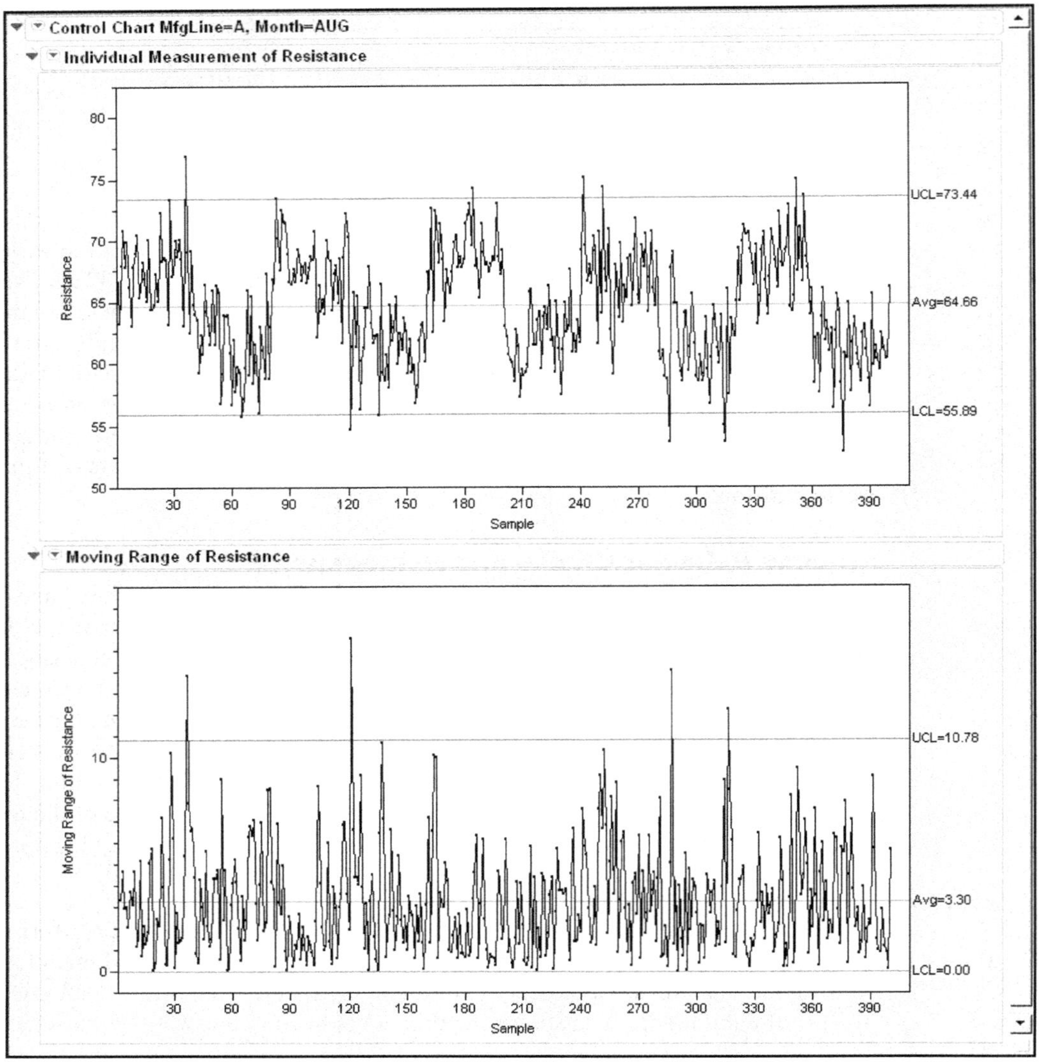

Where is this systematic pattern coming from? Since we have additional stratification variables (shift, hour, and so on) we can enhance the process behavior chart using different plotting symbols and colors to see whether we can gain more insight. Figure 3.8b shows the process behavior chart in Figure 3.8a with two different plotting symbols, a circle for shift 1, and a plus sign for shift 2. It is clear now that in August the lower resistance readings occurred on the 2nd shift. We should investigate to see whether this systematic pattern also occurs in other months and manufacturing lines.

Statistics Note 3.3: There are many different ways of setting up process behavior charts depending on the questions we want the charts to answer and how observations are arranged into subgroups. The charts, because they come in pairs, always look at the two components of variation between subgroups and within subgroups. For an in-depth discussion of process behavior charts and rational subgrouping see Wheeler and Chambers (1992) *Understanding Statistical Process Control.*

Figure 3.8b Process Behavior Chart with Different Plotting Symbols by Shift

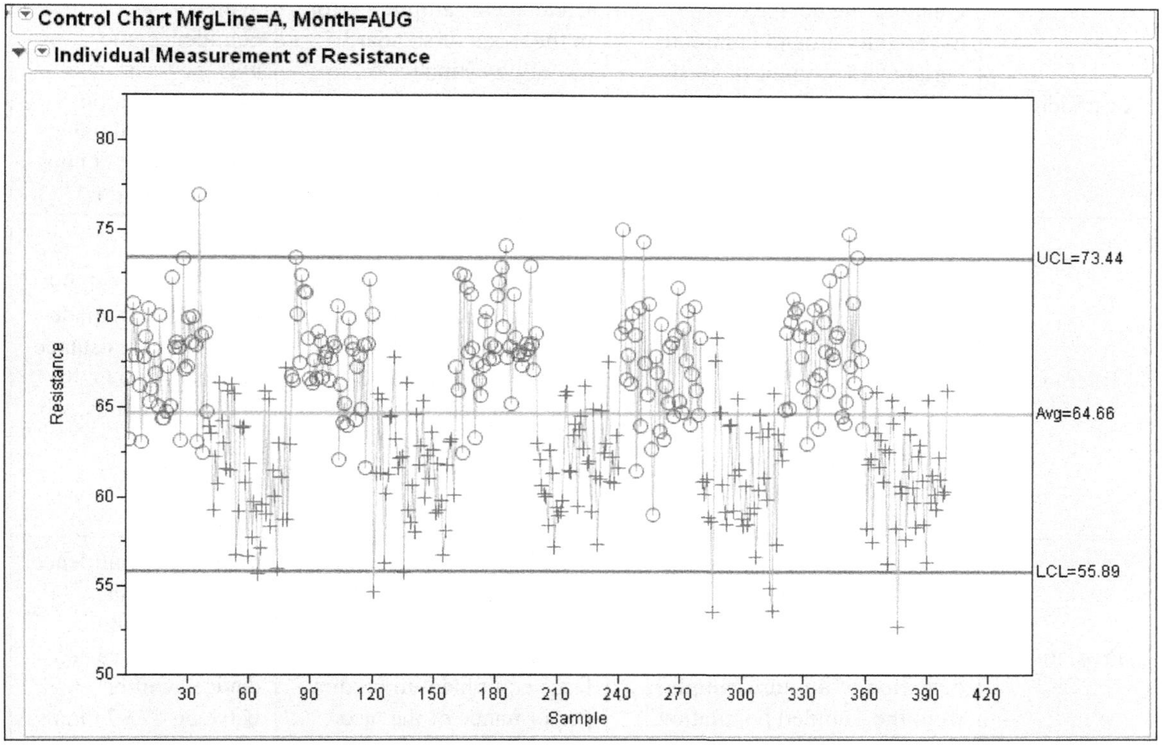

3.3.4 Statistical Intervals

In Chapter 2 we discussed how statistical intervals enable us to quantify the degree of uncertainty in our sample estimates and get an understanding of how much confidence we can place in them. For example, a statistical interval for the sample mean, $\overline{X}$, enables us to put a *margin of error* (MOE) around our estimate, $\overline{X} \pm$ MOE. This "plus or minus" MOE gives us a sense of the "goodness" of the estimate since the length of this interval is a reflection of the uncertainty in our estimate.

Table 3.4 provides some key information about the three types of statistical intervals (confidence, tolerance, and prediction) that are very useful in science and engineering. For more information about these types of intervals see Hahn and Meeker (1991) and Ramírez (2009).

Table 3.4 Statistical Intervals Definition, Uses, and Claims

Statistical Interval	Description	Uses	Claim
Confidence	Quantify the degree of uncertainty around common parameters of interest, such as the center of a sampled population, μ, or its spread, σ, or even statistics like C_{pk}, by adding a margin of error.	Understand sampling error in the mean or standard deviation estimates or, to conduct a more formal test of hypothesis.	With a 95% confidence, the average resistance of the population of devices will be between 49.23 Ohms and 50.49 Ohms.
Tolerance	An *enclosure* interval, based on a confidence level, for a specified proportion of the sampled population. For example, we might want to determine lower and upper bounds with a 95% confidence level such that 99% of the population is contained within them.	Set specification limits for individual values, determine whether specification limits can be met to achieve a specified yield, or to make a statement about an upper or lower performance bound of the population.	With a 95% confidence level, 99% of the devices made will have a resistance between 47.6 Ohms and 52.12 Ohms.
Prediction	Can be seen as an enclosure interval for one or more future observations, or as an interval for the mean or standard deviation of a future sample from the sampled population.	Predict performance bounds for a limited release of a new product to the market or, for a manufacturer of large equipment to predict performance of the next three to be made.	With 95% confidence, all the resistance values of a future sample of three devices will be between 47.97 Ohms and 51.75 Ohms.

Making Claims Using Statistical Intervals

Statistical intervals can be used to make claims about the measured performance of a material, process, or product. These claims are related to the type of inference that we want to make using the sample we drew from the population under study. In many instances we want to make claims pertaining to the unsampled population from our study. For example, if we collected resistance data for a random sample of electronic devices made in the month of August, we want to make a claim about the average performance of all of the electronic devices made in the month of August, including those not sampled. In other instances we want to make claims pertaining to a future sample that will be taken from a population. For example, if we qualify a new piece of equipment in manufacturing, we want to make a claim regarding upper and lower performance bounds for parts made in the future. The last column in Table 3.4 provides some examples of statistical claims that can be made for each type of interval.

Here are some specific examples of claims that can be made with the statistical intervals described in Table 3.4:

- claims about the average performance of a batch or lot of material
- claims about the performance variation of a batch or lot of material
- claims about a single future observation
- claims about a small lot
- claims about a percentage of the population
- claims about process performance based on C_{pk} or yield

As we see in later chapters, the validity of these claims depends on the following assumptions:

1. The data is a random and representative sample of the population.
2. The data are homogeneous.
3. The data can be approximated using a probability distribution.

For a great number of continuous data found in practice, the normal distribution is a good approximation. There are situations, however, where other types of probability distributions can be used to construct confidence intervals.

In addition to understanding the uncertainty in our estimates and making claims about the measured performance of a material, process, or product, statistical intervals can also be used to conduct tests of hypothesis, as was discussed in Chapter 2 and will be shown in Chapter 4.

Statistics Note 3.4: A confidence interval can be used to test a performance claim on a parameter of interest. For example, in Chapter 2 a 95% confidence interval for the mean resistance of an electronics device was calculated as 49.86 Ohms ± 0.63 Ohms = [49.23 Ohms, 50.49 Ohms]. Since this interval contains the value 50 Ohms, we would conclude, with 95% confidence, that there is not enough evidence to suggest that the average resistance is off from the 50 Ohms target. For the Socket qualification, we will use this approach to see whether the new raw material has a similar C_{pk} to the existing raw material and also to check the significance of stratification factors. For a refresher on the meaning of "confidence" see Section 2.6.4.

3.4 Step-by-Step JMP Analysis Instructions

For the raw material second source problem described in Section 3.1, we will walk through the steps necessary to characterize the sockets made in the injection molding process. We will use the seven steps shown in Table 3.5, which were first described in Chapter 1, and that form the basis for the analyses presented in the next chapters. Steps 1 and 2 should be stated before any data is collected or analyzed, and they do not necessarily require the use of JMP. However, Steps 3 through 6 will be conducted using JMP and the output from these steps will help us complete Step 7.

Table 3.5 Step-by-Step JMP Analysis for New Raw Material for Socket Manufacturing Process

Step	Objectives	JMP Platform
1. Clearly state the question or uncertainty.	We need to characterize the thickness performance of the sockets using the new raw material.	Not applicable.
2. Specify the expected performance of the material, product, or process.	We expect the thickness performance of the new material to be as good as or better than the old.	Not applicable.
3. Determine the appropriate sampling plan and collect the data.	Identify how many sockets will be manufactured with the new raw material and what sources of variation should be included in the study. How well do we want to estimate the average and standard deviation?	• **DOE > Sample Size and Power**
4. Prepare the data for analysis and conduct exploratory data analysis.	Make sure all stratification factors (sources of variation) are included in the JMP table, along with the measured response. Set column properties as needed. Check for any outliers or discrepancies with the data.	• **Cols > Column Info** • **Analyze > Distribution** • **Graph > Variability/Gauge Chart**
5. Characterize your data and answer the questions of interest.	Calculate appropriate descriptive statistics and statistical intervals to understand and quantify the sources of variation in the data. Check the analysis assumptions to make sure your results are valid.	• **Analyze > Distribution** • **Graph > Variability/Gauge Chart** • **Graph > Control Chart**
6. Summarize the results with key graphs and summary statistics.	Find the best summary measures and graphical representations of the results that give us insight into our uncertainty. Select key output for reports and presentations.	• **Analyze > Distribution** • **Graph > Variability/Gauge Chart**
7. Interpret the results and make recommendations.	Translate the statistical jargon into the context of the original problem. Assess the practical significance of your findings.	Not applicable.

Step 1: Clearly state the question or uncertainty

Due to an impending shortage for a key raw material used in sockets, you must qualify a second source supplier for this raw material. The most important performance measure of a socket is its thickness, which should be around 12 millimeters. However, since one side of the socket is convex (Figure 3.9a), a special gauge was made to measure not the thickness but the thickness in excess of 12 millimeters, or effective thickness. Customer specifications for this attribute require the effective thickness to fall within 0 and 15 hundredths millimeters. As a reference, one thousand inch (0.001 in) is about 2.54 hundredths millimeter (0.0254 mm).

Figure 3.9a Socket Cross Section

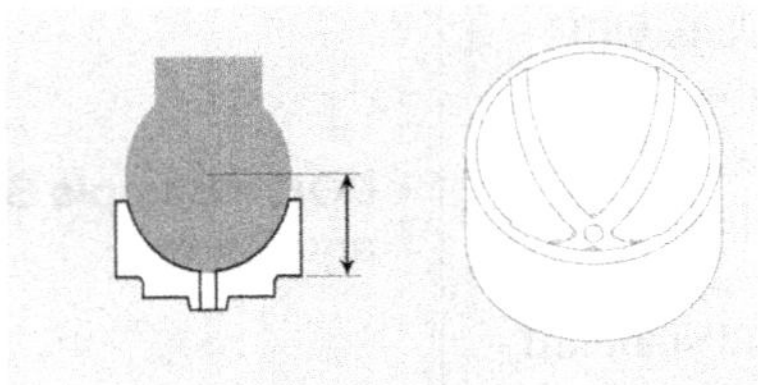

You want to show that the sockets produced with the new material are similar to the sockets that you have been producing using the old material. In other words, the business wants to make sure that the raw material coming from the new supplier results in a stable process that generates sockets with the same, or better, process capability as those produced with raw material from the current supplier as measured by $C_{pk} = 1$.

The injection molding operation is the manufacturing step that uses this particular raw material to make the sockets, and it is the focus of the qualification. The injection mold has four cavities (Figure 3.9b), and is able to produce four sockets at a time. One operator is in charge of the injection molder to make the sockets.

Figure 3.9b Injection Mold with Four Cavities

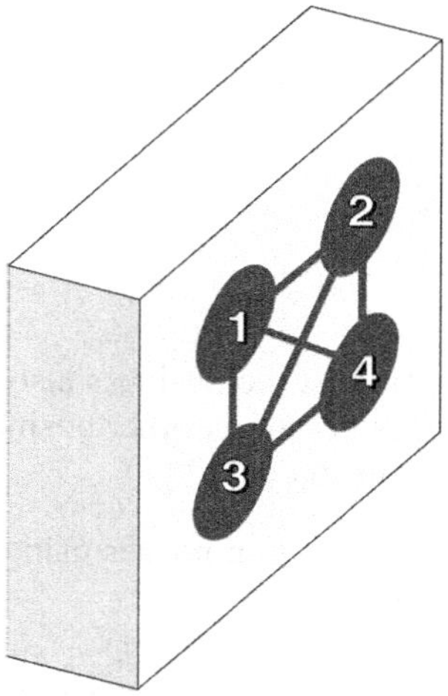

Figures 3.9a and 3.9b were taken from Section 5.6, Chapter 5 of Understanding Statistical Process Control (Wheeler and Chambers 1992). Used by permission of Dr. Donald J. Wheeler.

Step 2: Specify the expected performance of the material, product, or process

We are interested in characterizing the effective thickness performance of the sockets made in the injection molder with the new raw material. The information provided in Step 1 suggests that the performance of the sockets made with the new material must be similar to those made with the current raw material supplier. We need to characterize the thickness performance by estimating the following:

- average performance (μ) of the effective thickness
- performance variation (σ) of the effective thickness
- process capability (C_{pk}) of the effective thickness

In order for the raw material substitution to go undetected by our customers and the business, we must show that the mean, standard deviation, and process capability of the effective thickness of sockets made with the new raw material are similar to the current one. In other words, we must show that:

- $\mu = 7.5$ hundredths millimeters
- $\sigma \leq 2.5$ hundredths millimeters
- $C_{pk} \geq 1$

Apart from these performance and statistical considerations, the business also wants to make sure that the introduction of the raw material does not affect other business metrics such as profit margins, throughput, on-time delivery, and so on.

Note that in Chapter 4, we will show how to test this type of hypotheses using a more formal statistical framework; that is, tests of significance.

Step 3: Determine the appropriate sampling plan and collect the data

We will use the current socket sampling plan, but we need to clearly identify the experimental unit and the observational unit as was discussed in Section 2.3.2.

Determine the Experimental Unit (EU) and Observational Unit (OU)

One of the claims that we need to make with this study pertains to the process capability of the effective thickness, measured after injection molding. The injection molding is such that four sockets can be produced at the same time, and these constitute one experimental unit.

The specification limits for the effective thickness apply to one measurement made on each socket. The current measurement device enables us to take one measurement per socket, the observational unit (OU).

Select the Sources of the Variation

There are a number of things that contribute to the thickness variation in the injection molding process when manufacturing sockets. For example, the variation in the measured thicknesses can be due to differences in the daily set-ups of the equipment, wax build-up on the equipment over time, cavity differences, raw material batch changes, and humidity changes. Some of these sources of variation (such as cavity differences) can be included in the qualification, but others might not, such as humidity changes that might be difficult to measure.

Based on the current sampling plan, along with the quantity of raw material on-hand, four sources of variation will be included in this characterization. Sockets will be sampled from the population as is specified in Table 3.6. This sampling plan will result in the manufacturing of 400 sockets coming from 5 days x 4 times each day x 5 cycles x 4 cavities for this qualification.

Table 3.6 Sources of Variation and Sampling Frequency for Thickness Characterization

Source of Variation	Description	Sampling Frequency
Day-to-day	Represents variation that occurs from one day to the next, which can be caused by starting up the molder each day after the nightly shut down, differences in the temperature and humidity on the production floor, or wax build-up in the equipment.	Five consecutive days (Monday, Tuesday, Wednesday, Thursday, Friday)
Hour-to-hour	Variation in sockets can occur throughout the day due to factors such as controller drift on the equipment or raw material batch changes. Historical data suggests that these shifts can be detected within a two-hour time frame.	Every two hours throughout each day (10 a.m., 12 p.m., 2 p.m., 4 p.m.)
Cycle-to-cycle	This source of variation represents short-term variation within a two-hour interval, since each cycle takes minutes to run. Historically, there are not too many things that can result in large shifts within such a short time interval. However, this allows us to get a better estimate of the mean and standard deviation.	Five consecutive cycles (A, B, C, D, E)
Cavity-to-cavity	There are four different cavities on the mold used in the injection molding step. Variation can be due to inconsistent wear and tear of the press and inherent variation of the press itself.	Four cavities from molder (I, II, III, IV)

Estimating the Mean with a Given Margin of Error (MOE)

The sampling plan just described will result in 400 sockets and 400 effective thickness measurements. To determine how well we can estimate the average performance of the distribution of effective thickness measurements with a sample size, $n = 400$, we select

DOE > Sample Size and Power > One Sample Mean, which launches the **Sample Size and Power** dialog box that can be used for estimating the mean as shown in Figure 3.10.

Figure 3.10 One Sample Mean Sample Size and Power Dialog Box

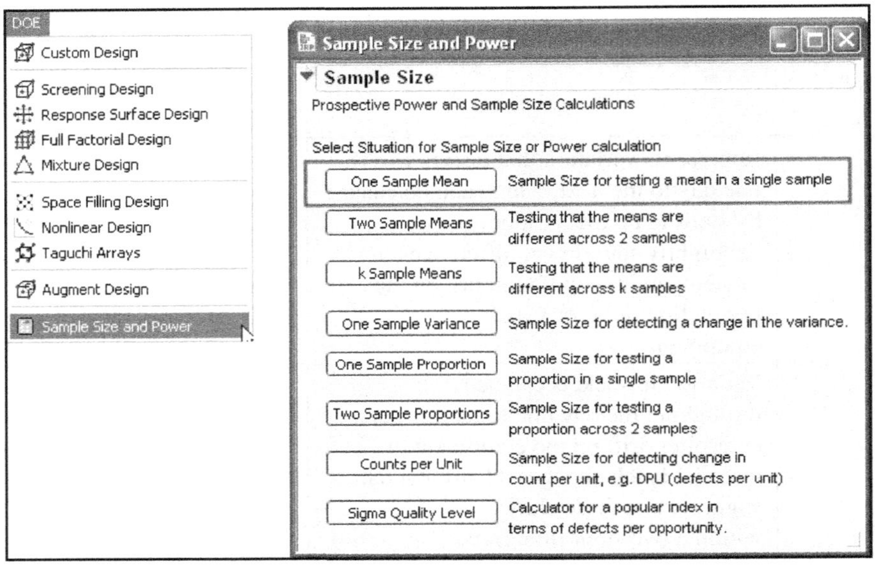

JMP 8 Note 3.1: In the **Sample Size and Power** dialog box in JMP 8 the **One Sample Variance** calculation is now called **One Sample Standard Deviation**.

Clicking **One Sample Mean** produces the dialog box on the left in Figure 3.11. This dialog box contains the following six fields:

a. **Alpha:** this is the risk we are willing to take that our estimate of the mean is off by *d* or more units (field 4). Alpha is also related to the confidence level that we want to use to establish a $100(1 - \alpha)\%$ confidence interval around our mean. For example, if we want to have a 95% confidence interval for the mean we enter 0.05 in this field.

b. **Error Std Dev:** this is an estimate of the standard deviation of our response. If we do not have an estimate of the standard deviation we can enter 1 into this field and obtain the **Difference to detect** in multiples of the standard deviation.

c. **Extra Params:** this field is used for multi-factor designs. Leave this field 0 for the applications described in this chapter.

d. **Difference to detect:** this represents the smallest acceptable difference (margin of error) for estimating the mean; that is, mean ± d. For example, a value of 2 Ohm would imply that we want to estimate the average resistance within ± 2 Ohms. Note that if a value of 1 was used for the **Error Std Dev** (b) then this value is given in multiples of the standard deviation, that is, mean ± ds. A value of s = 1.5 implies that we want to estimate the average resistance within ± 2(1.5) or ± 3 Ohms.

e. **Sample Size:** this is the total sample size (number of experiment units (EUs)) required in our study to be able to estimate the population mean with the inputs provided in (a), (b), (c), (d), and (f).

f. **Power:** this is a value between 0 and 1 that represents the probability of detecting that the mean differs by more than ± d units. Higher power values result in a higher sample size. Typical values are 0.8 or higher. For this chapter, we will set the power to equal 0.8.

Figure 3.11 One Sample Mean Sample Size Calculator

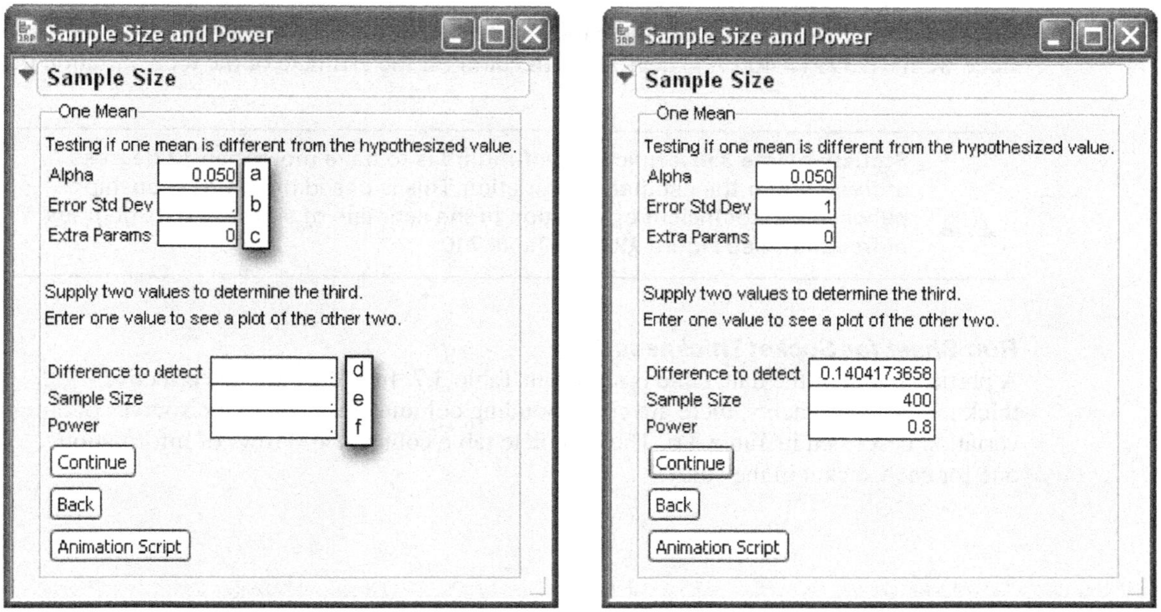

With inputs **Alpha** = 0.05, a 5% chance of not estimating the mean with the given margin of error, **Error Std Dev** = 1, **Sample Size** = 400, and **Power** = 0.8. The dialog box on the right side of Figure 3.11 shows the calculated margin of error, **Difference to detect**, as ± 0.14. With such a large sample size, 400 sockets, we are able to estimate the mean within a very small margin of error. Since we entered **Error Std Dev** = 1 the **Difference to detect** is given terms of units of the standard deviation for effective thickness, or ± 0.14σ. If, for example, σ = 0.025 mm, then our margin of error is ± 0.14(0.025) or ± 0.0035 mm.

Statistics Note 3.5: The calculations above assume that the observations are independent. For the socket data this might not be true since four sockets from the *same* mold are produced at a time. For correlated observations the estimate of the std dev is biased. However, the large sample size (400) brings the bias to 0.

JMP 8 Note 3.2: In JMP 8 the entry, **Error Std Dev**, is called **Std Dev**, and entry **Extra Params**, is called **Extra Parameters**.

What about our ability to estimate the total standard deviation? No worries here either, since we have 399 (= 400 – 1) degrees of freedom for the estimate of the total variation σ.

Statistics Note 3.6: A quick rule-of-thumb is to have more than 12 degrees of freedom for the estimate of variation. This is based on the relationship between the coefficient of variation of the estimate of sigma and its degrees of freedom. See Figure 7.20 and Table 7.10.

Run Sheet for Socket Thickness Data

A partial listing of the data table is shown in Table 3.7. In addition to the effective thickness measurements, there are corresponding columns for each of the sources of variation described in Table 3.6. The complete table contains 400 rows of information, one for each socket in the study.

Table 3.7 Run Sheet for Socket Thickness Characterization

Day	Hour	Cycle	Cavity	Socket	Effective Thickness	Comments
1	10:00 AM	A	I	1	13.773	Qual starts today
1	10:00 AM	A	II	2	9.259	
1	10:00 AM	A	III	3	8.465	
1	10:00 AM	A	IV	4	8.866	
1	10:00 AM	B	I	5	17.445	
1	10:00 AM	B	II	6	12.768	
1	10:00 AM	B	III	7	7.614	
1	10:00 AM	B	IV	8	9.994	
1	10:00 AM	C	I	9	16.667	
1	10:00 AM	C	II	10	11.784	
1	10:00 AM	C	III	11	9.396	
1	10:00 AM	C	IV	12	9.756	
1	10:00 AM	D	I	13	14.987	
1	10:00 AM	D	II	14	10.377	
1	10:00 AM	D	III	15	6.027	
1	10:00 AM	D	IV	16	9.397	
1	10:00 AM	E	I	17	17.644	
1	10:00 AM	E	II	18	9.914	
1	10:00 AM	E	III	19	8.415	
1	10:00 AM	E	IV	20	10.873	
…	…	…	…	…	…	…
5	4:00 PM	E	IV	400	6.393	

Data adapted from Table 5.2, Section 5.6, Chapter 5 of *Understanding Statistical Process Control* (Wheeler and Chambers 1992). Used by permission of Dr. Donald J. Wheeler.

Step 4: Prepare the data for analysis and conduct exploratory data analysis

Reading the Data File into JMP

Since the data was collected in an Excel file, we need to read it into JMP and create a JMP table before we can proceed with any analysis. From the main title bar in JMP, select **File > Open** or select the **Open Data Table** icon in the JMP Starter window. This launches the **Open Data File** dialog box, where we can select the location of our file, along with the filename. Make sure that the **Files of type** is set to *.XLS in order to view the Excel files in the chosen directory. Also, if there are multiple worksheets in the Excel file, you might want to check the box to **Allow individual worksheet selection**, otherwise a JMP table is created for each worksheet in the file. Figure 3.12 shows this dialog box, along with the appropriate choices to read in the socket qualification data. The resulting JMP table is shown in Figure 3.13.

Figure 3.12 Dialog Box to Read Excel Files into JMP

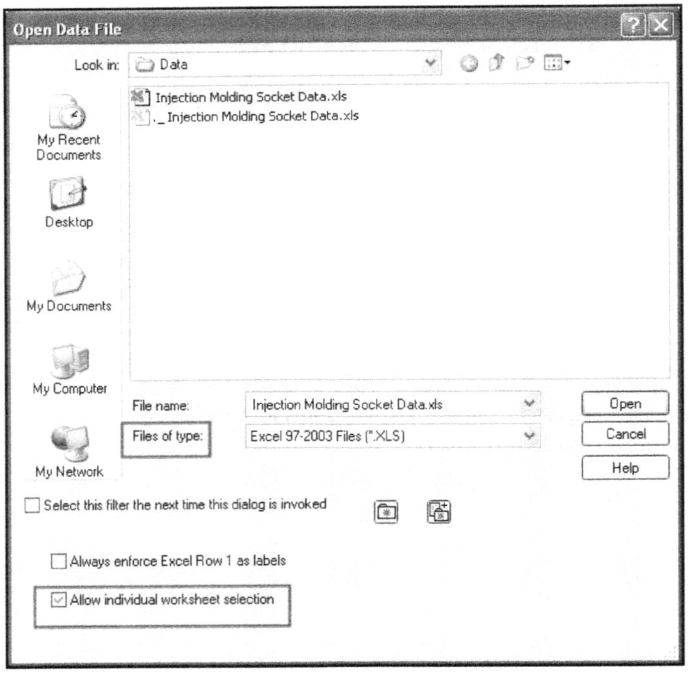

Figure 3.13 JMP Table Containing Data for Socket Thickness Characterization

Injection Molding Socket
Notes C:\Documents and Se

Columns (7/0)
Day
Hour
Cycle
Cavity
Socket
Effective Thickness
Comments

Rows
All rows 400
Selected 0
Excluded 0
Hidden 0
Labelled 0

	Day	Hour	Cycle	Cavity	Socket	Effective Thickness	Comments
1	1	10:00 AM	A	I	1	13.773	Qualification starts today
2	1	10:00 AM	A	II	2	9.259	
3	1	10:00 AM	A	III	3	8.465	
4	1	10:00 AM	A	IV	4	8.866	
5	1	10:00 AM	B	I	5	17.445	
6	1	10:00 AM	B	II	6	12.768	
7	1	10:00 AM	B	III	7	7.614	
8	1	10:00 AM	B	IV	8	9.994	
9	1	10:00 AM	C	I	9	16.667	
10	1	10:00 AM	C	II	10	11.784	
11	1	10:00 AM	C	III	11	9.396	
12	1	10:00 AM	C	IV	12	9.756	
13	1	10:00 AM	D	I	13	14.987	
14	1	10:00 AM	D	II	14	10.377	
15	1	10:00 AM	D	III	15	6.027	
16	1	10:00 AM	D	IV	16	9.397	
17	1	10:00 AM	E	I	17	17.644	
18	1	10:00 AM	E	II	18	9.914	
19	1	10:00 AM	E	III	19	8.415	
20	1	10:00 AM	E	IV	20	10.873	
21	1	12:00 PM	A	I	21	13.577	
22	1	12:00 PM	A	II	22	9.844	
23	1	12:00 PM	A	III	23	6.529	
24	1	12:00 PM	A	IV	24	9.368	
25	1	12:00 PM	B	I	25	18.272	
26	1	12:00 PM	B	II	26	9.193	
27	1	12:00 PM	B	III	27	10.466	
28	1	12:00 PM	B	IV	28	12.697	
29	1	12:00 PM	C	I	29	14.336	
30	1	12:00 PM	C	II	30	11.620	
31	1	12:00 PM	C	III	31	8.354	
32	1	12:00 PM	C	IV	32	13.652	
33	1	12:00 PM	D	I	33	15.573	

Altering JMP Table Properties

The first thing we notice about the JMP table shown in Figure 3.13 is the numeric values for Day. If we create any plots using Day, then they are labeled with the values 1 through 5 instead of Monday through Friday. We prefer to label these values with the names of the days of the weeks but we do not want to retype all the 400 values. Not only does JMP offer great statistical and visualization tools, but it also offers powerful data manipulation features. In order to change the Day values from 1–5 to Monday–Friday, all we need to do is change the column properties for the Day column. This can be accomplished by performing the following steps:.

1. Double-click the Day label in this column in order to launch the column information window. You can also right-click on the Day label and select **Column Info**.
2. Click **Column Properties** and select **Value Labels** from the drop-down list (see Figure 3.14).

Figure 3.14 Accessing Value Labels for a Day Column

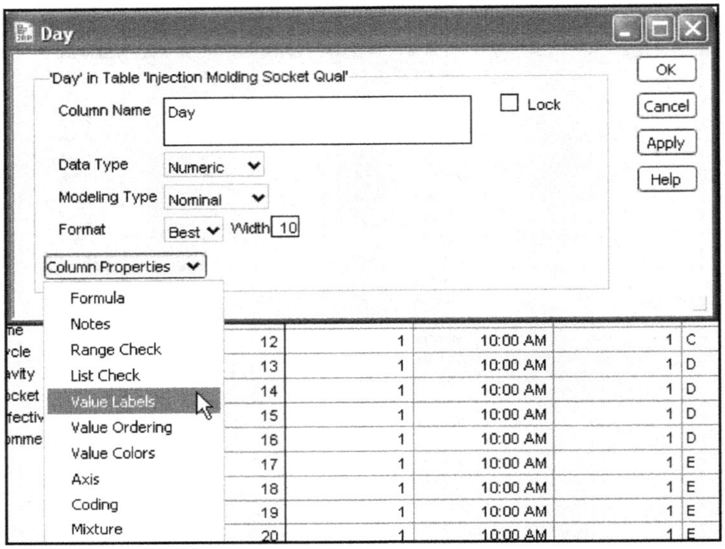

3. The **Value Labels** window appears to the right of the **Column Properties** drop-down list where we can enter the labels for Days 1 through 5, one at a time. First we enter Monday for Day = 1 by entering `1` into the **Value field** and `Monday` into the **Label field**. Clicking **Add** adds this assignment into the field directly above (see Figure 3.15) as well as clearing out the fields directly below it.

Figure 3.15 Adding Value Labels to Columns

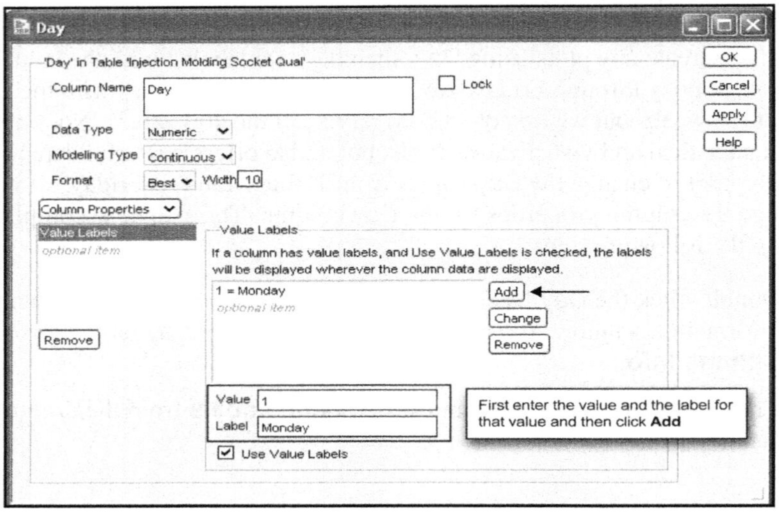

4. Continue to add labels for Days 2 through 5 by entering the following combinations into the **Value** and **Label fields** and clicking the **Add** after each one to enter them into the field above: Value = `2` and Label = `Tuesday`, Value = `3` and Label = `Wednesday`, Value = `4` and Label = `Thursday`, and Value = `5` and Label = `Friday`. Figure 3.16 shows all of these assignments.

Figure 3.16 All Value Labels for Day

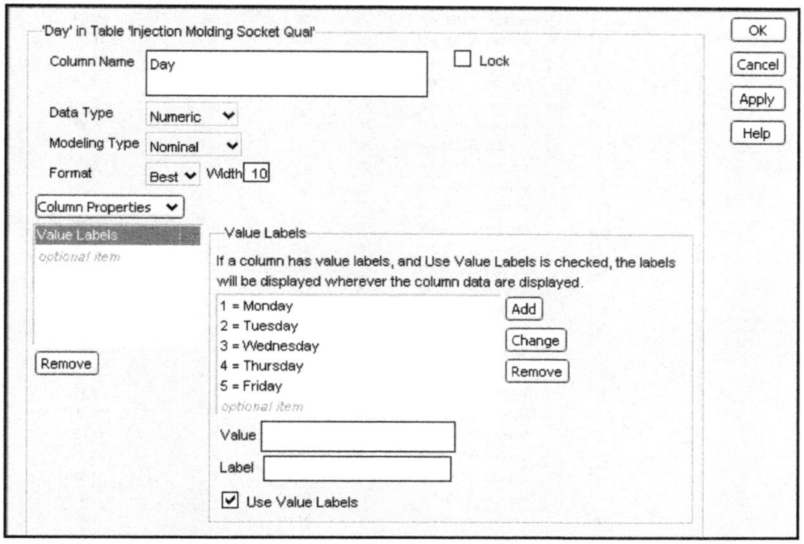

5. When finished making all assignments, click **OK** to update the JMP table, as is partially shown in Figure 3.17.

Figure 3.17 Partial Listing of Updated Labels for Day

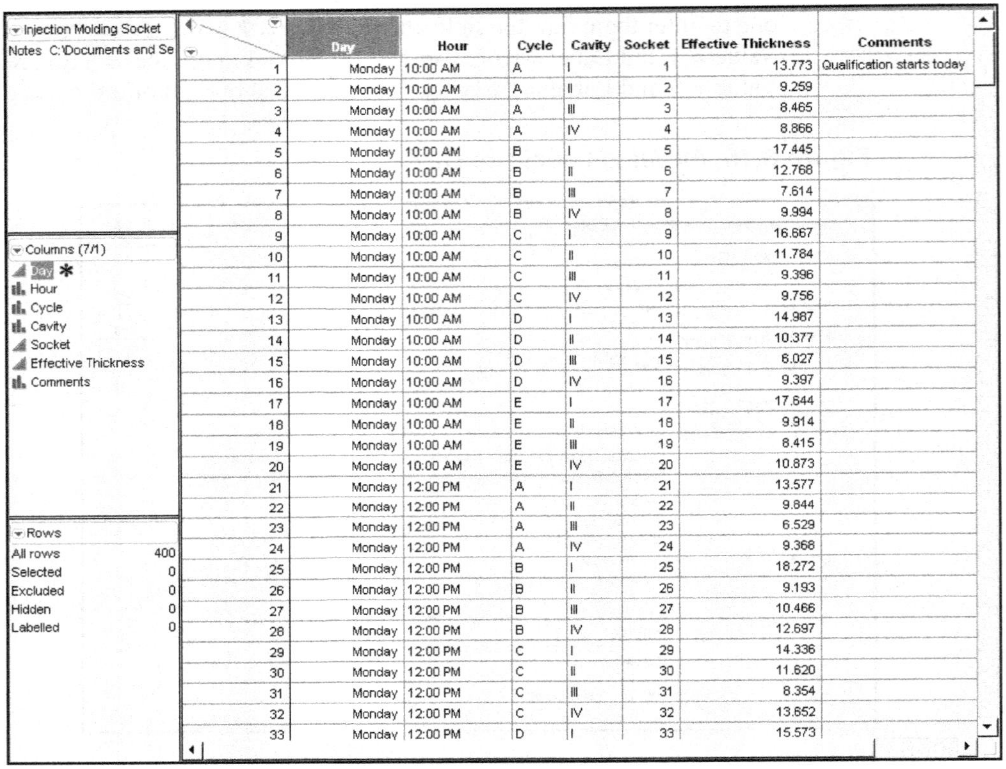

	Day	Hour	Cycle	Cavity	Socket	Effective Thickness	Comments
1	Monday	10:00 AM	A	I	1	13.773	Qualification starts today
2	Monday	10:00 AM	A	II	2	9.259	
3	Monday	10:00 AM	A	III	3	8.465	
4	Monday	10:00 AM	A	IV	4	8.866	
5	Monday	10:00 AM	B	I	5	17.445	
6	Monday	10:00 AM	B	II	6	12.768	
7	Monday	10:00 AM	B	III	7	7.614	
8	Monday	10:00 AM	B	IV	8	9.994	
9	Monday	10:00 AM	C	I	9	16.667	
10	Monday	10:00 AM	C	II	10	11.784	
11	Monday	10:00 AM	C	III	11	9.396	
12	Monday	10:00 AM	C	IV	12	9.756	
13	Monday	10:00 AM	D	I	13	14.987	
14	Monday	10:00 AM	D	II	14	10.377	
15	Monday	10:00 AM	D	III	15	6.027	
16	Monday	10:00 AM	D	IV	16	9.397	
17	Monday	10:00 AM	E	I	17	17.644	
18	Monday	10:00 AM	E	II	18	9.914	
19	Monday	10:00 AM	E	III	19	8.415	
20	Monday	10:00 AM	E	IV	20	10.873	
21	Monday	12:00 PM	A	I	21	13.577	
22	Monday	12:00 PM	A	II	22	9.844	
23	Monday	12:00 PM	A	III	23	6.529	
24	Monday	12:00 PM	A	IV	24	9.368	
25	Monday	12:00 PM	B	I	25	18.272	
26	Monday	12:00 PM	B	II	26	9.193	
27	Monday	12:00 PM	B	III	27	10.466	
28	Monday	12:00 PM	B	IV	28	12.697	
29	Monday	12:00 PM	C	I	29	14.336	
30	Monday	12:00 PM	C	II	30	11.620	
31	Monday	12:00 PM	C	III	31	8.354	
32	Monday	12:00 PM	C	IV	32	13.652	
33	Monday	12:00 PM	D	I	33	15.573	

JMP Note 3.2: The **Column Properties** in the Column Info window enable us to change the attributes of the variable in that column like **Data Type**, **Modeling Type** and **Format**, as well to add additional information for that variable like **Value Labels**, as we did for the **Day** variable; **Value Ordering** to modify the default alpha-numeric sorting (helpful to order months from January to December); **Value Colors** to define pre-specified graph colors for each level of the variable; **Units** to add the units of a particular variable; **Notes** to add any additional information about the variable; **Response Limits** to specify the range of values for the variable; **Spec Limits** to add specification limits (Figure 3.18), and so on.

Another thing that will make our analysis easier is to add the specification limits to the column properties for effective thickness. This automatically generates the process capability analysis every time we use the **Analyze > Distribution** platform. To accomplish this double-click on the column heading labeled Effective Thickness and the column info window will be displayed. From the **Column Properties** drop-down menu, we want to select **Spec Limits**, and then enter the values for the upper and lower specifications in the appropriate fields, as is shown in Figure 3.18. Click **OK** when finished.

Figure 3.18 Adding Specification Limits to Column Properties

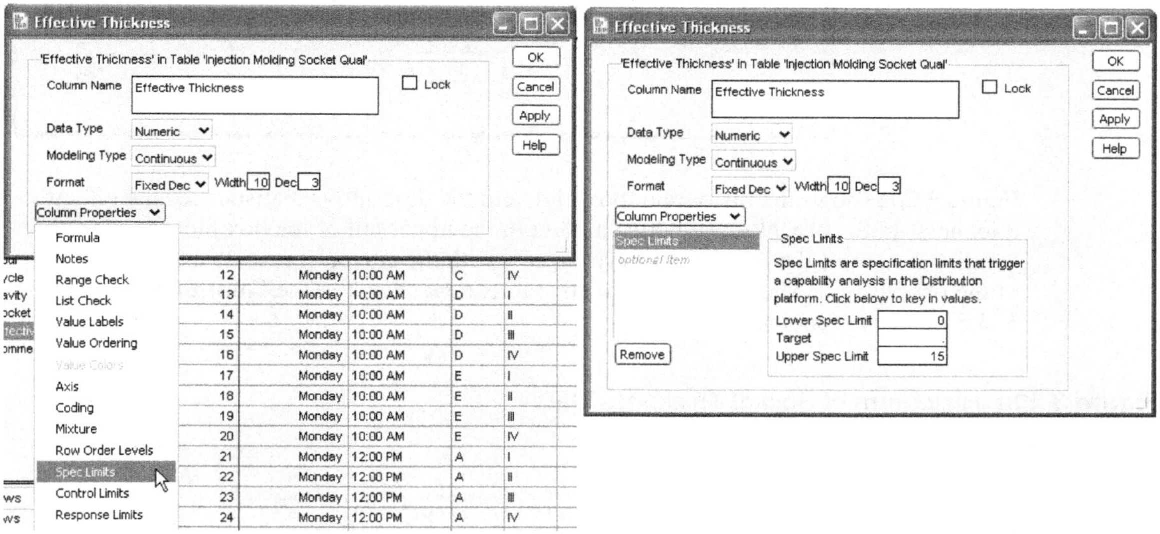

Checking for Outliers and Typos

Before moving on to Step 5, it is a good practice to examine the data for possible outliers, typographical errors, or any other unusual results. We can use the **Analyze > Distribution** platform to help visualize the data and check for anomalies in the data. From the primary toolbar, select **Analyze > Distribution** and a dialog box appears (see Figure 3.19) that enables us to make the appropriate choices for our data. Select Effective Thickness from the **Select Columns** window, and then click **Y, Columns**. This populates the window to the right of this button with the name of our response. Click **OK**.

Figure 3.19 Distribution Dialog Box for Effective Thickness

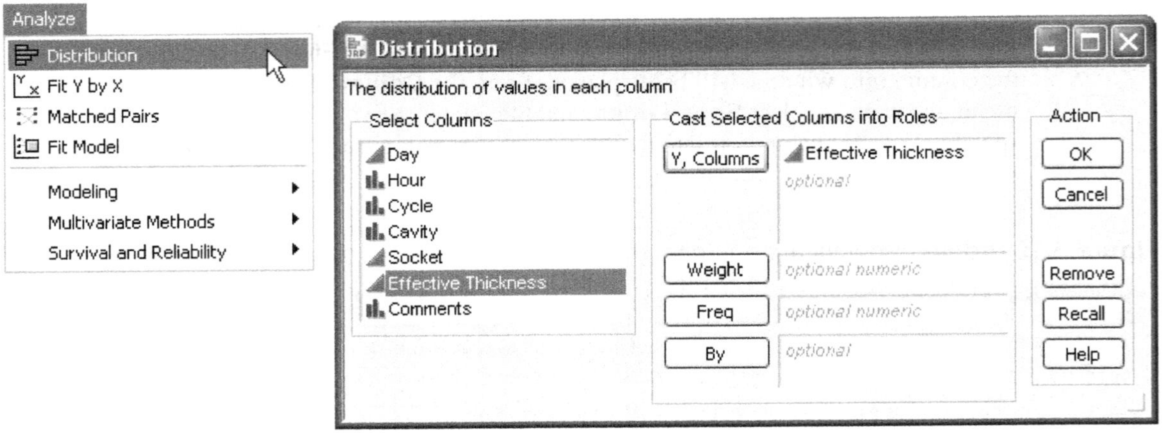

Figure 3.20a shows the histogram, box plot, and the descriptive statistics for the effective thickness. In the box plot, we notice a point in the upper tail of the box plot far away from the distribution of the data. If we hover over it with the cursor, it reveals the row number of the observation. The maximum effective thickness shown in the **Quantiles** output is 153.58, which seems odd.

Figure 3.20a Histogram of Socket Thickness Data

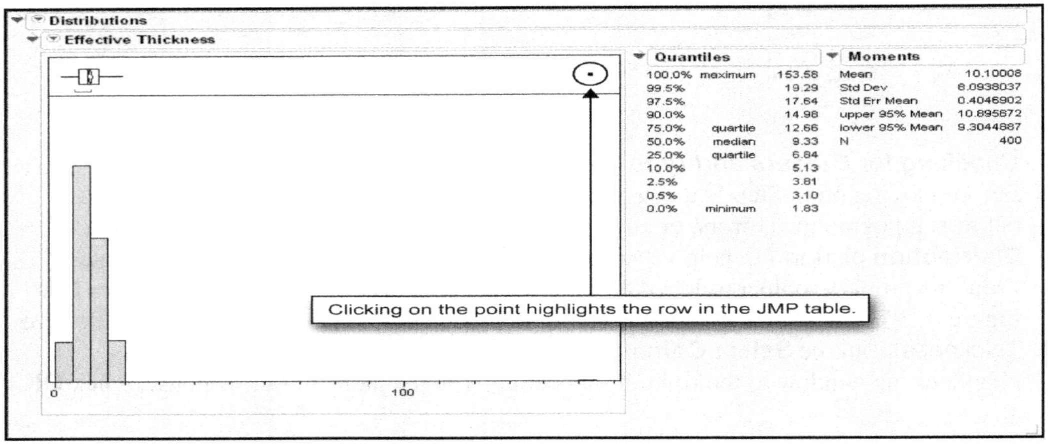

One of the powerful features of JMP is that all tables and graphs are linked. This means that if we click this far away point on the graph, the corresponding row in the JMP table is highlighted (see Figure 3.20b), which helps us to identify and locate this outlying observation. We can quickly see that that the value for Effective Thickness in row 309 is 153.58. After checking the notes, we realized that this number has the decimal point in the wrong place. This can be easily fixed by double-clicking in row 309 in the JMP table and re-entering the correct number, which is 15.358.

Figure 3.20b Outlier in Socket Thickness Data

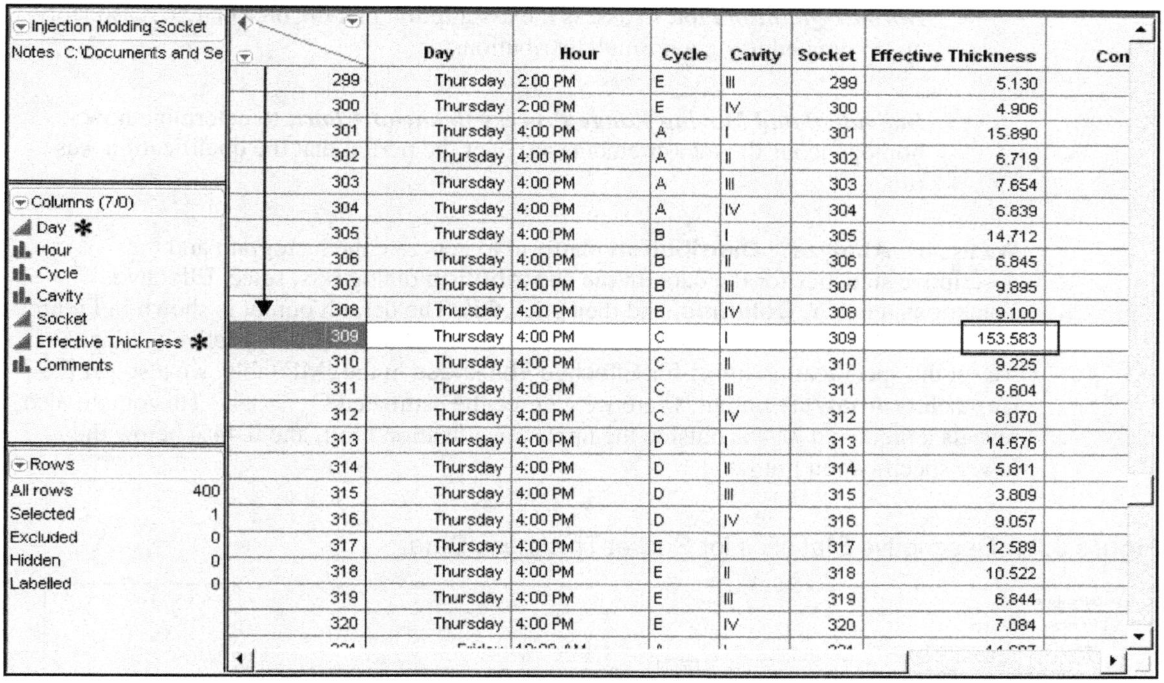

	Day	Hour	Cycle	Cavity	Socket	Effective Thickness	Con
299	Thursday	2:00 PM	E	III	299	5.130	
300	Thursday	2:00 PM	E	IV	300	4.906	
301	Thursday	4:00 PM	A	I	301	15.890	
302	Thursday	4:00 PM	A	II	302	6.719	
303	Thursday	4:00 PM	A	III	303	7.654	
304	Thursday	4:00 PM	A	IV	304	6.839	
305	Thursday	4:00 PM	B	I	305	14.712	
306	Thursday	4:00 PM	B	II	306	6.845	
307	Thursday	4:00 PM	B	III	307	9.895	
308	Thursday	4:00 PM	B	IV	308	9.100	
309	Thursday	4:00 PM	C	I	309	153.583	
310	Thursday	4:00 PM	C	II	310	9.225	
311	Thursday	4:00 PM	C	III	311	8.604	
312	Thursday	4:00 PM	C	IV	312	8.070	
313	Thursday	4:00 PM	D	I	313	14.676	
314	Thursday	4:00 PM	D	II	314	5.811	
315	Thursday	4:00 PM	D	III	315	3.809	
316	Thursday	4:00 PM	D	IV	316	9.057	
317	Thursday	4:00 PM	E	I	317	12.589	
318	Thursday	4:00 PM	E	II	318	10.522	
319	Thursday	4:00 PM	E	III	319	6.844	
320	Thursday	4:00 PM	E	IV	320	7.084	

Step 5: Characterize your data and answer the questions of interest

In Step 2 we specify the expected performance of the effective thickness in terms of the average (μ), variation (σ), and process performance (C_{pk}). Recall for the new material:

- $\mu = 7.5$ hundredths millimeter.
- $\sigma \leq 2.5$ hundredths millimeter.
- $C_{pk} \geq 1$.

High-Level Data Description

Let's start the analysis by looking at the overall performance of the qualification runs using the following:

- ***Histogram and Descriptive Statistics***: to look at the distribution of effective thickness measurements, obtain estimates for the mean and standard deviation, along with their confidence intervals, and check the performance to specifications using process capability indices. The estimates can be compared to the historical numbers from the previous supplier.

- ***Normal Quantile Plot***: to assess the assumption that the data can be reasonably approximated using a normal distribution.

- ***Individual and Moving Range Process Behavior Chart***: to determine how homogeneous the measurements are over the period that the qualification was run.

We use the **Analyze > Distribution** platform to generate the histogram and the descriptive statistics for the data. In the **Distribution** dialog box, select Effective Thickness, click **Y, Columns**, and then click **OK.** The default output is shown in Figure 3.21. The sample mean and standard deviation are 9.75 and 3.72, respectively. Since we set the specification limits for Effective Thickness in the JMP table, we also get the **Capability Analysis** output where we can see the estimated $C_{pk} = 0.47$. This output also reveals a predicted 7.94% outside the upper specification limit, and 0.44% below the lower specification limit.

Figure 3.21 Descriptive Statistics for Socket Thickness Data

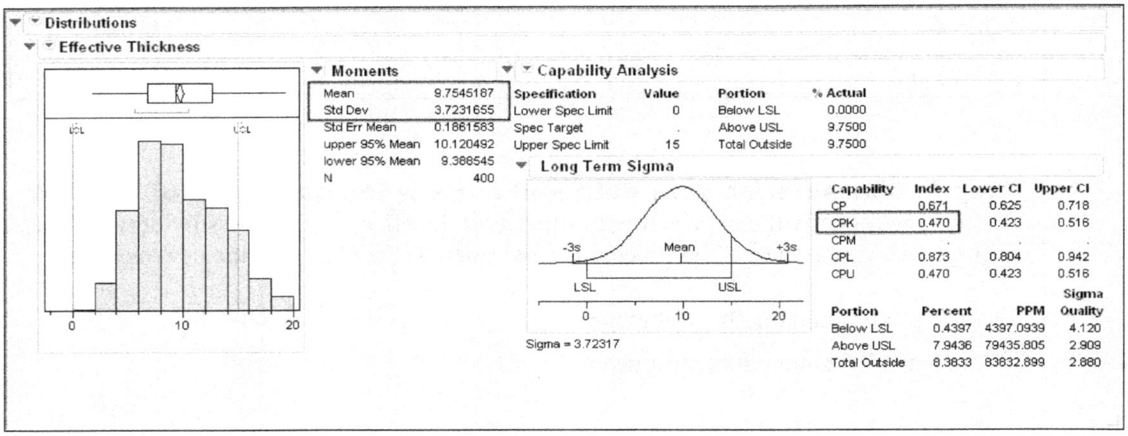

We can also see that our sample estimates are far off from their desired values. Our sample mean of 9.75 hundredths millimeters is 2.25 units higher than our target value of 7.5 hundredths millimeters; our sample standard deviation is almost 1.5 times larger than our target value of 2.5; and our process capability, 0.47, is significantly smaller than the desired $C_{pk} \geq 1$.

Additional output can be added to the default JMP Distributions output, shown in Figure 3.21, by clicking the red triangle next to the Effective Thickness label at the top of the window. The available options are shown in Figure 3.22, where we have selected the ones needed for our characterization: the **Normal Quantile Plot** and **Confidence Interval**.

Figure 3.22 Additional Distribution Options

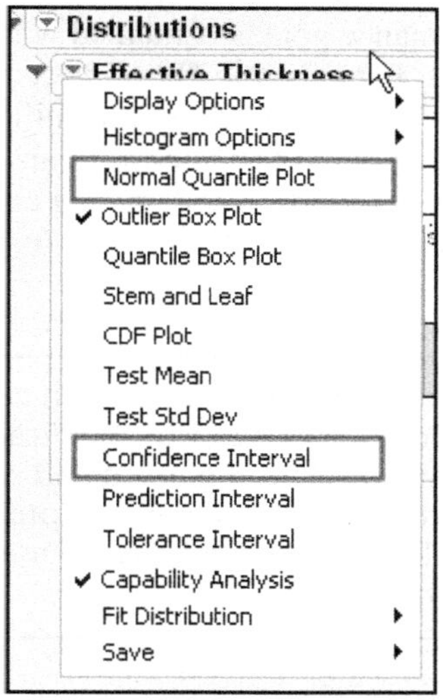

A confidence interval for the mean is automatically provided at the bottom of the **Moments** section of the default output, and the one for the C_{pk} is part of the **Capability Analysis** section. However, an interval for the standard deviation is not. When the **Confidence Interval** option is selected a window comes up that enables us to enter our choice for the confidence level = 1 – α, to select a one or two sided interval, and to enter a known value for the standard deviation (Sigma). Our choices for the socket thickness data, along with the resulting intervals, are provided in Figure 3.23.

Figure 3.23 95% Confidence Intervals for Mean and Standard Deviation

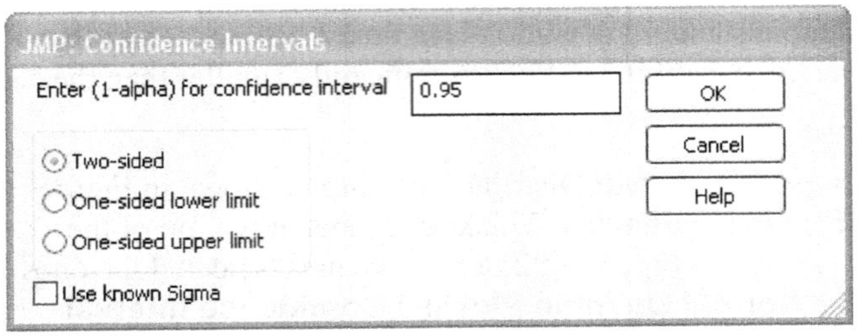

Confidence Intervals

Parameter	Estimate	Lower CI	Upper CI	1-Alpha
Mean	9.754519	9.388545	10.12049	0.950
Std Dev	3.723165	3.481801	4.000761	

Confidence intervals for the mean, standard deviation, and C_{pk} can provide us with an even stronger indication that the effective thickness data is not performing as expected. This is accomplished by determining if the performance goal is contained within the corresponding upper and lower confidence bounds. For example, the 95% confidence interval for mean effective thickness estimate goes from 9.39 hundredths millimeters to 10.12 hundredths millimeters, and we can see that the expected performance of 7.5 hundredths millimeters is not contained within these bounds. Similarly, the confidence intervals for the standard deviation and C_{pk} do not contain the expected performance values of 2.5 and 1, respectively.

Statistics Note 3.7: The goal in this chapter is to provide you with a quick way to explore and characterize your data. More formal tests of significance are introduced in Chapter 4 (using one-sample tests for the mean and standard deviation) to compare the measured performance of a material, product, or process with a standard. We also show the relationship between these tests and the corresponding confidence intervals.

When the **Normal Quantile Plot** option is selected, a normal quantile plot is added to the output directly above the histogram if the stacked option has been selected. The plot in Figure 3.24 shows the effective thickness data falling outside of the upper confidence band. The histogram shows a unimodal distribution with a longer tail to the right; that is, the distribution is skewed to the right, which is also suggested by the mean = 9.755 > median = 9.329.

Figure 3.24 Normal Quantile Plot Socket Thickness Data

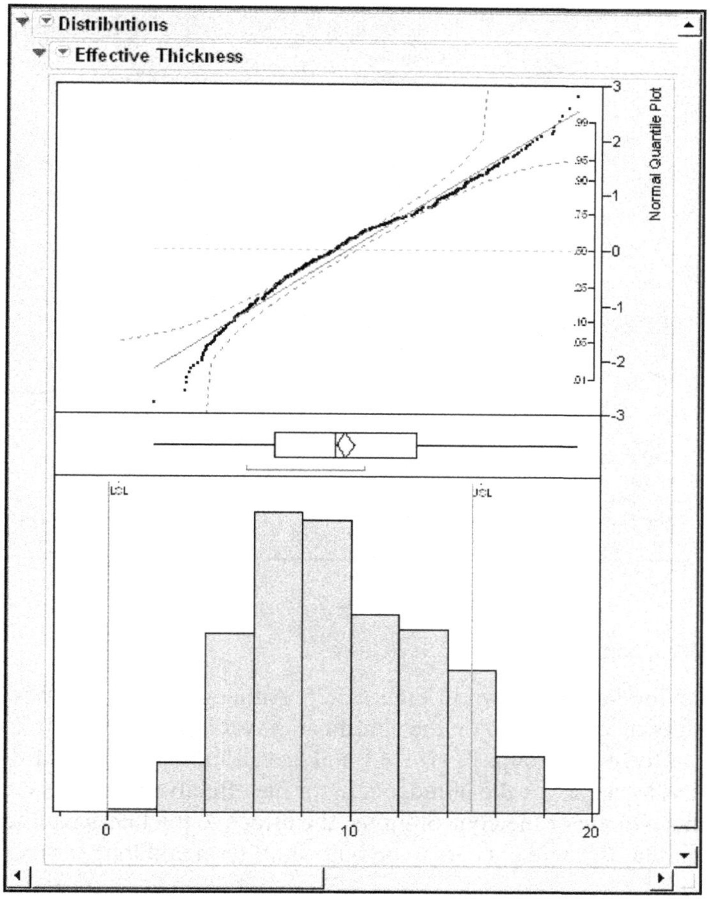

Finally, to check the homogeneous assumption, a process behavior chart for the data can be created using the **Graph > Control Chart** platform and following these steps:

1. Select **Graph > Control Chart > IR** from the main menu (Figure 3.25).
2. From within the **Control Chart** dialog box, select Effective Thickness under **Select Columns**, and click **Process** to assign it to the role of the process variable. Then select Socket and click **Sample Label** to assign this variable to the subgroup (Figure 3.25).

Figure 3.25 Control Chart Dialog Box for Socket Thickness Data

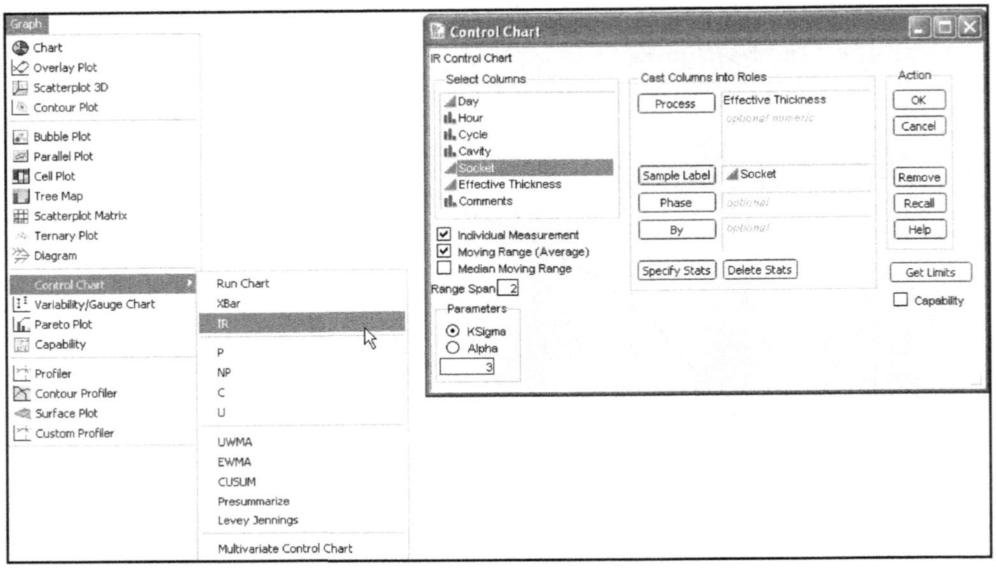

3. Click **OK** when complete.

The process behavior chart is shown in Figure 3.26. Although we can form subgroups in our data in a variety of ways based on the stratification variables (day, hour, cycle, cavity), we have chosen to form subgroups of size = 1 and use an Individual and Moving Range chart as a first pass to check for the homogeneity of the effective thickness data. While there are no points outside of the control limits, the effective thickness measurements do not seem random; that is, there appears to be both short term and long term cycles in the data.

Figure 3.26 Process Behavior Chart for Socket Thickness Data

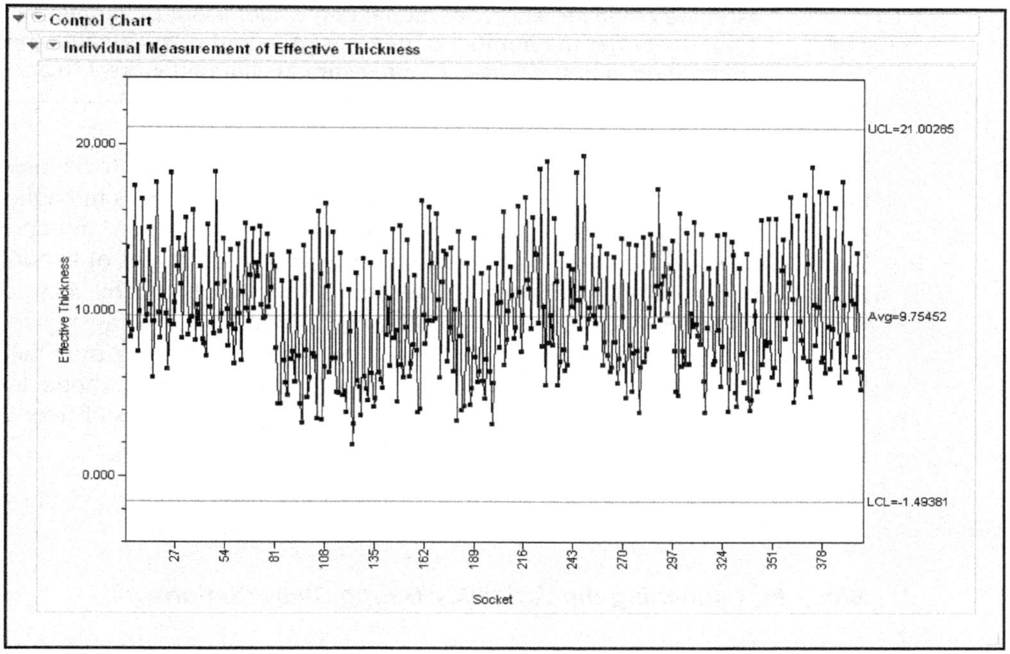

Understanding Sources of Variation

Now that we have discovered that the molding process with the raw material substitution is incapable of performing as expected, we must figure out how to reduce our variation and shift down the process average closer to 7.5. In other words, we would like to try to uncover any systematic trends or patterns in our data, using box plots, for example, and to quantify the sources of variation using Variance Components Analysis. The sources of variation in the data, Table 3.6, include day-to-day, hour-to-hour, cycle-to-cycle, and cavity-to-cavity. This ordering also reflects the hierarchy of the data, which is relevant in the subsequent steps used to graph and quantify variation. The day-to-day variation is at the top of the data hierarchy. Since only five days were used in the study, we have only 4 (5–1) degrees of freedom to estimate this variance component, which provides a "soft" estimate (See Table 7.10) of the day-to-day variation.

Statistics Note 3.8: When estimating variation, the degrees of freedom of the estimate are inversely proportional to the coefficient of variation (Equation 7.12). The larger the number of degrees of freedom, the smaller the coefficient of variation, and the better we can estimate sigma (Figure 7.20 and Table 7.10).

The second level in the data hierarchy is hour-to-hour variation, or to be more precise, the two-hour-to-two-hour variation, since data was taken four times throughout the day, or every two hours starting at 10:00 a.m. until 4:00 p.m. At this level of our hierarchy, hour within day, the hour-to-hour variation estimate has 15 degrees of freedom coming from the 5 days × (4−1) hour periods. The third level in the hierarchy is cycle-to-cycle variation, which consists of five consecutive cycles of the press within a given hour. This source of variation is estimated with 80 degrees of freedom coming from the 5 days × 4 hour periods × (5−1) cycles. Finally, the cavity-to-cavity variation is at the lowest level of the data hierarchy and is estimated with a whopping 300 degrees of freedom, coming from the 5 days × 4 hours × 5 cycles × (4−1) cavities.

The best tool within JMP for visualizing and estimating the different variance components is the **Graph > Variability/Gauge Chart** platform.

Steps for Launching the Variability/Gauge Chart Platform

1. From the primary toolbar, select **Graph > Variability/Gauge Chart**. A dialog box appears that enables us to make the appropriate choices for our data.
2. Select Effective Thickness from the **Select Columns** window, and then click **Y, Response**. This populates the window to the right of this button with the name of our response.
3. Select Day from the **Select Columns** window, and then click **X, Grouping** in the right-hand side of the window. This populates the window to the right of this button with the name of the factor under investigation.
4. Click **OK**. The default JMP output for the socket data is shown in Figure 3.27.

Figure 3.27 Default Variability Chart for Socket Thickness Data: Two Sources of Variation

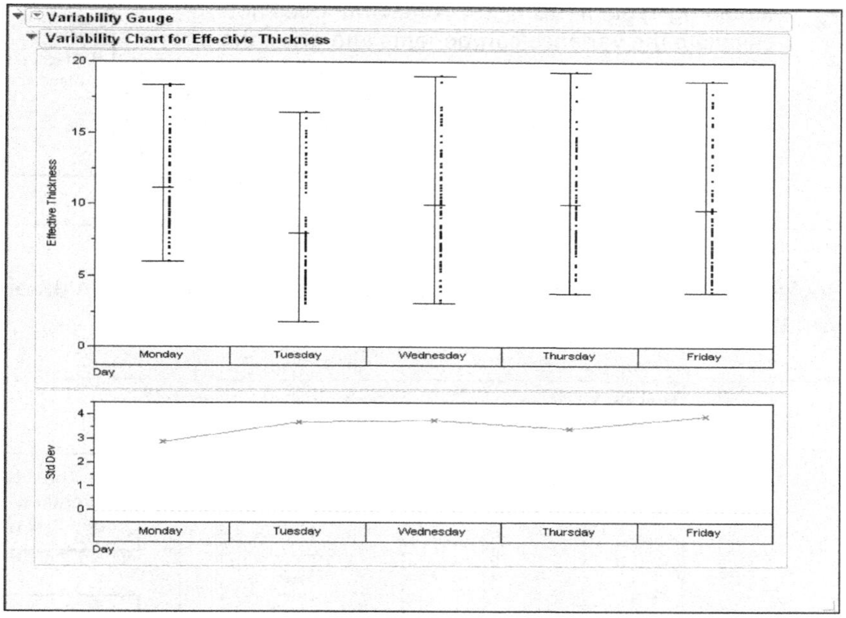

5. To enhance the output, click the red triangle at the top of the window. Turn off the range bars by selecting the label **Show Range Bars**, and add box plots and connect the cell means by selecting **Show Box Plots** and **Connect Cell Means**, respectively. Finally, variance components are obtained by selecting **Variance Components**. Figure 3.28 shows theses selections and the resulting output.

JMP Note 3.3: In order to run the variance components analysis we need to make sure that the factors in the **X, Grouping** area have the **Nominal Modeling Type**. If this is not done, JMP does not calculate the variance components and gives the error message "Grouping columns cannot be Continuous." This can be accomplished by double-clicking Day and selecting **Nominal** from the drop-down menu of **Modeling Type**. On the data table, you can also click the blue triangle next to the Day name, and select **Nominal** from the drop-down menu.

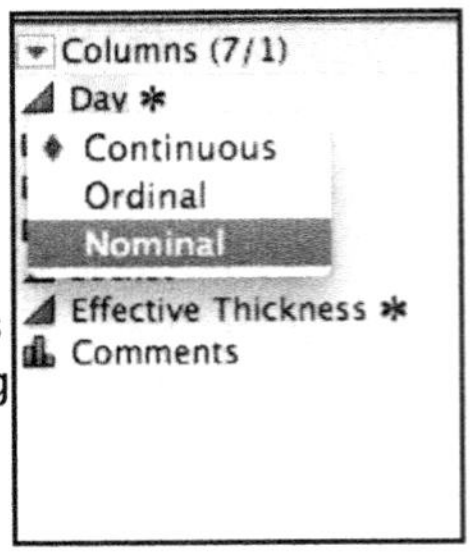

Figure 3.28 Revised Variability Chart for Socket Thickness Data: Two Sources of Variation

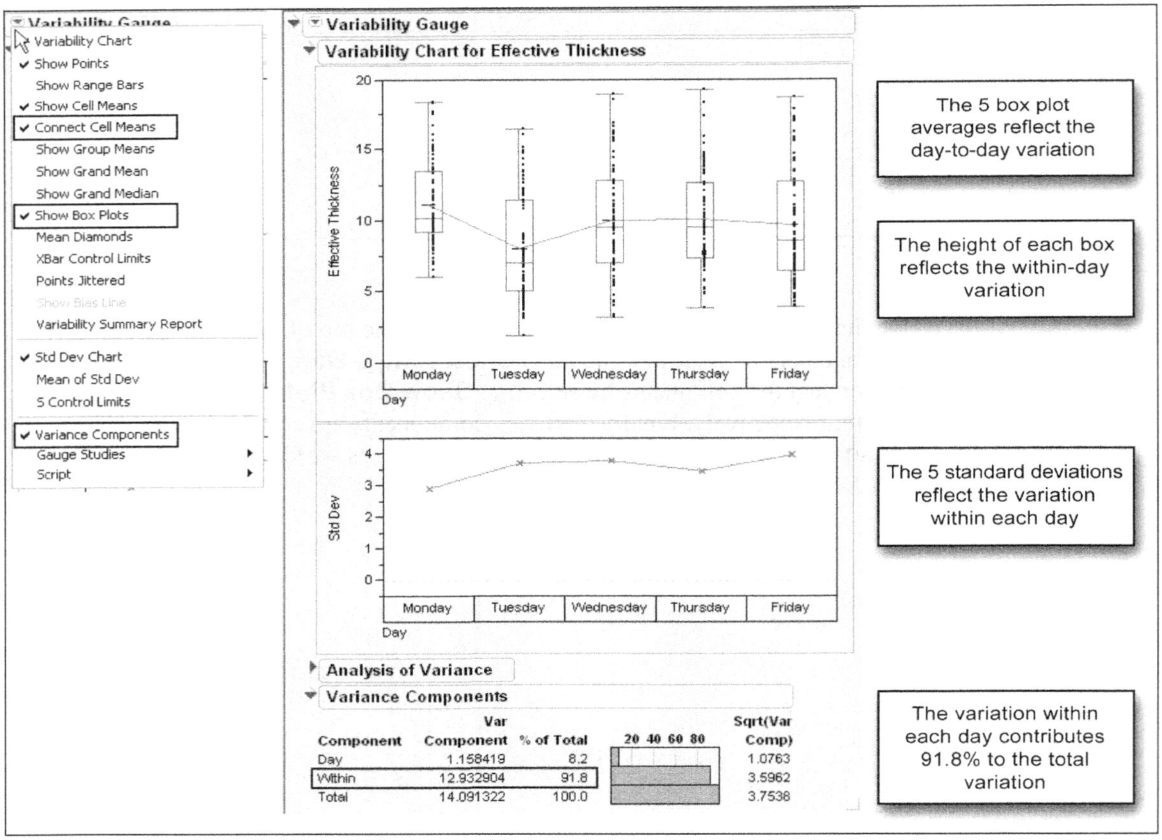

The first stratification of our data enables us to compare the day-to-day variation with the within-day variation. The daily variation is observed by looking at the differences in the five averages representing the five days of the qualification. For example, the average effective thickness on Monday appears to be close to 11, Tuesday is close to 8, and, the averages for Wednesday through Friday are closer to 10, giving a difference around 3 units between Tuesday and the other days. The height of each box in Figure 3.28 gives us a sense of the variation within each day. For the Friday measurements, we see a large range with the lowest effective thickness close to 4 and a highest effective thickness around 19. In addition to comparing the height of the five boxes to see how consistent the within-day variation is throughout the week, the standard deviation plot (directly below) shows the standard deviations to be hovering between three- and four-hundredths of a millimeter.

The box plots in Figure 3.28 suggest that the within-day variation is larger than the day-to-day variation. What is the actual contribution of each variance component? The **Variance Components** analysis at the bottom of the output shows that the total variation in the effective thickness data ($s^2 = 14.09$) consists of the day-to-day variation (1.158 [8.2% of the total]) and the within-day variation (12.93 [91.8% of the total]). As we saw in the plot, the majority of the variation is coming from the variation present within each day, which consists of the variation coming from hours, cycles, and cavities within the mold. We need to break out the within-day variation into these sources of variation in order to discover where the largest contributor to the within-day variation is.

> **JMP Note 3.4:** It is important how the variables are entered in the **X, Grouping** field because how they are entered determines how the variables are displayed in the plot, and how the variance components are calculated. The ordering should reflect the hierarchy in the data.

The next stratification of our data pulls out the hour-to-hour variation from the within-day variation shown in Figure 3.28. This is accomplished by adding the **Hour** variable to the **X, Grouping** role in the **Variability/Gauge** dialog box, as is shown in Figure 3.29. In order to calculate the variance components, on the bottom left-hand side of the dialog box, select **Model Type** and then **Nested** from the drop-down menu. The JMP output is shown in Figure 3.30. The same options were chosen to alter the default output as was shown in Figure 3.28 except with the addition of **Show Group Means**, **Mean of Std Dev**, and **S Control Limits**.

Figure 3.29 Variability/Gauge Dialog Box: Three Sources of Variation

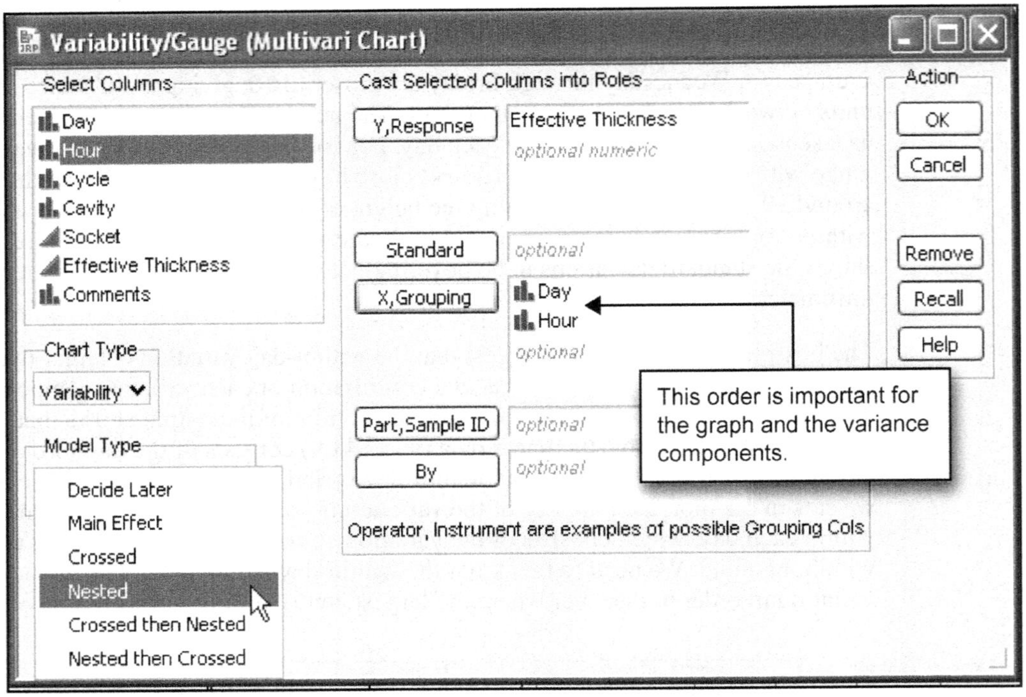

Figure 3.30 now contains 20 box plots, 5 days × 4 hour periods within a day, and the differences between the 20 box plot averages now reflects the daily and hourly variation from the socket study. The five straight lines shown in the plots reflect the same five daily averages described previously, while the jagged lines that connect the box plots averages within each day reflect the hour-to-hour variation. Finally, the height of each box now consists of the variation of the effective thickness measurements coming from the four sockets from the four cavities, for each of the five consecutive cycles every two hours. Looking closely at the box plots, it looks like the variation within each box (as represented by the height of each box) is larger than the variation between the boxes.

To confirm our observations, we rely on the variance components analysis at the bottom of Figure 3.30. The total variation in the effective thickness measurements ($s^2 = 14.09$) consists of the day-to-day variation (1.04 [7.4% of the total]), the hour-to-hour variation (0.5 [3.5% of the total]), and the within-day variation (12.56 [89.1% of the total]). As suspected, the majority of the variation is now coming from variation observed within the hour period, consisting of five cycles within an hour period, and the four cavities within the mold.

Figure 3.30 Variability Chart for Socket Thickness Data: Three Sources of Variation

Variability Gauge

Variability Chart for Effective Thickness

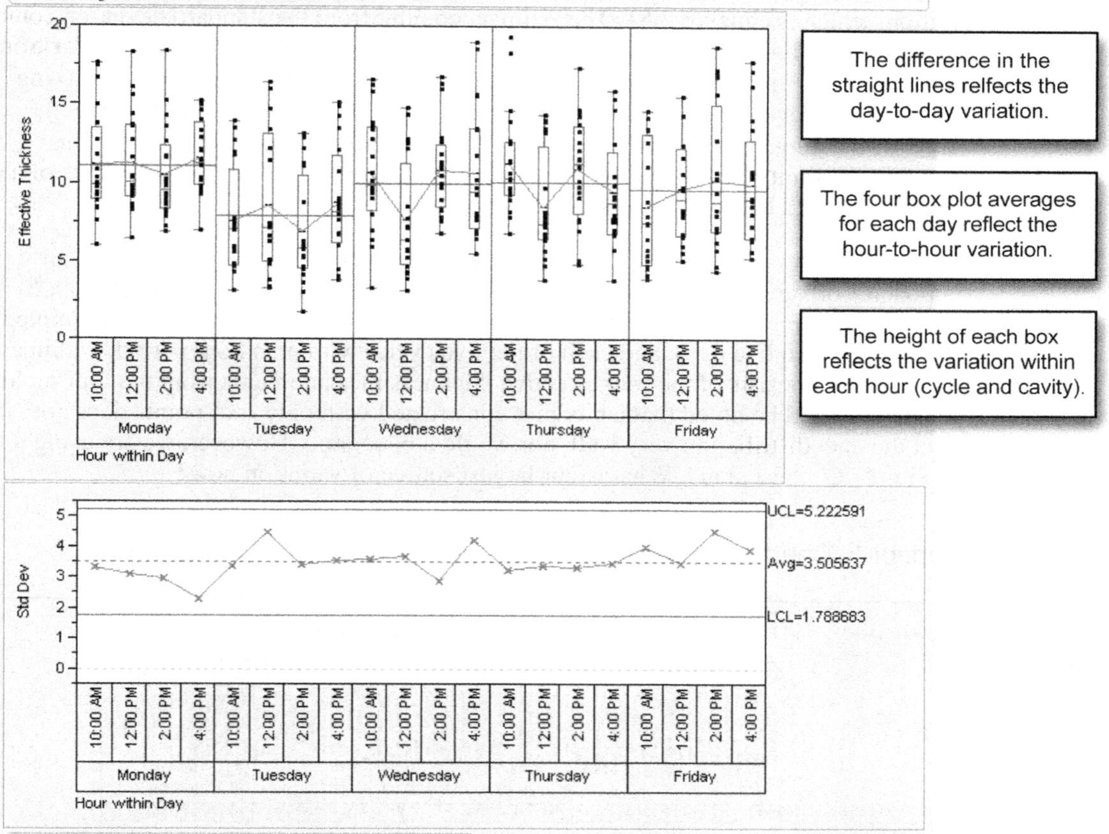

Analysis of Variance

Source	DF	SS	Mean Square	F Ratio	Prob > F
Day	4	422.4256	105.606	4.70751	0.0116*
Hour[Day]	15	336.504	22.4336	1.78642	0.0347*
Within	380	4771.993	12.5579		
Total	399	5530.922	13.862		

Variance Components

Component	Var Component	% of Total	20 40 60 80	Sqrt(Var Comp)
Day	1.039660	7.4		1.0196
Hour[Day]	0.493786	3.5		0.7027
Within	12.557876	89.1		3.5437
Total	14.091322	100.0		3.7538

The Within (cycle and cavity) variation contributes 89.1% to the total variation.

The plot below the box plot chart in Figure 3.30 displays 20 standard deviations, one for each of the day-hour combinations. This plot has been enhanced by selecting **Mean of Std Dev** and **S Contol Limits** from the options menu to the left of the **Variability Gauge title** (Figure 3.28). The centerline is an estimate of the average within-day-hour variation, which is equal to 3.51. The estimate coming from the standard deviation control chart is very similar to the within-variance component estimate (3.54) from the **Variance Components** section of the output. The control limits serve as a guide for identifying those standard deviations that are smaller or larger than the average, and they give us a sense of the range of the within-day-hour variation. All the 20 standard deviations are within the limits, and the within-day-hour variation goes from 1.79 hundredths of a millimeter to 5.22 hundredths of a millimeter.

Figure 3.31 shows the variability chart and variance component analysis for the four components of variation in the injection molding study. The output has 100 (5 x 4 x 5) box plots, one for each of the five days of the week, each the four hour periods sampled within each of the five days, and each of the five cycles within an hour period, within a day. Each box consists of the four effective thickness measurements from the four mold cavities. It should be noted that a box plot constructed with only four points does not reflect the data distribution very well, nor would a histogram. However, we are using it here simply as a visual aid. What is our largest source of variation now?

Figure 3.31 Variability Chart for Socket Thickness Data: Four Sources of Variation

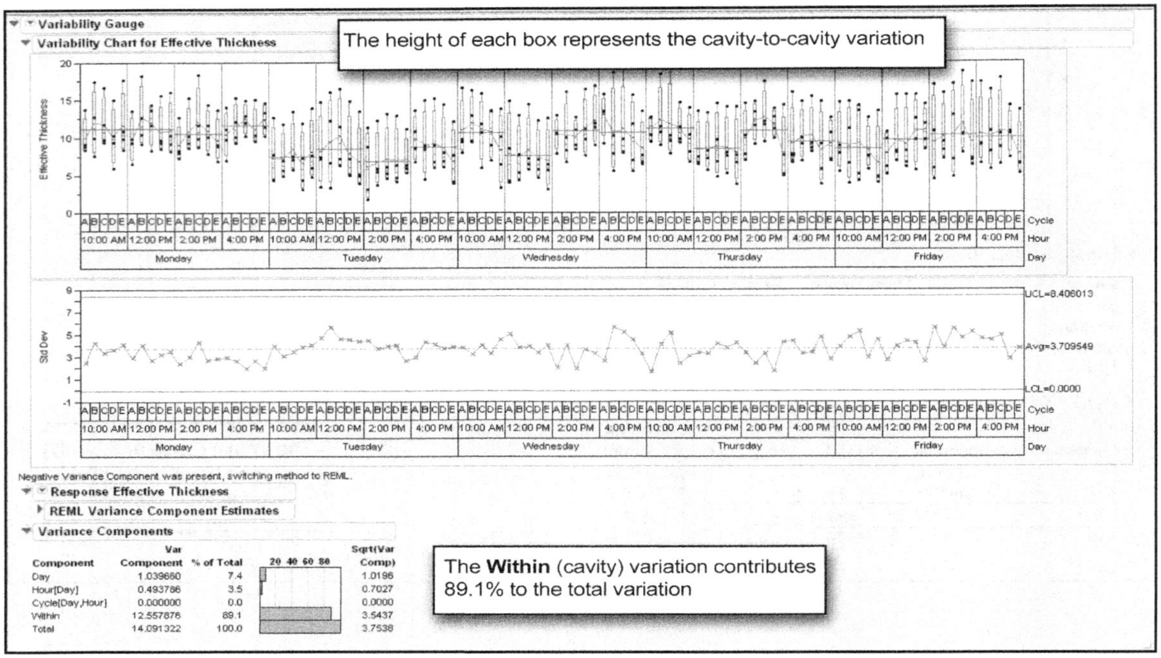

The **Variance Components** analysis at the bottom of the variability chart output shows that our total variation ($s^2 = 14.09$) consists of the day-to-day variation (1.04 [7.4% of the total]), the hour-to-hour variation (0.5 [3.5% of the total]), cycle-to-cycle variation (0 [0% of the total]), and cavity-to-cavity within variation (12.56 [89.1% of the total]). It is clear now that the largest source of variation in the effective thickness data is due to differences between the four cavities. The plot of the cavity standard deviations in the JMP output suggests that the cavity-to-cavity variation is pretty consistent from day to day, hour to hour, and cycle to cycle. In the standard deviation chart no standard deviations fall outside of the control limits, which also supports the claim that cavity variation is fairly consistent.

Statistics Note 3.9: Why is the estimate for the cycle-to-cycle variation zero? Does that mean that there is no variation across the five consecutive cycles, A, B, C, D, and E? If we examine the five cycle averages sampled every two hours that are shown in Figure 3.31, we see that there are small differences between the means, so the cycle-to-cycle variation is definitively not zero. Whenever we see a variance component equal to zero, it is an indication that the variance components below it in the hierarchy are large, which in turn drives the variance component to zero. In this case the variance component within Cycle (the cavity-to-cavity variation) is really large and accounts for almost 90% of the total variation. The cycle-to-cycle variation, being so small as compared to the cavity-to-cavity variation, is driven to zero. Here it is obvious that the first order of business is to decrease the cavity-to-cavity variation. Once this is done the cycle-to-cycle variation will surface.

Understanding the Cavity-to-Cavity Variation

The easiest way to see what is going on between the cavities is to add different plotting colors and symbols to the box plot to see whether there is a systematic pattern between the four cavities. To accomplish this, right-click in the top plot shown in Figure 3.31, choose **Row Legend**, and then selecting Cavity from the dialog box. Also select the Set **Marker by Value** check box so we can get different plotting symbols and colors for each value of Cavity (see Figure 3.32). The updated variability chart is shown in Figure 3.33a. In order to show more of the detail, only the Monday data is shown in Figure 3.33b.

Figure 3.32 Add Plotting Symbols Using Row Legend

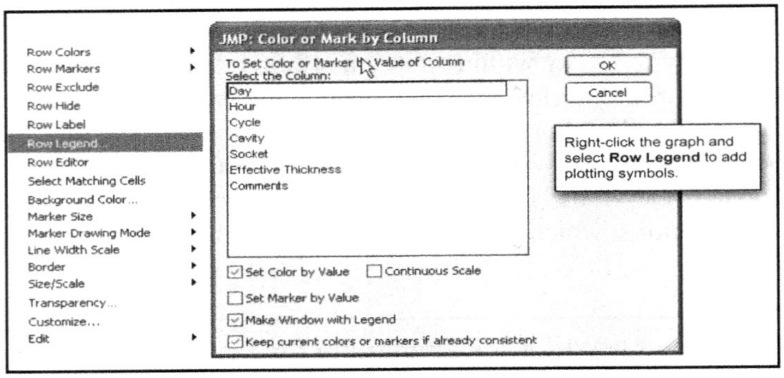

Figure 3.33a Variability Chart for Socket Thickness Data with Different Plotting Symbols

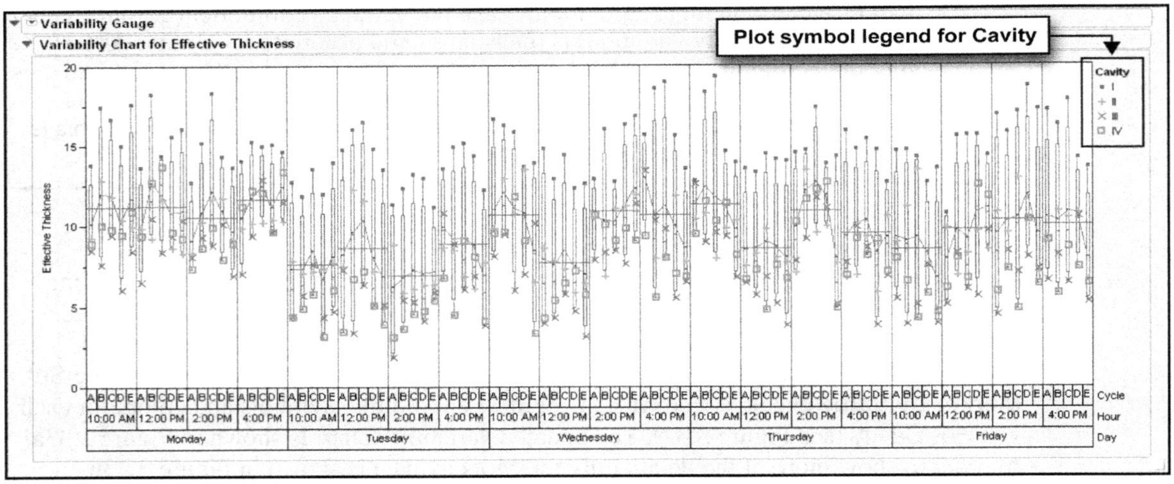

Figure 3.33b Variability Chart for Socket Thickness Data (Monday Only)

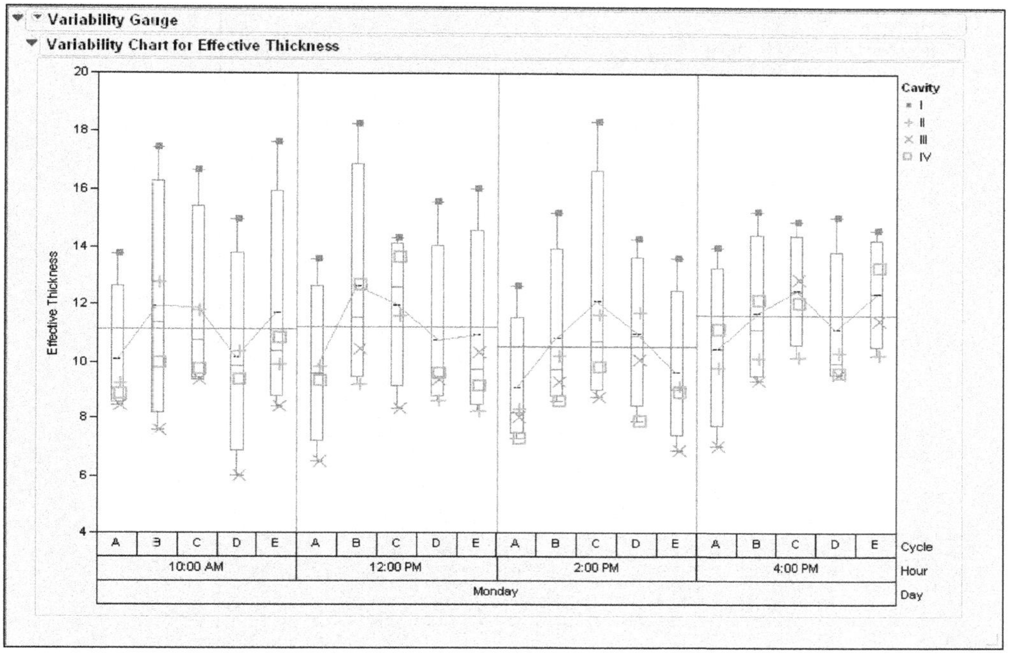

We can now clearly see why cavity is contributing so much variation to the total effective thickness variation. For every cycle of the press, Cavity I is making sockets with a higher effective thickness than the other three cavities. For completeness, we will now examine the cavity-to-cavity component using the **Variability/Gauge** platform.

In the dialog box for the **Variability/Gauge** platform, we select Cavity for the **X**, **Grouping** factor, Effective Thickness for the **Y**, **Response field**, and click **OK**. The JMP output is shown in Figure 3.34 where we see that Cavity I has an average Effective Thickness of 14.7687 hundredths millimeter, while Cavities II, III, and IV, have an average Effective Thickness of 9.07, 7.32, and 7.86 hundredths of a millimeter, respectively. It is also interesting to note that the standard deviations for each cavity seem to be increasing by about 0.2 hundredths of a millimeter as a function of Cavity. The **Variability Summary Report** option was used to get the sample means and standard deviations shown underneath the box plot in Figure 3.34.

Figure 3.34 Variability Chart for Socket Thickness Data by Cavity

Variability Gauge

Variability Chart for Effective Thickness

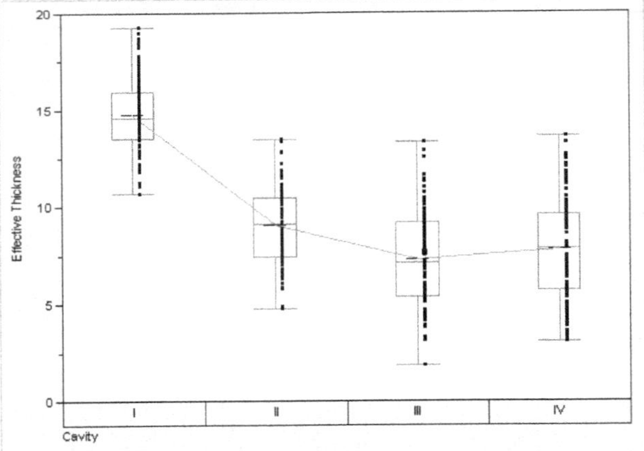

Variability Summary for Effective Thickness

	Mean	Std Dev	Std Err Mean	Lower 95%	Upper 95%	Minimum	Maximum	Observations
Effective Thickness	9.754519	3.723165	0.186158	9.388545	10.12049	1.833202	19.29565	400
Cavity[I]	14.7687	1.857236	0.185724	14.40019	15.13722	10.68982	19.29565	100
Cavity[II]	9.067703	2.084421	0.208442	8.654109	9.481297	4.77889	13.48978	100
Cavity[III]	7.323191	2.393551	0.239355	6.848259	7.798124	1.833202	13.37937	100
Cavity[IV]	7.858479	2.621044	0.262104	7.338407	8.378551	3.103183	13.65152	100

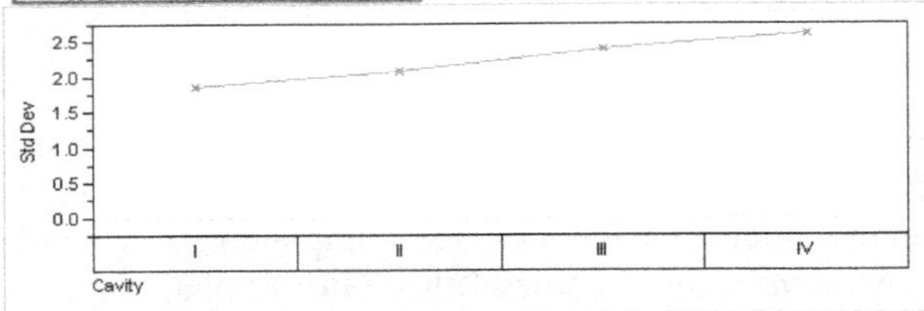

Process Performance by Cavity

The process capability results using all of the data gave a C_{pk} equal to 0.47, a mean equal to 9.75, and a standard deviation equal to 3.72, which fell short of our expected performance goals of a C_{pk} equal to 1, a mean equal to 7.5, and a standard deviation equal to 2.5. As we saw in Section 3.3, when we combine different populations together we can get a distorted picture because the variation could be inflated and the mean biased by other components in the data. Based on what we learned about the cavity-to-cavity differences, we should perform the capability analysis for each cavity and see how it compares to our goals.

An analysis by cavity is easily done by taking advantage of the **By** option in the **Analyze > Distribution** platform dialog box. For this we put Cavity in the **By** window, as is shown in Figure 3.35, in order to calculate the capability indices for each cavity.

Figure 3.35 Dialog Box for Analyze > Distribution Using By Statement

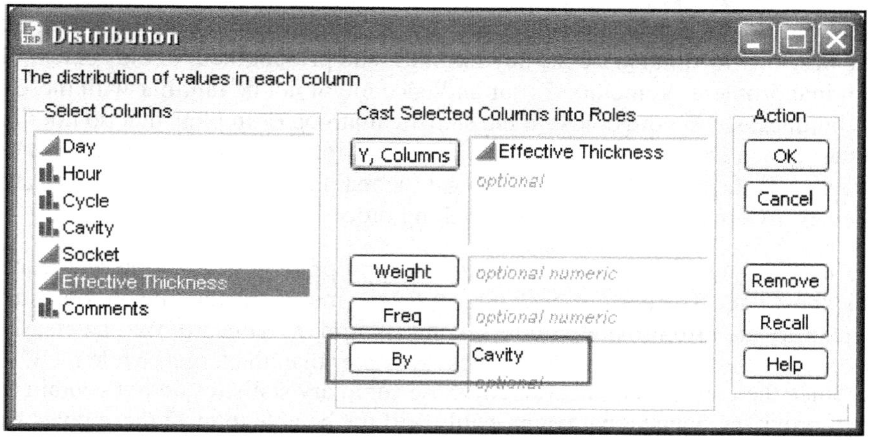

Figure 3.36 shows the C_{pk} values for each of the four cavities: Cavity I = 0.042, Cavity II = 0.949, Cavity III = 1.020, and Cavity IV = 0.908. Note that the process capability for Cavity I is very low while for Cavities II, III, and IV the process capability is close to 1.

Figure 3.36 Process Capability of Socket Thickness Data by Cavity

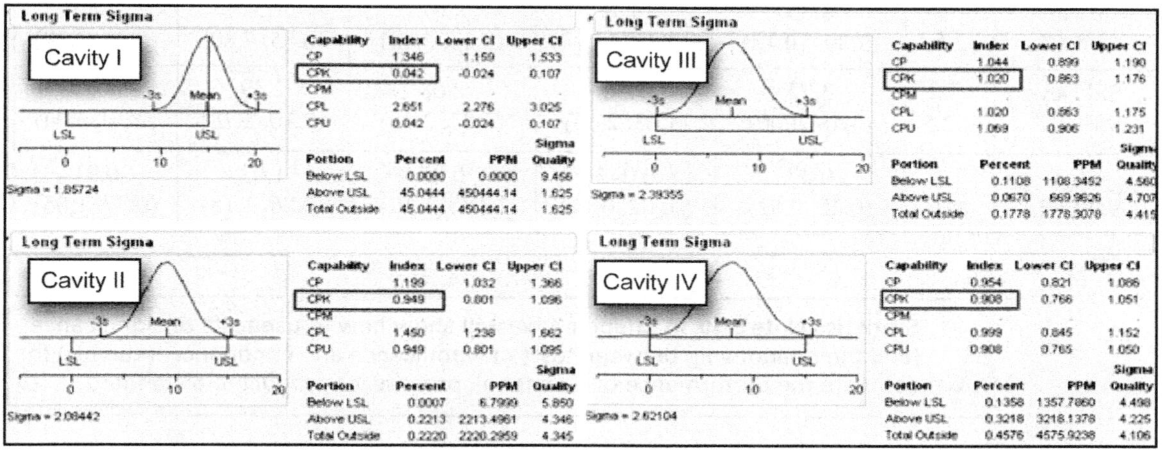

Step 6: Summarize your results with key graphs and summary statistics

We need to sort through all of the output and analysis that we did in Step 5 and determine the best way to summarize the key findings, and present them in the context of the original problem. Remember, your audience might not be familiar with these tools and techniques, so do not get lost in the statistical jargon or in plots that do not illustrate the issues at hand. A combination of the right descriptive statistics, statistical intervals, and graphics is the most effective way to get the main points across for the characterization of the new raw material in the socket-making process.

Table 3.8 provides summary statistics and appropriate statistical intervals for the socket thickness data. The performance targets for the mean, standard deviation and process capability are also provided in the second column of Table 3.8. We can clearly see that for Cavity I the average effective thickness is higher than the target, while the C_{pk} is less than 1. Since the confidence intervals for these summary statistics do not contain the targets, they provide evidence that we probably will not be able to meet these goals with Cavity I. For Cavity II the average effective thickness is also higher than the target.

Table 3.8 Performance Characterization of Socket Thickness Data

Summary Measure	Goal	All Cavities	Cavity I	Cavity II	Cavity III	Cavity IV
Mean 95% CI	7.5	9.75 (9.39, 10.12)	14.77 (14.40, 15.14)	9.07 (8.65, 9.48)	7.32 (6.85, 7.80)	7.86 (7.34, 8.38)
Std Dev 95% CI	≤ 2.5	3.72 (3.48, 4.00)	1.86 (1.63, 2.16)	2.08 (1.83, 2.42)	2.39 (2.10, 2.78)	2.62 (2.30, 3.04)
C_{pk} 95% CI	≥ 1	0.47 (0.42, 0.52)	0.042 (-0.02, 0.11)	0.95 (0.80, 1.10)	1.02 (0.86, 1.18)	0.91 (0.77, 1.05)

Statistics Note 3.10: In Chapter 4 we will show how to use tests of significance (and the relationship between tests of significance and confidence intervals) to compare the performance of a material, process, or product to a standard.

We are evaluating the process performance by using the confidence intervals in the following way:

- Mean, $\mu = 7.5$: if the 95% confidence interval for the mean includes the target value of 7.5, then we conclude that we have met the goal. However, if the target of 7.5 is outside of this interval, then we conclude that we have not met the goal.
- Standard Deviation, $\sigma \leq 2.5$: if the 95% confidence interval includes the target value of 2.5 or if the upper confidence limit is less than 2.5, then we conclude that we have met the goal. However, the second situation makes a stronger statement about the population standard deviation because we have shown that even the largest possible value is still less than the goal of 2.5. Finally, if the target 2.5 is not included in the interval and the lower confidence limit is greater than 2.5, then we conclude that we have not met the goal.
- $C_{pk} \geq 1$: if the 95% confidence interval includes the target value of 1 or if the lower confidence limit is greater than 1, then we conclude that we have met the goal. However, the second situation makes a stronger statement because we have shown that even the smallest possible value is still greater than 1 and therefore, we have shown that we have exceeded the goal of 1. Finally, if the target 1 is not included in the interval and the upper confidence limit is less than 1 then we conclude that we have not met the goal.

Based on these comparisons, we see that Cavities III and IV have met the goals for all three summary measures, while Cavities I and II have not met all three goals. Cavity I has a mean that is off target on the high side, contributing to the low C_{pk}, but its standard deviation is the lowest between the three cavities. Centering the Cavity I mean between the specification limits will improve the C_{pk}. Cavity II also has a desirable standard deviation and an acceptable C_{pk}, but its mean is also on the high side.

Tolerance Intervals

A tolerance interval (as described in Section 2.6.3 and Table 3.4), is a very useful interval for yield predictions because it gives bounds that contain a specified yield (as given by a percentage of the sampled population). They can be generated using the **Analyze > Distribution** platform in JMP, and are shown in Table 3.9 along with the actual and estimated percentage out of specification. For example, a 95% (confidence) tolerance interval for the total population (all cavities) was calculated by selecting **Tolerance Interval** from the drop-down menu that is displayed when you click the red triangle at the top of the JMP output window (Figure 3.37). The dialog box for this interval requires us to specify the confidence level and the containment (or proportion) for which we want the bounds calculated. We specified 99.73% in the **Specify Proportion to cover** field in this dialog box, since this is equivalent to ± 3sigma, as shown in Figure 3.37.

Figure 3.37 Tolerance Interval for All Cavities

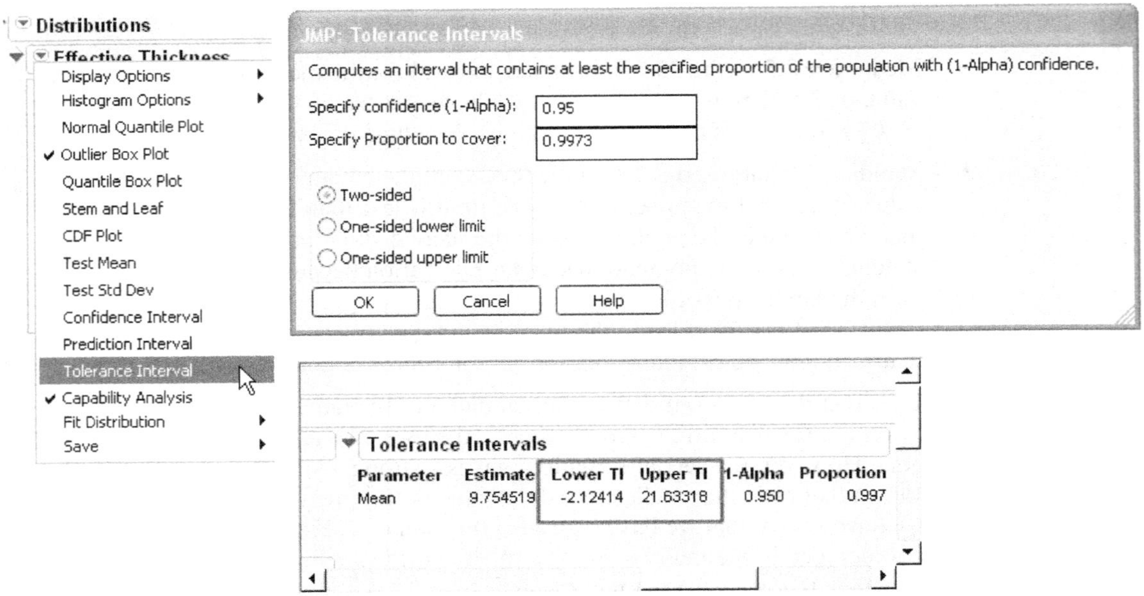

Statistics Note 3.11: The confidence intervals for the mean, standard deviation, and C_{pk}, and the tolerance interval for All Cavities in Tables 3.8 and 3.9 should be used with caution since the assumption of independence might not hold.

The estimated percentage out of specification for Cavity I is 45%, and close to 0% for the other 3 cavities. Comparing the tolerance interval for Cavity I with the specifications limits (0, 15) tells the same story. The tolerance interval indicates that 99.73% of the sockets coming from Cavity I will have an effective thickness between 8.42 and 21.12 hundredths of a millimeter; the upper bound being six units away from the upper specification of 15 hundredths of a millimeter. For the other three cavities the upper bounds of the 99.73% tolerance intervals are only 0.5 to 1.8 units above the upper specification of 15 hundredths millimeter.

Table 3.9 Performance Characterization of Socket Thickness Data

Summary Measure	All Cavities	Cavity I	Cavity II	Cavity III	Cavity IV
Actual % Out of Spec	9.75	39	0	0	0
Estimated % Out of Spec	8.38	45.04	0.22	0.18	0.46
99.73% Tolerance Interval	(-2.12, 21.63)	(8.42, 21.12)	(1.94, 16.19)	(-0.86, 15.51)	(-1.1, 16.82)

Graphical Displays

A variability chart with the specification limits LSL = 0 and the USL = 15 added as reference lines is the most effective graphical display to complement the tabular results in Tables 3.8 and 3.9. You can add reference lines to the plot by double-clicking in the y-axis when the cursor turns into a hand in the plot, entering `15` in the empty box at the bottom of the dialog box, and clicking **Add Ref Line**. One can now clearly see in Figure 3.38 that Cavities II-IV are within specifications, but about half of the Cavity I population is outside the upper specification limit.

Figure 3.38 Variability Chart for Socket Thickness Data with Reference Line

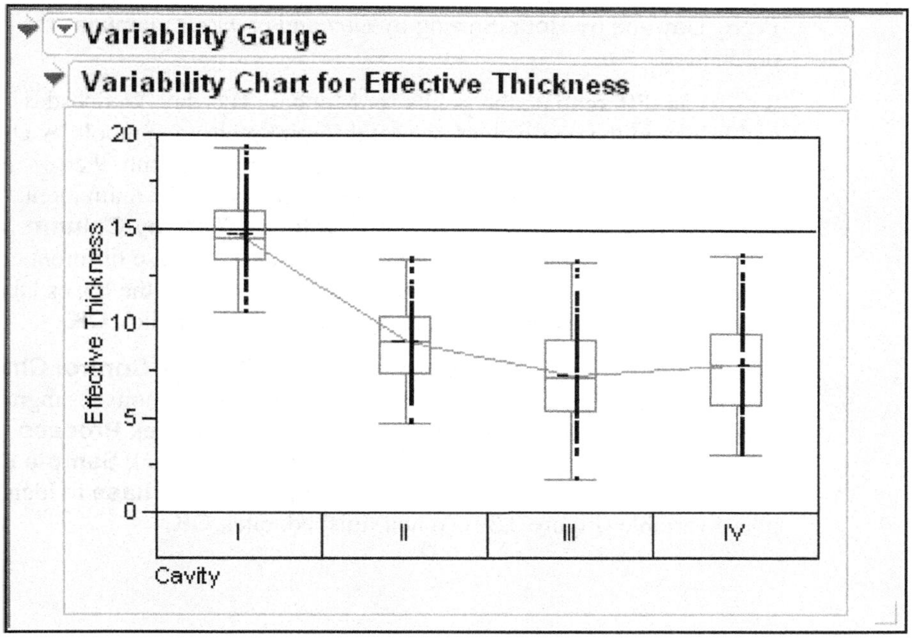

> **JMP 8 Note 3.3:** In JMP 8 the **Y Axis Specification** window has changed. The **Add Ref Line** button is now called **Add**, and is located inside the **Reference Lines** section of the window.

Phase Charts for Process Stability

Although Figure 3.38 reveals the problem with Cavity I, the time dimension is lost. A phase chart can be used to show the stability of the molding step process throughout the week that the qualification was run. A phase chart is a process behavior chart that has been broken down based on a stratification factor in our data. The charts are placed side by side on the same axis to facilitate comparisons.

To create a phase chart we need to identify our phase variable (cavity in our case), our subgroup, and type of process behavior chart we want to create. For each cavity, we define the *subgroup* as the five consecutive cycles of the press within a two-hour period each day. Since we have four two-hour periods in a day, and we ran the process for five days, each cavity chart has 20 points (or subgroups).

In JMP the $\overline{X}$ phase chart is generated as follows:

1. To sort the JMP table by Cavity, Day, and Hour, first select **Tables > Sort**, select Cavity, Day, and Hour from the list of column names in the left-hand side of the dialog box, and then click **By** to enter it in the field on the right. Then click **OK**. The JMP table will now be sorted by Cavity, from I through IV, by Day and by Hour. Sorting by Day and by Hour maintains the order of production.
2. So we can differentiate the points for Monday, Tuesday, Wednesday, Thursday, and Friday in the phase chart, we set different plotting symbols by Day. If the plotting symbols were already set using a different column, then we need to reset these by selecting **Rows > Clear Row States** from the main menu. To set Day as the plotting symbol, select **Rows > Color or Mark by Column**, and then select Day from the list of columns. Since we want to use different colors and different plotting symbols, we make sure that we check the boxes labeled **Set Color by Value** and **Set Marker by Value**, and then click **OK**.
3. To launch the **Control Chart** platform, select **Graph > Control Chart > XBar**. This opens a dialog box that enables us to select the response, subgroup, and phase variable. Select Effective Thickness, and then click **Process** to let JMP know that this is our response. Then select **Hour** and click **Sample Label** to identify our subgroup. Finally, select Cavity and click **Phase** to identify our phase variable (Figure 3.39). When finished, click **OK**.

Figure 3.39 Creating a $\overline{X}$R Phase Chart for Socket Thickness Data

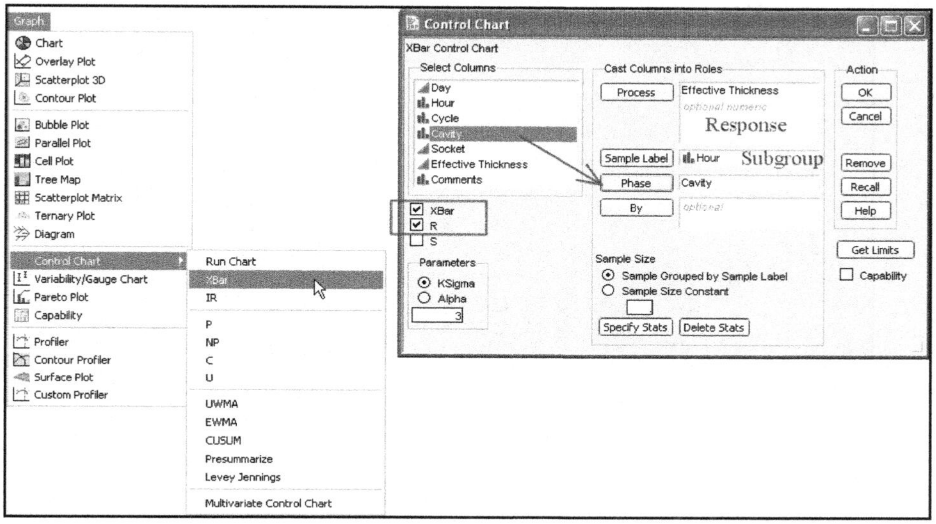

> **JMP 8 Note 3.4**: The Color or Mark by Column window has changed in JMP 8. The new window has different options for coloring and marking the plots, including different color scales and marker choices.

The phase chart is shown in Figure 3.40. For each cavity, there are 20 subgroups that correspond to the 5 days × 4 2-hour periods per day. On the $\overline{X}$ chart, each symbol represents the average of the five effective thickness measurements coming from the five cycles within a two-hour period. On the R chart, each symbol represents the range of the five effective thickness measurements coming from the five cycles within a two-hour period.

For each cavity, each $\overline{X}$ chart helps us answer the question: *Are there large differences between the five-cycle means for the two-hour-to-two-hour periods?* For each cavity, each R chart helps us answer the question: *Are the ranges of the effective thickness of the five cycles consistent within a two-hour-to-two-hour period?*

Figure 3.40 Phase Chart for Effective Thickness Data

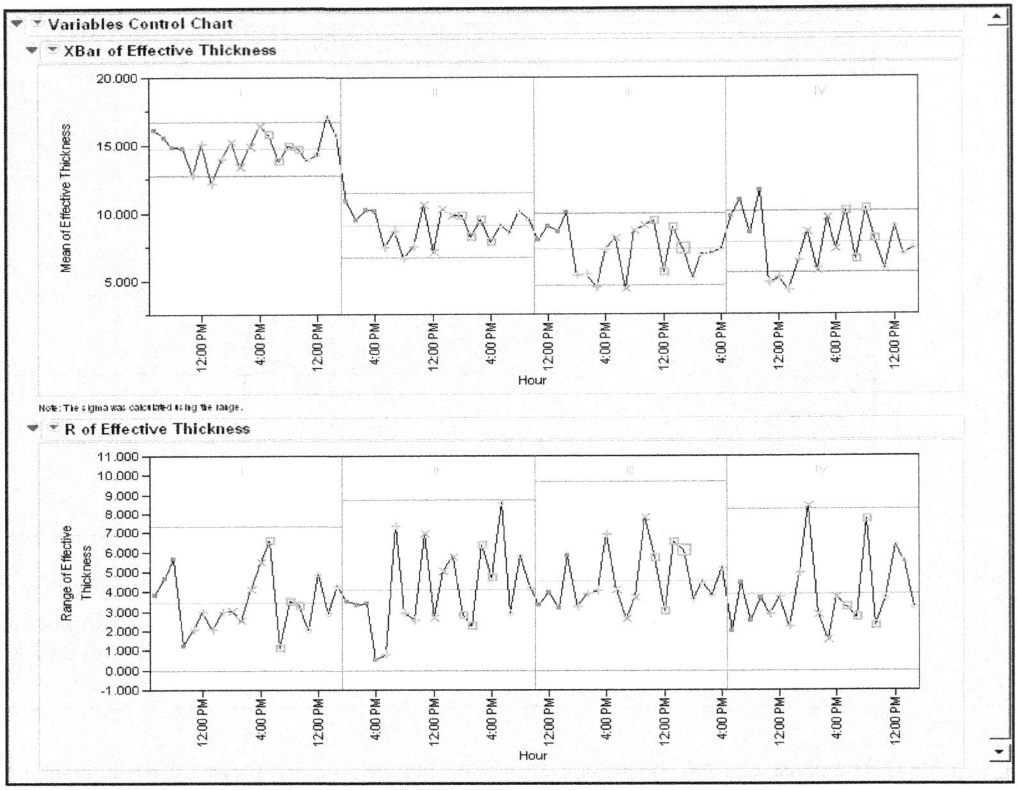

Each of the $\overline{X}$ charts corresponds to the four cavities, and has at least one point outside either the lower or upper control limit. They also appear to have some unusual patterns or trends within the control limits. For example, Cavity IV has a run of points above the centerline for Day 1, the first four points, and then drops below the centerline for Day 2, the next four points. By comparing the centerlines in the $\overline{X}$ chart we get another view of the differences between the cavities, with Cavity I having a higher average than the other three.

The range chart (the bottom plot in Figure 3.40) displays the within-subgroup variation, or the variation between the sockets for the five consecutive cycles within a two-hour period. There is only one point above the upper control limit for Cavity IV, indicating a range that is higher than the rest and perhaps unusual. If we click this point and then go to the JMP table, we can see that the five measurements in this subgroup occurred on Wednesday at 10:00 a.m. with values 9.45, 9.56, 11.75, 9.02, and 3.31. The range for this subgroup is then 11.75–3.31 = 8.44, which is the larget range for Cavity IV in the R chart. We can see that for this particular time period there was a difference of as much as 8.44 hundredths of a millimeter in the effective thickness measurements.

Step 7: Interpret the results and make recommendations

It is easy to get lost in all the statistical output and lose sight of the original questions that motivated the study. It is important to always go back to the problem statement in Step 1 and the expected performance we defined in Step 2 to make sure we answer the key uncertainties.

Recall from Step 1 that due to an impending shortage of a key raw material that is used in the molding step of a process for manufacturing sockets, we needed to characterize a new raw material from a second supplier. The most important performance characteristic of a socket is its effective thickness. In order for the raw material substitution to go undetected by our customers and the business, we must show that the mean, standard deviation, and process capability of sockets made with the new raw material are similar to those made with the current raw material (other business metrics like cost, throughput, on-time delivery, and so on were not included in this analysis). In other words, we must show that this new raw material has the following characteristics:

- The average effective thickness is about 7.5 hundredths of a millimeter.
- The effective thickness variation is no more than 2.5 hundredths of a millimeter.
- The process capability $C_{pk} \geq 1$.

If we look at the overall performance based on the 400 effective thickness measurements, we would have to conclude that we did not meet our criteria, and that we should not qualify this supplier. This is because the overall average effective thickness is 9.75 hundredths of a millimeter, the effective thickness standard deviation is 3.72 hundredths of a millimeter, and the process capability $C_{pk} = 0.47$. In addition, 9.75% of the sockets fall above the upper specification limit of 15 hundredths millimeter. However, as we examined the data in more detail, using our stratification factors, we discovered that the sockets made in Cavity I were too thick, and that Cavity I was responsible for producing all of the sockets that had effective thickness above the upper specification limit. The sockets made in Cavities II, III, and IV all had more favorable results relative to our goals. In fact, the sockets from Cavities III and IV met our target values for the mean, standard deviation, and C_{pk} and Cavity II missed only the target for the mean but met the targets for the standard deviation, and C_{pk}.

The above analysis indicates that the problem seems to be with the injection mold, Cavity I being different from the rest, and not the raw material. In other words, it is difficult to blame the higher effective thicknesses coming from Cavity I on the raw material.

We can assure the business that the sockets from Cavities II, III, and IV made with the new raw material will achieve the expected performance targets. However, further investigation is needed in order to understand why Cavity I is producing sockets that are too thick. In the interim, we recommend that if production volume is not a pressing issue, we can either not make sockets in Cavity I, or we can implement a more rigorous sampling plan to make sure that unacceptable sockets do not slip through to the customer.

The Rest of the Story

Wheeler and Chambers (1992, pp. 109–10) describe how the engineer in charge of the molding process sent the mold to the tool shop to have a three-thousandth-inch shim put behind Cavity I to solve the thickness problem. After he got the mold back he asked the toolmaker if he had done anything else to the mold, and the toolmaker said, "I did clean it up real good—there was a wax build-up of the face of the mold. I cleaned that off for you." It was then that the process engineer realized that the reason the process was out of control was because the operators were not cleaning off the wax build-up often enough. After changing the cleaning interval for the molding step, the process became capable and stable with the raw material from the new supplier.

3.5 Summary

In this chapter we learned how to characterize the measured performance of a material, product, or process in terms of average, variation, and process capability. We showed how the use of descriptive statistics helps us to efficiently summarize both small and large quantities of data using common measures such as the sample mean, standard deviation, and process capability indices like C_{pk}. We discussed how exploratory data analysis (EDA) helps us gain insight into the data, and how, by leveraging graphical and visualization tools, we can reveal many interesting or unexpected features, such as different relationships and sources of variation, that go beyond what we can do with descriptive statistics alone. In addition, we characterized a new supplier's raw material that is used to manufacture sockets in order to show the relevance of these techniques within engineering and science.

Among the graphical tools, we used histograms and normal quantile plots to look at the overall shape of the distribution of data, identify outliers, examine the spread or variation, and assess normality. In order to look for time-related patterns and trends and to check the homogeneity of our data, we used process behavior charts. We used box plots extensively, in conjunction with variability charts, to examine the variation coming from multiple stratification factors such as different machines, operators, production days, or mold cavities.

We also reviewed some useful statistical intervals, including confidence, prediction, and tolerance intervals, and discussed their differences. We used tolerance intervals for the socket data to show how we can identify the lower and upper bounds where some stated proportion of our population will fall.

We provided step-by-step instructions for characterizing a second source raw material for making sockets. These instructions included how to use JMP to help us conduct the corresponding statistical analysis and interpret the output, how to translate our findings in the context of the problem, and how to make appropriate decisions.

Some of the key concepts and practices that we should remember from this and previous chapters include the following:

- Use the appropriate combination of descriptive statistics and graphs to discover and interpret unexpected and exciting new findings about your study.
- Be proactive when identifying possible sources of variation in the system generating the data, so we can include them in our data collection efforts in order to use them as stratification factors in our analysis.
- Remember to clearly define the statistical claims that you want to make, the population of interest, the experimental unit, the observational unit, and how the study is executed in order to to conduct the appropriate analysis.
- Select key graphs and analysis output to facilitate the presentation of results.
- *Always translate the statistical results back into the context of the original question.*

3.6 References

Box, G.E.P., J.S. Hunter, and W.G. Hunter. 2005. *Statistics for Experimenters: Design, Innovation, and Discovery*. 2d ed. New York, NY: John Wiley & Sons.

Greene, B. 2009. "Questions, Not Answers, Make Science the Ultimate Adventure." *Wired Magazine*, 17 May 2009. Available http://www.wired.com/culture/culturereviews/magazine/17-05/st_essay.

Hahn, Gerald J., and William Q. Meeker. 1991. *Statistical Intervals: A Guide for Practitioners.* New York, NY: John Wiley & Sons.

Natrella, M.G. 1963. *Experimental Statistics, NBS Handbook 91*. Washington, D.C.: U.S. Department of Commerce, National Bureau of Standards. Reprinted 1966.

Ramírez, José G. 2009. "Statistical Intervals: Confidence, Prediction, Enclosure." Cary, NC: SAS Institute Inc. Available http://www.sas.com/apps/whitepaper/index.jsp?cid=4430.

Velleman, P.F., and D.C. Hoaglin. 1981. *Applications, Basics, and Computing of Exploratory Data Analysis*. Boston, MA: Duxbury Press.

Wheeler, D.J., and D.S. Chambers. 1992. *Understanding Statistical Process Control.* 2d ed. Knoxville, TN: SPC Press.

Chapter 4

Comparing the Measured Performance of a Material, Process, or Product to a Standard

Chapter Goals

- Become familiar with the terminology and concepts needed to compare average performance and performance variation to a given standard.
- Use JMP to compare average performance and performance variation to a given standard.
- Translate JMP statistical output in the context of a problem statement.
- Present the findings with the appropriate graphs and summary statistics.

4.1 Problem Description

What do we know about it?

You are an engineer working in the semiconductor industry and your primary area of expertise is in thin film deposition. Your team has just installed a three-zone, temperature-controlled vertical furnace in order to grow a 90 Angstrom (Å) layer of silicon dioxide on 200 mm silicon wafers. The oxidation of silicon in a well-controlled and repeatable manner is a critical step in the fabrication of modern integrated circuits. This thin layer of silicon dioxide is obtained by flowing oxygen gas into the chamber, where a quartz boat containing 100 wafers has been loaded. Virgin wafers, without any previous processing, are also included in each quartz boat in order to check the final thickness of the silicon dioxide layer when the oxidation is completed.

What is the question or the uncertainty to be answered?

After spending many months installing the furnace, running a series of experiments to understand cause-and-effect relationships, and establishing the standard operating procedures (SOP) to be used in manufacturing, you are finally ready to qualify this new piece of equipment. The primary objective of the qualification is to demonstrate that the new vertical furnace is capable of consistently producing wafers with a silicon dioxide layer of 90 Angstrom, using the SOP settings and a reliable measurement technique. Operations would also like to get an estimate of any potential scrap that will result if the thickness of the silicon dioxide layer is above or below the specification limits. The specification limits for the virgin wafer average thickness are LSL = 87 Angstrom (Å) and USL = 93 Angstrom (Å). The furnace that is being replaced by the new furnace has a yield loss of about 1%.

How do we get started?

Our goal is to demonstrate that the performance of the new furnace consistently meets the target (standard) of 90 Å, with a yield loss of no more than 1%. We begin by doing the following:

1. Formulate our requirements in terms of a statistical framework that enables us to demonstrate, with a certain level of confidence, that the performance of the new furnace meets the standard.

2. Translate the statistical findings in a way that clearly conveys the message without getting lost in the statistical jargon.

What is an appropriate statistical technique to use?

The new furnace will produce a sample of thickness readings that we can use to check its average performance against a standard of 90 Å and a yield loss of less than 1%. One-sample tests of significance are best suited for comparing the average performance and performance variation to a standard. In Section 4.3 of this chapter we review the key concepts introduced in Chapter 2 that are relevant to a one-sample test of significance, and in Section 4.4 we introduce a step-by-step approach for performing a one-sample test of significance using JMP.

4.2 Key Questions, Concepts, and Tools

Several statistical concepts and tools that were introduced in Chapter 2 are relevant when comparing the measured performance of a material, process, or product to a standard. Table 4.1 outlines these concepts and tools and the Problem Applicability column helps to relate these to the qualification of the new vertical oxidation furnace.

Table 4.1 Key Questions, Concepts, and Tools

Key Questions	Key Concepts	Problem Applicability	Key Tools
Is our sample meaningful?	Inference from a random and representative sample to a population	There are several sources of variation that can impact the performance of the furnace, such as the temperature zones within the furnace, the orientation of the wafer within the boat, and the set-ups for each new run of wafers. These sources should be taken into consideration when establishing sampling plans for the qualification. Future inferences depend upon the stability of the process over time. If the process is not stable, then the results of our study are limited to the sample collected. How many samples do we need to conclude that the new furnace is acceptable?	Sampling schemes, sample size calculators
Are we using the right techniques?	Probability model using the normal distribution	There is no reason to suspect that oxide thickness is not symmetric about an average value. There are no boundary conditions that would truncate the response values. The normal distribution should provide an adequate description of the data.	Descriptive statistics using sample mean and sample variance, normal quantile plots to check distributional assumptions
What risk can we live with?	Decision theory	The silicon dioxide layer is part of the gate of each transistor and a high quality thin layer is critical to its performance. Therefore, we would like to have a good chance of detecting a difference in performance from the standard, and to have a high level of confidence that we are not going to make the wrong decision.	Type I and Type II errors, statistical power, statistical confidence

(continued)

Table 4.1 (*continued*)

Key Questions	Key Concepts	Problem Applicability	Key Tools
Are we answering the right question?	One-sample test of significance	We need to demonstrate that the new furnace is capable of consistently producing wafers with a 90 Angstrom layer of oxide. Oxide layers that are too thin or too thick must be scrapped according to our upper and lower specification limits. These criteria suggest that we need to conduct a statistical test of significance for the average performance, as well as the performance variation of oxide thickness.	Student's *t*-test for the mean, Chi-square test for the variance
How do we best communicate the results?	Visualizing the results	We need to demonstrate that our population of wafers being processed by this new furnace meets our goals for the mean and variance. This can be visualized very effectively with a histogram that is overlaid with specification limits and descriptive statistics. In addition, we want to examine the different sources of variation that we described previously by using box plots.	Histograms, variability charts, confidence interval for the mean, confidence interval for the variance, tolerance interval

4.3 Overview of One-Sample Tests of Significance

4.3.1 Description and Some Applications

When comparing the measured performance to a standard we collect data under similar conditions from one material, one process, or one product, measure some attribute of interest, and then decide whether it meets our goals. For this scenario, we are not intentionally altering the causal factors in order to try to create a signal in our responses but are most likely trying to verify an SOP or optimal conditions of our products or processes. This will result in just one sample of measurements taken from the population of interest and the comparison is done using a one-sample test of significance.

Some examples from science and engineering situations where we would want to conduct one-sample tests of significance include:

- Determining if a product meets or exceeds its stated claims of performance.
- Commercializing a new product to determine whether it meets fitness-for-use criteria.
- Qualifying a new supplier and deciding whether we should place them on a ship-to-stock program in order to minimize resources for incoming inspection activities.
- Making sure that a process enhancement associated with efficiency gains does not alter the product performance in any way.
- Verifying the optimal process settings for a piece of equipment based on the output from a statistically designed experiment.

As we can see, one-sample tests of significance are important for scientific and engineering explorations because they can be used to answer numerous inquiries that arise throughout a product's life cycle. All these situations imply something about the performance as compared to a standard, k. But where does this standard value come from? This value must be specified ahead of time and can be obtained from a variety of sources, depending upon the context of the inquiry. For example, if we are commercializing a new product and are interested in a key fitness-for-use parameter, like the average resistance of a cable, then an appropriate value of k is given by the middle of a specification window or the target value. Typically, k comes from a desired target value for a key product attribute; is estimated from historical performance of a similar process, product, or material; is specified by a performance claim; or can even come from a competitive option.

In sections 4.3.2 and 4.3.3 we discuss how a one-sample test of significance can be used to compare measured performance to a standard with respect to average performance and performance variation. In section 4.5, we show how to carry out this type of comparison for the new oxidation furnace using JMP, including how we derived our choice for the standard value k.

4.3.2 Comparing Average Performance to a Standard

In statistics, tests for comparing a population mean (or the average performance) to a standard are referred to as one-sample tests for the mean, or more commonly, one-sample *t*-tests. When we carry out a test for the mean we are interested in drawing a conclusion about the measured performance of our material, process, or product, with respect to a standard. This measured average performance can be equal to the standard, less than the standard, or greater than the standard. In statistical jargon these types of comparisons (equal, greater than, or less than) can be described using a null (H_0) and an alternative (H_1) statistical hypothesis that contain statements about the average performance (population mean, μ) as compared with the standard k.

Statistics Note 4.1: Tests of significance start with a null hypothesis (H_0) that represents our current beliefs that we *assume* to be true but cannot be proven, and an alternative hypothesis (H_1) that is a statement of what the statistical test of significance is set out to *prove* by rejecting the null hypothesis. In what follows we will label the null hypothesis *Assume,* and the alternative hypothesis *Prove* to remind us of what the test of significance sets out to do.

The three ways that we can write the null (H_0) and alternative (H_1) hypothesis are as follows:

1. The average performance is different from the standard.

   H_0: Average Performance (μ) = Standard (k) (Assume)
   H_1: Average Performance (μ) $\neq$ Standard (k) (Prove)

   This is a *two-sided* alternative since we do not care about the average performance being less or greater than the standard, just that it is different from it. The null hypothesis (H_0) *assumes* that the average performance (population mean, μ) is equal to our standard value, k. On the other hand, the alternative hypothesis (H_1) states that our average performance is not equal to our standard value, k. If our test of significance favors the alternative, then we have proven with statistical rigor that our average performance is different from the standard, k, without making any statements regarding the direction of departure from k.

   > *Two-sided alternative hypotheses do not make any claims about the direction of departure of the mean from our standard.*

2. The average performance is greater than the standard.

   H_0: Average Performance (μ) $\leq$ Standard (k) (Assume)
   H_1: Average Performance (μ) $>$ Standard (k) (Prove)

   In this one-sided alternative hypothesis, H_1, we are interested in knowing whether our average performance is greater than k. Here, a rejection of the null hypothesis H_0 will show, with statistical rigor, that the average performance is greater than *k*. This set-up might be appropriate if we let *k* equal a lower limit of some type and want to demonstrate that the population mean exceeds this lower limit. For example, this alternative hypothesis might be useful if we need to demonstrate that the average performance of a bond strength measurement exceeds a lower limit k.

3. The average performance is lower than the standard.

   H_0: Average Performance (μ) $\geq$ Standard (k) (Assume)
   H_1: Average Performance (μ) $<$ Standard (k) (Prove)

   Finally, for this one-sided alternative we are interested in knowing whether our average performance is less than the standard, *k*. If we do not have enough evidence to reject H_0 then we assume that our population mean is greater than or equal to *k*. However, if we reject the null in favor of H_1 then we have shown, with statistical rigor, that our average performance is less than the standard value of *k*. This set-up might be appropriate if we let *k* equal an upper limit of some type and want to demonstrate that the population mean does not exceed this upper limit. For example, this alternative hypothesis could be useful if we need to demonstrate that the average performance of a particle measurement does not exceed an upper limit, *k*.

> *Use a one-sided alternative hypothesis if you want to make a strong statement about the direction of the outcome.*

The *t*-statistic

Most test statistics that we encounter in practice represent a signal-to-noise ratio. If we know our performance variation, then a test for the average performance can be done using the Z-score described in Chapter 2:

$$Z_{score} = \frac{Signal}{Noise} = \frac{\text{(Estimated Average Performance} - \text{Standard)}}{\text{Known Performance Variation}}.$$

Recall from Chapter 2 that a value from the standard normal outside the ±2 standard deviations band occurs about 5% of the time, while a value outside the ±1 standard deviations band occurs 32% of the time, as is shown in Figure 4.1. Therefore, values of the Z-score less than -2 or greater than 2 provide evidence that the average performance is different from the standard.

Figure 4.1 Areas under the Standard Normal Distribution

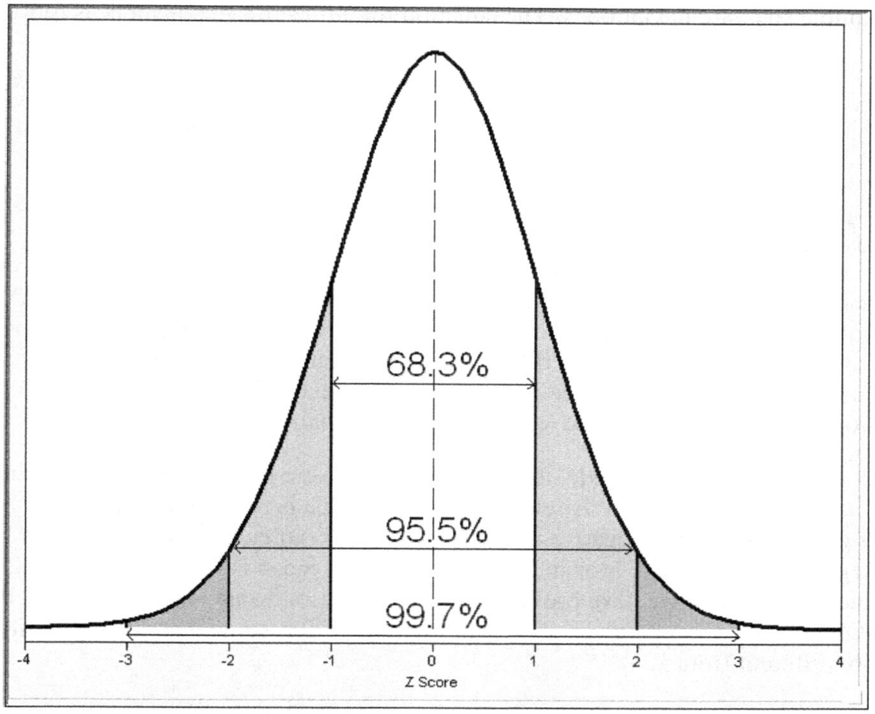

In real life, however, we seldom, if ever, know the true value of the performance variation, so it needs to be estimated from the data. The resulting signal-to-noise ratio is the *t*-statistic that can be used to test the three hypotheses described above: equal to the standard, lower than the standard, or greater than the standard. This test statistic can be written as:

$$t = \frac{\text{Signal}}{\text{Noise}} = \frac{(\text{Estimated Average Performance} - \text{Standard})}{\text{Estimated Performance Variation}} \qquad (4.1)$$

Note that the numerator is measuring the difference between the average performance and the standard; that is, the signal we are interested in, while the denominator is the yardstick to decide whether the signal is worth worrying about. In order to carry out these tests of significance for a given sample of *n* measurements, we need a measure of the average performance given by $\overline{X}$, and a measure of the performance variation given by the sample standard deviation, *s*. The equation for a *t*-test for the mean is as follows:

$$t = \frac{\text{Signal}}{\text{Noise}} = \frac{(\text{Performanc Average} - \text{Standard})}{\text{Performance Standard Deviation}/\sqrt{\text{sample size}}} = \frac{(\overline{X} - k)}{\text{s}/\sqrt{\text{n}}} \qquad (4.2)$$

This signal-to-noise ratio resembles a Z-score, but it depends on the sample size via $\sqrt{n}$. As with the normal Z-score, the magnitude of the *t*-statistic gives us a sense of how big the signal is with respect to the noise, and whether we have enough evidence to reject the null hypothesis. For example, if the calculated *t*-statistic is ≤ 1, then it is highly unlikely that we will reject the null and conclude that the average performance is different from the standard. However, if the calculated *t*-statistic is ≥ 2.5, then there is a good chance that we will reject the null and conclude that the average performance is different from our standard, since the signal is two-and-a-half times the noise.

In addition to its magnitude, the sign of the *t*-statistic gives us insight regarding where our average performance is relative to *k*. This value is positive if the average performance is greater than *k* and negative if it is less than *k*. If our calculated *t*-statistic is large and negative then we have proven that our population mean is less than *k*, and if it is large and positive then we have proven that our population mean is greater than *k*. If our test statistic is small, regardless of its sign, then we will assume that our population mean is not different from *k*.

As an illustration, Figure 4.2a shows the partial JMP output for a *t*-test that compares the average performance to a standard (Hypothesized Value) of 0. The alternative hypothesis of interest is a two-sided alternative hypothesis with H_1: $\mu \neq 0$. The average performance, calculated from the sampled data, is −1.2554 and the sample standard deviation is 7.14. We calculate the *t*-statistic using Formula 4.2 (−1.2554 – 0) / (7.00113 / √51) = −1.2805. Note: The *t*-statistic equals the estimated average because the estimated standard deviation, 7.00113, happens to be very close to the square root of the sample size, √51. We can see by the sign of this statistic that the average performance is less than our standard value, *k* = 0. We can also see by the magnitude of this statistic that the signal is only 1.28 times larger than the noise and, as compared with the empirical rule represented in Figure 4.1, is probably not large enough to be significant. The magnitude of the *t*-statistic is related to the area under the Student's *t*-distribution curve, the *p*-value, as explained next.

Figure 4.2a Partial JMP Output: *t*-Test for a Sample of 51 Observations

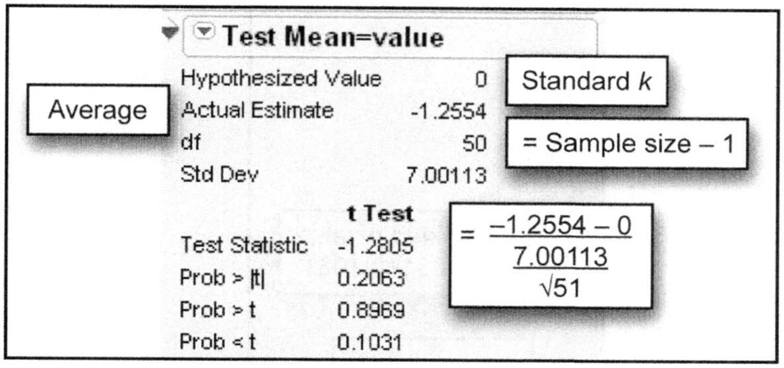

However, what conclusions can we draw with *t*-statistics that are in the "gray" zone, that is, slightly greater than 1 but less than 2.5? The question is how large is large? Unfortunately, the answer depends on the way the sample was collected and the sample size. What we need is a measure of the probability of observing a *t*-statistic as large as, or larger than, the one obtained from the data at hand, assuming that the population mean is equal to the specified standard. This is done in JMP by calculating the area under the Student's *t*-distribution curve with a mean equal to the specified standard and sample size. This calculated area is called the *p-value* and is dependent on the type of comparison that we want to test. For example, if our alternative hypothesis is that the average performance is greater than the standard (H_1: $\mu > k$), then the *p*-value is the area under the Student's *t*-distribution curve, to the right of the calculated *t*-statistic. Similarly, if our alternative hypothesis is that the average performance is less than the standard (H_1: $\mu < k$), the *p*-value is the area to the left of the calculated *t*-statistic. If the alternative hypothesis is that the average performance is different from the standard, then the *p*-value is the area to the right of the calculated *t*-statistic plus the area to the left of the calculated *t*-statistic.

> ***A p-value is an area under a probability curve that quantifies the likelihood of observing a test statistic as large as, or larger than, the one obtained from the data.***

Figure 4.2b shows the complete JMP output for the one-sample test for the average performance shown in Figure 4.2a. We see that the shaded dark blue in the plot represents the area to the left of −1.2805, while the shaded light blue represents the area to the right of 1.2805. Both of these areas are equal to 0.10315 and their sum, 0.2062, is the two-sided *p*-value (H_1: $\mu \neq 0$). This is the probability, assuming that the population mean is 0, of observing by chance a value less than −1.2805 or greater than 1.2805. Since the *t*-distribution is symmetric, the *p*-value $0.2062 = 2\times0.10315$ can be calculated as twice the area to the left of −1.2805.

Figure 4.2b Areas under the *t*-Distribution

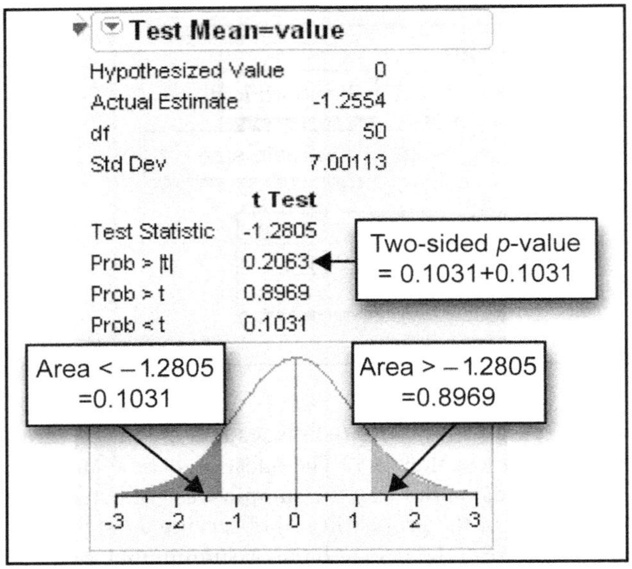

> *The p-value is compared with α, the probability of committing a Type I error. If p-value < α, then reject H_0.*

How do we weigh the evidence given by the *p*-value? Recall from Section 2.7.1 that when testing a hypothesis we are allowing ourselves to a "make a mistake" a certain percent of the time. That is, a certain percent of the time we are going to say that the average performance is not equal to the standard when, in fact, it is. This error rate (significance level), α, which is pre-specified *before* we perform the test, serves as the cutoff point for our significance test decisions. If the *p*-value < α, then we have enough evidence to prove our alternative hypothesis; but, if the *p*-value ≥ α, then we do not have enough evidence to say that our average performance is not equal to the standard *k*. The value of α is typically selected to be 0.05. For the example in Figure 4.2b the two-sided *p*-value = 0.2063 > 0.05, and, therefore, we do not reject the null hypothesis that the average performance is equal to 0. Can we say that the average is equal to 0? Not really. All we can say is that we do not have enough evidence to say that the "true" population mean is not 0.

> *If we do not reject the null hypothesis, all we can say is that we do not have enough evidence to say that the "true" population mean is not 0.*

Statistics Note 4.2: The validity of this procedure (for determining if the average performance is equal to a standard) depends upon the assumptions that were discussed in Chapter 2 and are reiterated here. The underlying population must be normally distributed, or close to normally distributed, the data are homogeneous, and the experimental units must be independent from each other.

JMP Note 4.1: Since JMP does not know which type of hypothesis we are testing (equality, greater than, or less than), JMP will output the *p*-values associated with all three alternative hypotheses.

Figure 4.3 shows the JMP output for a second example of a one-sample *t*-test, which is testing that the mean is equal to a standard value *k* of 49.5 which JMP labels **Hypothesized Value**. The average estimated from our sample (labeled **Actual Estimate**) is 50.4103, our sample size is 51, and our sample standard deviation (labeled **Std Dev**) is 1.92264. The label **t Test** indicates that JMP is using the Student's *t*-distribution for the *p*-value calculations. The calculated *t*-statistic is 3.3814. The graph at the bottom of Figure 4.3 provides a visual representation of the quantitative results provided above it. The *t*-distribution, or blue line, is centered at the standard value of 49.5, and the sample average is plotted to the right of the distribution as a red line located at 50.4103.

Figure 4.3 JMP Output for One-Sample Test for the Mean

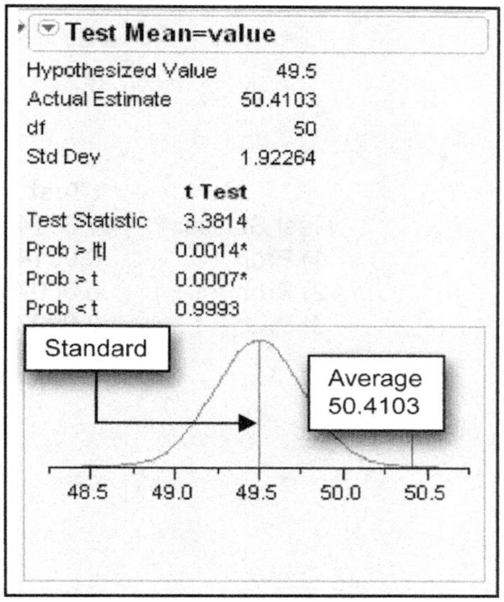

The shaded area to the right of the red line is the area under the curve that represents the probability of getting a sample mean equal to or larger than the one we obtained in our study, 50.4103, assuming that the actual mean is 49.5 (H_0). This area is 0.0007 and is labeled **Prob > t**. (See Figure 4.4.)

Figure 4.4 *p*-Value for the *t*-Statistic

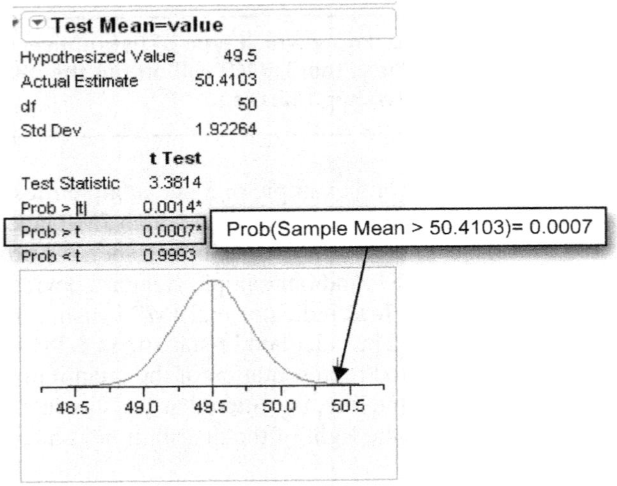

The *p*-value of the test depends on the type of hypothesis we are testing and is given by one of the three *p*-values shown in the **Test Statistic** section of the JMP output. The **Test Statistic** section indicates the following:

1. H_0: Average Performance = 49.5
   H_1: Average Performance ≠ 49.5
   p-value = 0.0014*

2. H_0: Average Performance ≤ 49.5
   H_1: Average Performance > 49.5
   p-value = 0.0007*

3. H_0: Average Performance ≥ 49.5
   H_1: Average Performance < 49.5
   p-value = 0.9993

	t Test
Test Statistic	3.3814
1) Prob > |t|	0.0014*
2) Prob > t	0.0007*
3) Prob < t	0.9993

> *The p-value depends upon the type of comparison we are making.*

Note that the symbols (> |t|, > t, < t) correspond to the alternative hypothesis, and that JMP puts an * next to those *p*-values that are < 0.05, which is the default Probability (Type I error) cut-off value. If we are testing the null hypothesis stating that the average performance is ≤ 49.5, as is shown in (2), then there is only a 0.07% chance of observing a value equal to or larger than the sample average of 50.4103. This means that the data supports that the average performance is greater than 49.5. If, on the other hand, we are testing the null hypothesis stating that the average performance is ≥ 49.5, as is shown in (3), then the probability of observing a value of 50.4103 or larger, is 99.93%, a very likely event, meaning that we do not have enough evidence to say that the average performance is not greater than 49.5.

> *These three p-values are related to each other:*
>
> Prob > t + Prob < t = 0.0007 + 0.9993 = 1
> Prob > |t| = 0.00014 = 2 × 0.0007 = 2 × Prob > t

Our one-sample test shows a statistically significant difference: however, is this difference practically significant? For this example, using $\alpha = 0.05$ we would conclude that the average performance is greater than our standard value, 49.5. While the actual mean of 50.4103 is greater than 49.5, this difference is only 0.9103 (50.4103–49.5) units. This might not have any practical significance or relevance in the context of the problem addressed with the study. Remember, that we are using statistics to help us decide whether we have enough evidence to reject our null hypothesis, or current beliefs, and we cannot forget to translate the statistical results into a practical meaning.

4.3.3 Comparing Performance Variation to a Standard

In statistics, tests for comparing a population variance (that is, performance variation) to a standard are referred to as one-sample tests for the variance. When we carry out a test for the variance, we are interested in making inferences about the population standard deviation, or the population spread around the mean, of the performance of our material, process, or product with respect to a standard. The performance variation can be equal to the standard, less than the standard, or greater than the standard. These tests can provide valuable information about our ability to meet specification limits or, if there are no specification limits, help us to understand the range of values that our customers might get over time.

As was the case for the test for the average performance, we begin by specifying the null hypothesis (H_0), which is what we believe is true, and the alternative hypothesis (H_1), which is what we set out to *prove* by rejecting the null hypothesis. There are three ways that we can write the null (H_0) and alternative (H_1) hypotheses for studying the standard deviation:

1. The performance variation is different from the standard.

   H_0: Performance Variation (σ) = Standard (k) (Assume)
   H_1: Performance Variation (σ) $\neq$ Standard (k) (Prove)

   This is a *two-sided* alternative since we do not care about the performance variation being less than or greater than the standard—just that is different from it. The null hypothesis (H_0) states our belief (our assumption) that the population standard deviation, σ, is equal to our standard value, k. On the other hand, the alternative hypothesis (H_1) states that our performance variation is not equal to our standard value, k. If our test of significance favors the alternative hypothesis, then we have proven, with statistical rigor, that our standard deviation is different from k, but we have made no statements regarding its direction of departure from k.

2. The performance variation is greater than the standard.

   H_0: Performance Variation (σ) $\leq$ Standard (k) (Assume)
   H_1: Performance Variation (σ) > Standard (k) (Prove)

   In this one-sided alternative hypothesis, we are interested in knowing whether our performance variation is greater than k. Here, a rejection of the null hypothesis, H_0, will show (with statistical rigor) that the performance variation is greater than k. This form of the hypotheses is useful if we want to demonstrate that the standard deviation is greater than the standard. For example, we might want to determine whether the measurement is incapable of meeting a yield goal based on a set of specification limits and a standard value that, if exceeded, would produce a higher yield loss than desired and result in a rejection of the null hypothesis. For this situation, the null hypothesis assumes that the product, process, or material is capable of meeting the standard value unless proven otherwise.

3. The performance variation is smaller than the standard.

   H_0: Performance Variation (σ) $\geq$ Standard (k) (Assume)
   H_1: Performance Variation (σ) < Standard (k) (Prove)

   Finally, for this one-sided alternative we are interested in knowing whether our performance variation is less than the standard, k. If we do not have enough evidence to reject H_0, then we assume that our population standard deviation, σ, is greater or equal to k. However, if we reject the null hypothesis in favor of H_1, then we have shown, with statistical rigor, that our population standard deviation is less than the standard value of k. This form of the hypotheses is useful if we want to demonstrate that the standard deviation is less than the standard. For example, we may want to demonstrate that the measurement is capable of meeting a yield goal based upon a set of specification limits and a standard value that, if not exceeded, would produce a lower yield loss than expected and result in a rejection of the null hypothesis. For this situation, the null hypothesis assumes that the product, process, or material meets or exceeds the standard value unless proven otherwise.

The χ^2-statistic

For testing performance variation we need another "signal-to-noise" ratio. In this case the signal is given by the sample standard deviation, *s*, and the noise is given by the standard value we want to test against, *k*. The test statistic is the ratio of the sample variation, s^2, to the square of the standard value, k^2, and is adjusted by the degrees-of-freedom = sample size – 1 as follows:

$$\chi^2 = (n-1)\frac{s^2}{k^2} \tag{4.3}$$

If the sample standard deviation, *s*, is close to our standard value, *k*, then their ratio will be close to 1 and the test statistic will be close to (n–1). If the sample standard deviation, *s*, is much larger than *k*, then the ratio will be larger than 1, and the test statistic will be larger than (n – 1). Finally, if the sample standard deviation, *s*, is much smaller than *k*, then the ratio will be smaller than 1, and the test statistic will be smaller than (n – 1). Unlike the *t*-statistic for testing the average performance, the Chi-square test statistic cannot be negative because it is the ratio of positive quantities.

As before, we need a yardstick to decide whether our calculated test statistic is "large"; or, in other words, to decide whether the signal is larger than the noise. The *p*-value is again a calculated area under a probability distribution, but this time we use the Chi-square distribution rather than the Student's *t*-distribution. That is, the *p*-value for a performance variation test is the area under the Chi-square curve to either side of our calculated test statistic. The alternative hypothesis will determine whether we want the area to the left or right of our calculated test statistic.

For the data in section 4.3.2, let our standard value for the performance variation, *k*, be 2.25 units. The JMP output in Figure 4.5 shows the results for this one-sample test of our performance variation. The **Hypothesized Value** represents our choice for *k* of 2.25, and the **Actual Estimate** of 1.92264 is the sample standard deviation. The label above the calculated test statistics is **ChiSquare** because JMP is using the Chi-square distribution to conduct a one-sample test on the standard deviation. The value of the **Test Statistic** is 36.5089, corresponding to 50 degrees-of-freedom, or the number of observations (51) minus 1. The magnitude of the calculated test statistic as compared with (n – 1), or 50, might suggest that our performance variation is less than the standard value, but it certainly would not suggest that it is larger than our standard.

Figure 4.5 JMP Output for One-Sample Test for σ

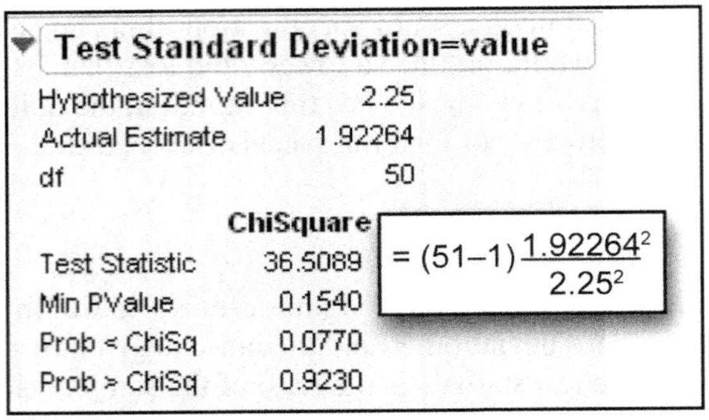

However, to know for sure, we need to seek out the *p*-values in this output. The JMP output in Figure 4.5 contains, once again, all three *p*-values corresponding to the three alternative hypotheses of not equal to, less than, and greater than the standard. The labels for the *p*-values are a bit different from the ones we observed with the one-sample *t*-test for the mean. The label **Min P value** for the first *p*-value in this column is the *p*-value that corresponds to the two-sided alternative hypothesis. The other two *p*-values correspond to the "less than" and "greater than" alternative hypotheses, respectively. As was stated previously, the *p*-value is an area under a probability density curve. The area to the left of the calculated statistic of 36.5089 is 0.077, while the area to the right of the calculated statistic of 36.5089 is 0.923.

How do we know if we can reject the null hypothesis? We need to compare the *p*-value that we obtained from JMP with our significance level, Probability(Type I error), α. If *p*-value < α, then we reject the null hypothesis in favor of the alternative, and we have enough evidence to state that the performance variance is different from our standard. If the *p*-value ≥ α, then we do not have enough evidence to reject the null hypothesis, and, therefore, we can assume that the performance variation is not different from the standard value.

JMP Note 4.2: Unlike the case for testing a mean, we do not get the graphical representation of the Chi-square distribution at the bottom of this output to help us visualize the results. A Chi-square distribution based on a sample size of (50 = 51 – 1) is shown in Figure 4.6. We can see that the calculated test statistic from the last example of 36.5089 contains an area of 0.077 to its left and is located to the left of 50. This *p*-value means that there is a 7.7% probability of obtaining, by chance, a value of the test statistic less than the one obtained in this sampled data. The area to the right of the test statistic of 36.5089 is 0.923 and implies that there is a 92.3% probability of obtaining, by chance, a test statistic greater than the one obtained in this sampled data.

Figure 4.6 *p*-Value for the Chi-Square Test

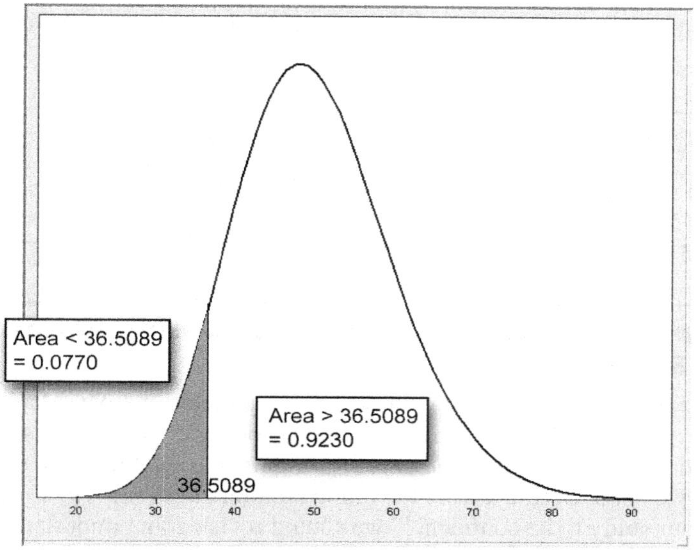

Just like the test for the population mean, the *p*-value of the test for the population standard deviation depends on the type of hypothesis we are testing and is given by one of the three *p*-values shown in the Test Statistic section of the output. If we are testing

1. H_0: Performance Variation (σ) = 2.25 (*k*)
   H_1: Performance Variation (σ) ≠ 2.25 (*k*)
   p-value = 0.1540

2. H_0: Performance Variation (σ) ≥ 2.25 (*k*)
   H_1: Performance Variation (σ) < 2.25 (*k*)
   p-value = 0.0770

3. H_0: Performance Variation (σ) ≤ 2.25 (*k*)
   H_1: Performance Variation (σ) > 2.25 (*k*)
   p-value = 0.9230

If we are interested in a two-sided alternative hypothesis (1), since the Min Pvalue = 0.154 > 0.05 we do not have enough evidence to say that the performance variation is different from 2.25 units. In fact, for $\alpha = 0.05$, we would not reject any of these hypotheses.

> *These three p-values are related to each other:*
>
> Prob < ChiSq + Prob > ChiSq = 0.077 + 0.923 = 1
> Min PValue > = 0.154 = 2 × 0.077 = 2 × Prob < ChiSq

4.3.4 Sample Size Calculations for Comparing Performance to a Standard

Data does not exist in isolation, and therefore all the statistics that we calculate are meaningful *only* in the context of the data. In other words, for any analysis to give useful results, we need to understand how the data is going to be collected, what they represent, and how many we need. Before asking the quintessential question: "How many samples do we need in our study to be confident?" we should revisit some important concepts discussed in Chapter 2. Recall that the experimental unit (EU) is the smallest unit that is affected by the "treatment" we want to study. The output from the sample size calculations specifies how many experimental units (n) we need in our study. Note that this is not necessarily the same as the number of measurements that we will have at the end of our study. If we decide to measure each experimental unit two times, then we will end up with twice as many readings as we have experimental units. For example, several thickness readings might be taken on a single wafer (the experimental unit), giving several "repeated measurements" on a single wafer.

The *sampling scheme* should describe how we will sample our data from the population under investigation. Recall from Chapter 2 that two popular sampling schemes are the "simple random sample" and "stratified simple random sample." In stratified simple random sampling we first subset our population into different strata, and then we take a random sample from each of the strata. The use of randomization in selecting our sample helps us prevent bias from creeping into our studies. we also need to ensure that our sample is representative of the population that we are interested in studying.

Our knowledge of the *sources of variation* that are present in our products, processes, and materials, is also relevant in how we sample our data. This is particularly true of manufacturing-based studies, such as the one described in the beginning of this chapter. Just as we have a choice about the type of sampling scheme that we select for our study, we also have a choice about which sources of variation we want to include. For example, in a batched-based operation in the chemical industry, we might include in our study sources of variation, such as batch-to-batch, within batch, different tanks, and different shifts. Could these different sources of variation represent the different strata of our population that could be used in a stratified simple random sample?

Finally, how many samples do we need to be relatively certain about the outcome?

> **JMP Note 4.3:** JMP software has functionality for determining sample sizes for one-sample tests for both the mean and the variance. These can be found under **DOE -> Sample Size and Power**, as is shown in Figure 4.7. The sample size calculator for testing the average performance is labeled **One Sample Mean** in Figure 4.7, and the sample size calculator for testing the performance variation is labeled **One Sample Variance**. Several inputs are required for these calculations, as shown in Figure 4.8 and Figure 4.13a.

Figure 4.7 Accessing Sample Size and Power Calculations in JMP

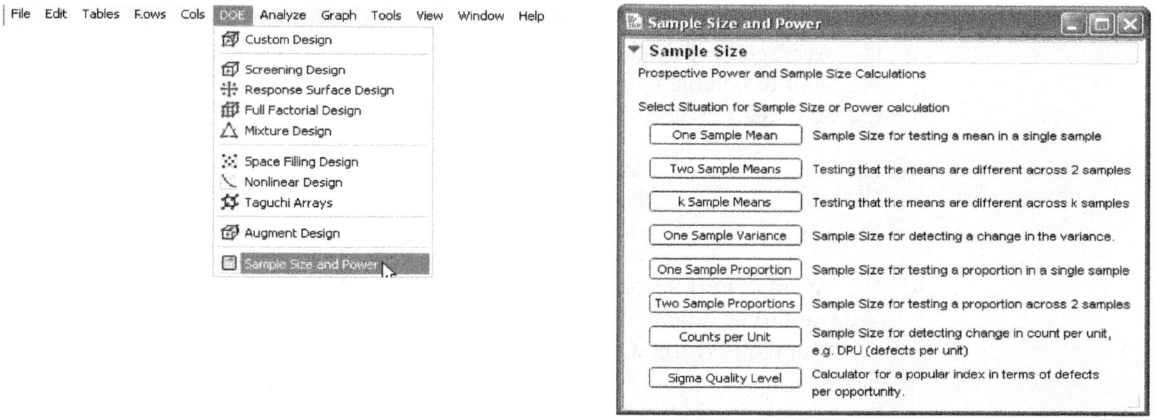

One Sample Mean (Comparing Average Performance to a Standard)

JMP provides an easy to use sample size and power calculator. Below we discuss the inputs needed to use the calculator for testing a one-sample mean. Note that JMP help can be obtained for any field by selecting the **?** icon in the toolbar and clicking in the desired field.

1. **Alpha**: This is the Probability (Type I error) or significance level that was introduced in Chapter 2 and discussed previously. This defines the amount of risk that we are willing to live with for incorrectly stating that our mean is different from our standard value, *k*, when, in fact, it is not. An alpha of 0.05 is the standard choice and is the default in JMP.

2. **Error Std Deviation**: This is the noise in our data that is usually described by the standard deviation, σ. Historical data is helpful for providing an estimate of the standard deviation. Otherwise, we need to input our best guess at the standard deviation. If this is not possible because we have no historical data, or if, for example, we are dealing with a new process or material, then we can enter a value of 1 in this field. If we do this, then we will need to make sure that we specify the **Difference to detect** in terms of multiples of this unknown standard deviation.

3. **Extra Params**: This field is used for studies involving more than one factor (multi-factor experiments), which is a topic not covered in this book. We will leave this blank for one-sample tests on the mean.

4. **Difference to detect**: This is the smallest deviation from the standard performance, *k*, that we would like to detect, that is, the smallest practical difference that we would like to detect. This value must be positive and larger than 0.

   For example, a difference to detect of 2 implies the following when testing:

   a. H_0: Average Performance $= k$ vs. H_1: Average Performance $\neq k$.
   We want to be able to detect a difference a large percent of the time whenever the $|$Average Performance $- k| > 2$.

   b. H_0: Average Performance $\leq k$ vs. H_1: Average Performance $> k$.
   We want to be able to detect a difference a large percent of the time whenever the Average Performance $> 2k$.

   c. H_0: Average Performance $\geq k$ vs. H_1: Average Performance $< k$.
   We want to be able to detect a difference a large percent of the time whenever the Average Performance $< 2k$.

Statistics Note 4.3: If we specify our error standard deviation as 1, then the difference to detect must be interpreted in numbers of standard deviations. For example, if we set **Error Std. Dev**= 1 and **Difference to detect** = 0.5, this implies that we would like to detect a statistical difference if the average performance is 0.5σ units away from our standard value, *k*. This solves the problem of not knowing the noise in the system. However, once we get an estimate of noise from the data, the difference that we are able to detect is a function of this estimated noise, which might be larger than anticipated.

---

5. **Sample Size**: The sample size is the minimum sample size that is recommended for the study. We need to collect measurements on this many experimental units in order to be able to detect our practical Difference (**Difference to detect**) with a specified power.

6. **Power**: The power of a test is the probability of detecting a difference between the average performance and the standard, when in fact there is a difference. Since it is a probability, the power takes on values between 0 and 1 and can be converted to a percentage by multiplying it by 100. Generally speaking, it is nice to have a higher power for our one-sample test of significance. We recommend starting with a power of at least 0.8 and adjusting it up or down as needed. Don't be too greedy because the more power you want, the more samples you need!

There are several ways that we can use the **JMP Sample Size and Power** calculator. We can input values for **Alpha**, **Error Std. Dev.**, **Difference to detect**, and **Power**, and let JMP calculate the sample size. For example, let's say we need to figure out how many samples we need to test that the average performance is equal to a standard. We enter **Alpha** = `0.05`, **Error Std. Dev.** = `1`, **Difference to detect** = `0.5`, and **Power** = `0.8` (see Figure 4.8). In other words, we want to be able to detect a difference of half the noise (0.5), or half the standard deviation, 80% of the time, when in fact the difference is half the noise or more.

Figure 4.8 Sample Size Calculation for One-Sample Mean

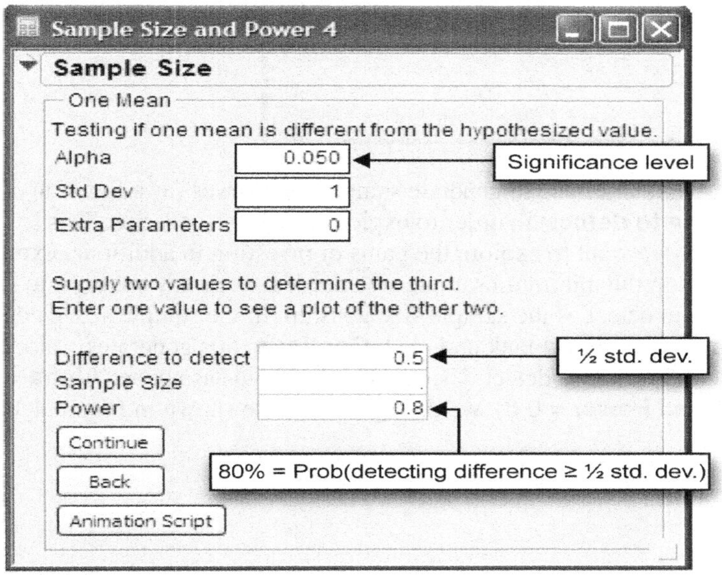

When we click **Continue** using the parameters outlined previously (Figure 4.8), we get a sample size of 33.367 (Figure 4.9). Yes, it is normal to have fractional sample sizes. We recommend that you round up since this always increases the power, albeit by only a little bit. For this example, we will need 34 experimental units in order to be able to detect a difference in average performance that is more than 0.5 standard deviations away from our standard, *k*, with a power of 80%.

Figure 4.9 Sample Size Output for One-Sample Mean

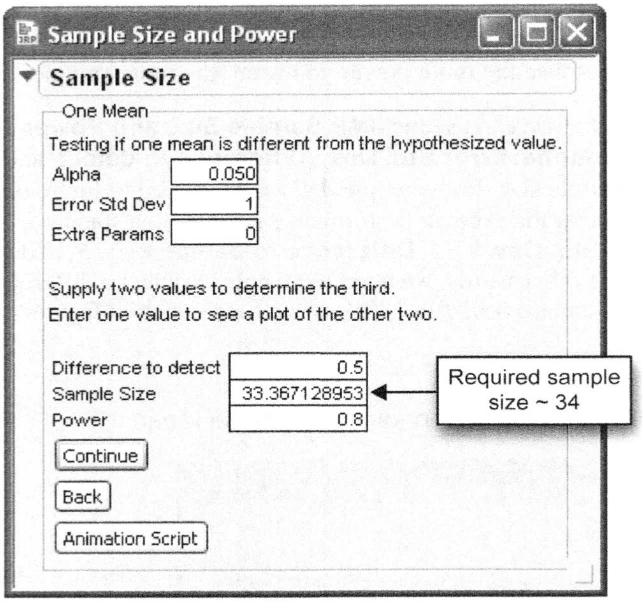

We can also use this interface to generate sample size curves (as a function of **Power** or the **Difference to detect**) in order to explore different scenarios. This is very, very useful when we want to explore the gains of investing in additional experimental units or in selecting the minimum sample size needed. Let's say we want to see what differences we can detect as the sample size goes up. In the sample size calculator we leave **Difference to detect** blank and click **Continue**; this generates a plot of sample size versus the difference to detect. Using the same inputs as above (**Alpha** = 0.05, **Error Std. Dev.** = 1, and **Power** = 0.8), we then get the curve shown in Figure 4.10.

Figure 4.10 Sample Size vs. Difference Curve for One-Sample Mean

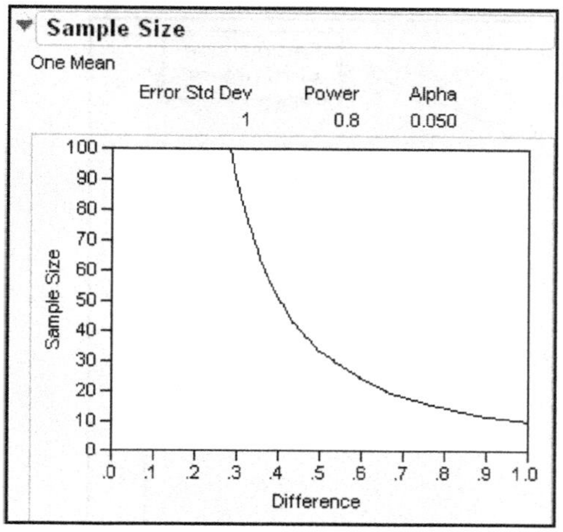

To get a better view of the curve at larger sample sizes, expand the y-axis to reach 300. Move the cursor over the y-axis and, when the cursor turns into a hand, right-click over the y-axis and select **Axis Settings**. The Y Axis Specification dialog box will appear. Just type over the entry in the **Maximum** field and change the value to `300` (see Figure 4.11), and then click **OK** and this will change the scale on the y-axis in the plot.

Figure 4.11 Y Axis Specification Dialog Box for Changing the Y-axis to 300

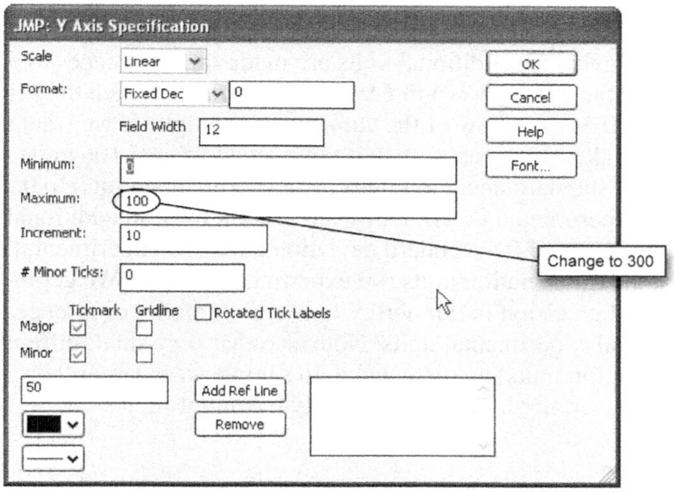

Figure 4.12 Sample Size vs. Difference Curve for One-Sample Mean

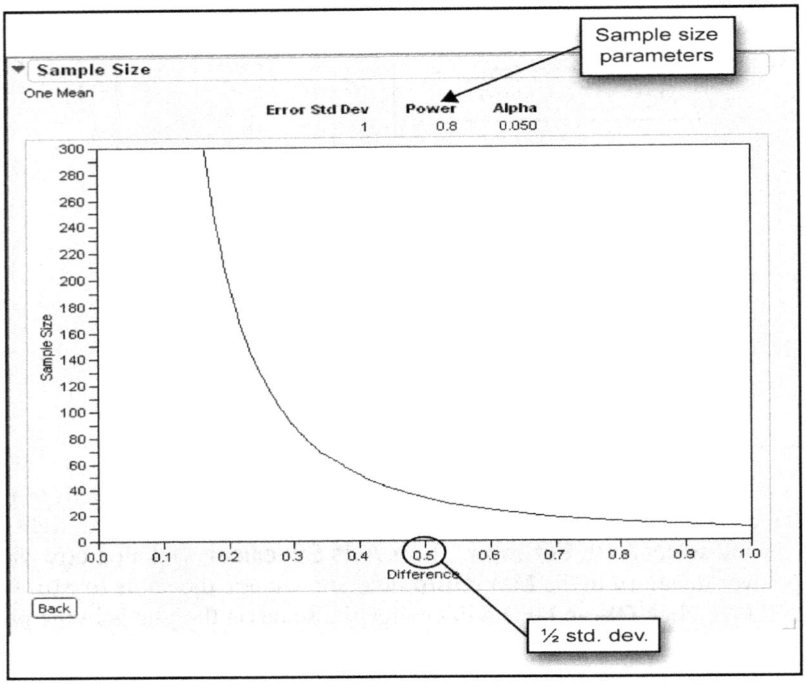

Now the plot clearly shows the point of diminishing returns for adding sampling units in our study. The biggest gains for additional units are made for differences to detect greater than or equal to where the curve shown in Figure 4.12 begins to reach the asymptote, at around a difference of 0.3 (the elbow of the curve). For example, if we want to increase our ability to detect smaller differences, then it takes an additional 106 units to move us from a detection of 0.3 standard deviation units (90 experimental units) to 0.2 standard deviation units (196 experimental units). However, it takes only an additional 40 units to move us from a detection of 0.4 standard deviation units (50 experimental units) to a detection of 0.3 standard deviation units (90 experimental units). We get the same magnitude of increased precision in our ability to detect a smaller difference but at half the amount of additional experimental units. Note also that very small differences (less than 0.1 standard deviation units) are detected with sample sizes larger than 300. These small differences, however, might not be practically meaningful.

One Sample Variance

The calculations for performance variation are performed using the variance or the square of the standard deviation. That is, we need to convert the standard deviation, σ, to the variance, σ^2. The inputs needed for the sample size calculator include the following:

1. **Alpha**: This is the Probability (Type I error) or significance level that we discussed in Chapter 2. This defines the amount of risk that we are willing to live with for incorrectly stating that our standard deviation is different from our standard value, k, when in fact it is not. An alpha of 0.05 is the standard choice and the default in JMP.

2. **Baseline Variance**: This is the square of our standard value for the standard deviation and is equal to k^2. For example, if we want to test whether our standard deviation is close to 4 Ohms, then 16 (= 4^2) would be the value that is entered in this field.

3. **Guarding a change**: This is the direction of change that we want to detect in our true standard deviation, σ from our standard value, k. Only two choices are available in this calculator and depend on which alternative hypothesis you choose:

   If H_1: $\sigma > k$, then select the **Larger** option.

   If H_1: $\sigma < k$, then select the **Smaller** option.

4. **Difference to detect**: This is the increase or decrease in the variance that we would like to detect from our baseline variance as described in (2).

   For example, consider the situation where the baseline standard deviation is 4 Ohms and we would like to detect a change of at least 1 Ohm. In other words, we will detect a difference if our standard deviation increases at least from 4 to 5 or, the variance increases at least from 16 to 25. Suppose we are testing the following one-sided alternative hypothesis:

   H_0: Performance Variation (σ) ≤ 4.

   H_1: Performance Variation (σ) > 4.

   We would enter $(4 + 1)^2 - 4^2 = 25 - 16 = 9$ in this field. Note that this is a 25% increase in the standard deviation, but a 64% increase in the variance.

   We can flip the sign in the hypotheses to the following:

   H_0: Performance Variation (σ) ≥ 4.

   H_1: Performance Variation (σ) < 4.

If we still want to detect a change of at least 1 Ohm, then we would enter $(4 - 1)^2 - 4^2 = 9 - 16 = -7$ in this field. In this scenario, we will detect a difference if our standard deviation decreases at least from 4 to 3 or the variance decreases at least from 16 to 9. Note that this is a 25% decrease in the standard deviation, but around a 56% decrease in the variance.

5. **Sample Size**: This is the sample size that is recommended for the study. We need to collect measurements on this many experimental units in order to be able to detect our practical Difference (**Difference to detect**) with a specified power.

6. **Power**: The power of a test is the probability of detecting a difference between the performance variation and the standard, when in fact there is a difference. The power is a probability and therefore takes on values between 0 and 1, which can be converted to a percentage by multiplying it by 100. Generally speaking, it is nice to have a high power for our one-sample test of significance. We recommend starting with a power of at least 0.8 and adjusting it up or down as needed. Again, don't be too greedy because the more power you want, the more samples you need!

Unlike the sample size calculator for the mean, we cannot generate sample size curves by leaving two fields blank. Also, sample size calculations for the variance are not as intuitive as sample size calculations for the mean and can generate results that do not make sense at first glance. This is because the Chi-square probability distribution for the variance is not symmetrical like the Normal or Student's *t* probability distributions for the mean.

For example, if we select values for the input fields that were discussed previously and are shown in Figure 4.13a (that is, **Alpha** = 0.05, **Baseline Variance** = 16, **Guarding a change** = Larger, **Difference to detect** = 9, and **Power** = 0.8), then JMP will output a **Sample Size** = 60.8486. For practical purposes we round this number up to 61 units.

Figure 4.13a Sample Size for One-Sample Test for Variance (Larger Than)

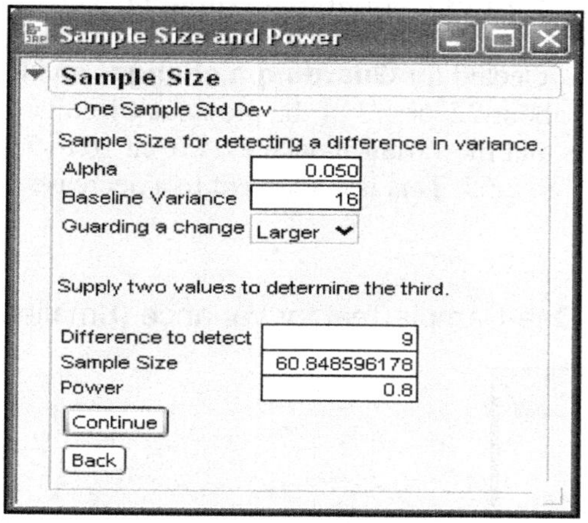

What is this telling us? Suppose we are testing the one-sided alternative hypothesis:

H_0: Performance Variation (σ) ≤ 4.

H_1: Performance Variation (σ) > 4.

A sample size of 61 units and $\alpha = 0.05$ enables us to detect a statistical difference 80% of the time (power) when our standard deviation, as compared with our standard value of 4, is at least 5 or more.

What if we are testing the reverse of the one-sided alternative hypothesis shown below and we still want to detect a difference of 1 unit in our standard deviation? Why can't we just use the same inputs and results that we obtained in Figure 4.13a? Here are the hypotheses:

H_0: Performance Variation (σ) ≥ 4.

H_1: Performance Variation (σ) < 4.

As we previously described in step 4, if we want to detect a decrease in the standard deviation of a least 1 unit, then the difference to detect in this case is $(4 - 1)^2 - 4^2 = 9 - 16 = -7$ and not -9 because the underlying Chi-square distribution is not symmetric (we are dealing with square quantities that are always positive).

To see this lack of symmetry, let's select all of the same parameters settings shown in Figure 4.13a and just change the **Guarding a change** to **Smaller**. This change is shown in Figure 4.13b, with the **Difference to detect** = -9, **Alpha** = 0.05; **Baseline Variance** = 16; the **Smaller** option selected for **Guarding a change**, and **Power** = 0.8. JMP will output a **Sample Size** of about 22, or 1/3 of the previous 61 units. Setting the **Difference to detect** to –9 implies that the variance decreases from 16 to 7, or that the standard deviation decreases from 4 to 2.65. This is equivalent to a decrease of 1.35 units and not a decrease of 1 unit.

Figure 4.13b Sample Size for One-Sample Test for Variance (Smaller Than)

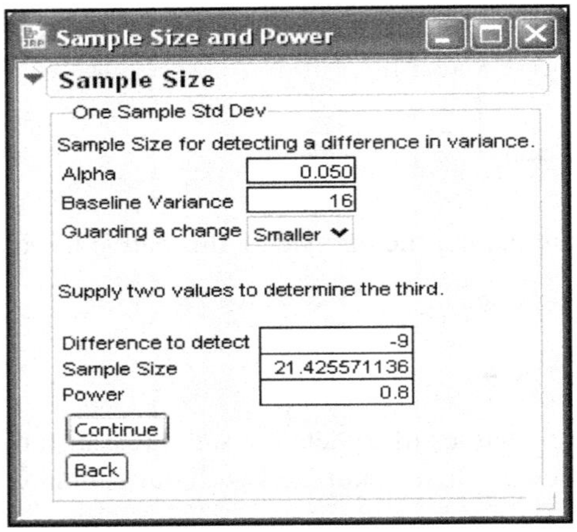

Caution: Sample size calculations do not guarantee that the outcomes from our studies exactly meet the **Difference to detect**, **Power**, and **Alpha** values that we specified. The values that we input into the various fields of the JMP sample size calculator give us a rough idea of the number of units that we will need in order to make an informed decision with some degree of confidence.

4.4 Step-by-Step JMP Analysis Instructions

We will illustrate the furnace qualification example using the seven steps shown in Table 4.2. The first two steps should be stated before any data is collected or analyzed, and they do not necessarily require the use of JMP. However, steps 3 through 6 will be conducted using JMP and the output from these steps will help us complete step 7.

Table 4.2 Step-by-Step JMP Analysis for Furnace Qualification

Step	Objectives	JMP Platform
1. Clearly state the question or uncertainty.	Make sure that we are attempting to answer the right question with the right data.	Not applicable.
2. Specify the hypotheses of interest.	Demonstrate that the new vertical furnace is capable of consistently producing wafers with a silicon dioxide layer of 90 Angstrom.	Not applicable.
3. Determine the appropriate sampling plan and collect the data.	Decide how many samples to take and identify the experimental and observational units. State the significance level (α), that is, the risk we are willing to take.	• **DOE > Sample Size and Power**
4. Prepare the data for analysis and conduct exploratory data analysis.	Visually compare the oxide distribution to the standard. Look for outliers and possible data entry mistakes. Do a preliminary check of assumptions.	• **Analyze > Distribution**
5. Perform the analysis to verify your hypotheses and to answer the questions of interest.	Are the observed differences between the data and the standard real or due to chance? We want to be able to detect these differences and estimate their size. Check the analysis assumptions to make sure our results are valid.	• **Analyze > Distribution > Test Mean, Std Dev** • **Graph > Control Chart > IR Chart**

(continued)

Table 4.2 *(continued)*

Step	Objectives	JMP Platform
6. Summarize the results with key graphs and summary statistics.	Find the best graphical representation of the results that give us insight into our uncertainty. Select key output for reports and presentations.	• **Graph > Variability/Gauge Chart**
7. Interpret the results and make recommendations.	Translate the statistical jargon into information that is meaningful in the context of the problem. Assess the practical significance of your findings.	Not applicable.

Step 1: Clearly state the question or uncertainty

We are finally ready to qualify the new vertical furnace that is used to oxidize silicon wafers. The primary objective of the qualification is to demonstrate that the vertical furnace is capable of consistently producing wafers with an average silicon dioxide layer of 90 Angstrom (Å), using the SOP (standard operating procedure) settings. Operations would also like to get an estimate of any potential scrap that will result if the silicon dioxide layer is too thick or too thin. The upper and lower specification limits for the oxide layer are LSL = 87 Å and USL = 93 Å. We need to demonstrate with a high degree of confidence what this equipment is capable of outputting.

Step 2: Specify the hypotheses of interest

For this example, there are two sets of hypothesis that are needed. We need to make a statement that pertains to the average performance of the vertical oven, as well as the variation in oxide thickness of our vertical oven. For the average performance the standard value that we will use for oxide thickness is 90 Å. The null and alternative hypotheses for the average performance are shown below:

H_0: Average Performance (μ) = 90 Angstrom (Assume)

H_1: Average Performance (μ) $\neq$ 90 Angstrom (Prove)

Why have we chosen a two-sided alternative hypothesis? The wafers should have a target oxide thickness of 90 Å, and there is a concern if the oxide layer is too thick or too thin. Also, both upper and lower specification limits are used to scrap wafers at this step. There is no concern regarding the direction of departure from the standard that would lend itself to a one-sided alternative hypothesis.

Statistics Note 4.4: We "prove" the alternative hypothesis in the sense of *establishing by evidence a fact or the truth of a statement* (The Compact Oxford English Dictionary).

We also need to test that the performance variation (for the wafers) is less than or equal to a standard value, but where do we get this standard value for the wafer variation? Let's review the rational for selecting a standard value for the performance variation. We are given the values of the lower specification limit, 87 Å, and the upper specification limit, 93 Å. These specification limits apply to the wafer averages since we are interested in centering the average wafer thickness within these limits. We will use this information to back calculate a value for our standard deviation. If the business can tolerate a 1% yield loss, assuming that it occurs equally on either side of the specification limit, then we can determine the value for our standard deviation that "guarantees" the 1% yield loss, as long as the distribution is centered on the target value of 90 Å. In order to do this we find the normal distribution quantile, z, such that the area between $\pm z$ is 1 – Loss = 1 – 0.01 = 0.99 (99%). This corresponds to a 0.5% yield loss on either side of our specification limits. The normal quantile function in the Formula Editor easily provides this value, which is 2.5758. The JMP steps to find this value are shown below and highlighted in Figure 4.14.

1. Create a new JMP table by pressing CTRL+N.
2. Add one row to the table by double-clicking in the first row.
3. Right-click the column name and select **Formula**.
4. Select **Probability** under the **Functions (grouped)** menu.
5. Select **Normal Quantile**.
6. Enter a probability—a number between 0 and 1.

For our problem, we need to type `0.005 (0.5%)` in the box to the right of **Normal Quantile** and click **OK**. This will output the value -2.5758 in the first row of the newly created JMP table. Since the normal distribution is symmetric, we know that the area to the right of 2.5758 is also equal to 0.005, that is, the area under the normal density function between ±2.5758 is 0.99.

Figure 4.14 Formula Editor for the Normal Quantile Function

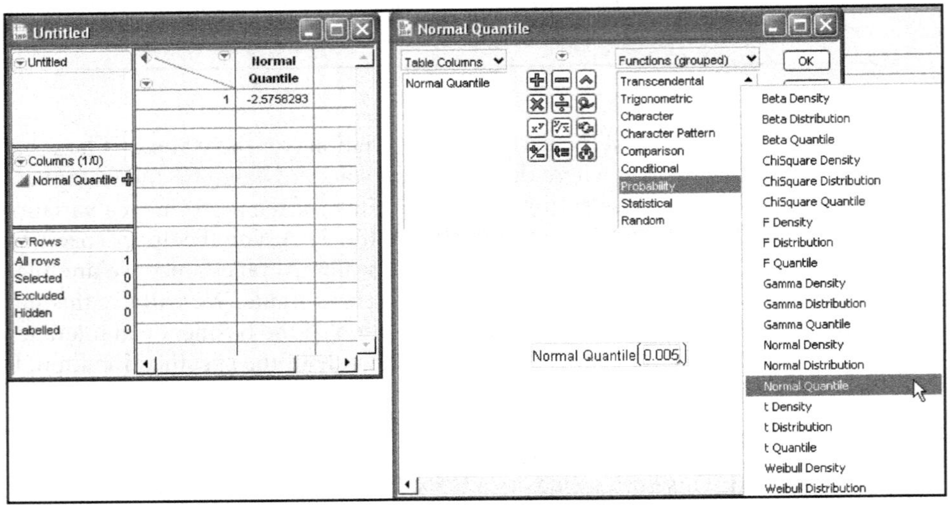

JMP Note 4.4: The Normal Quantile function produces a quantile from the Normal distribution such that the area to its *left* is equal to the specified probability.

In order to complete this calculation, we must solve the following equation for the standard deviation: Z_{score} = (USL – target) / (standard deviation). In other words, standard deviation = (USL – μ) / Z_{score} = (93 – 90)/ 2.5758 = 3/2.5758 = 1.165 that we round up to 1.17. We can now write the hypotheses for the performance variation as:

H_0: Performance Variation (σ) $\leq$ 1.17 Angstrom (Assume)

H_1: Performance Variation (σ) > 1.17 Angstrom (Prove)

Note that the null hypothesis (H_0) specifies that the performance variation is less than or equal to 1.17. This assumes that the yield will be at least 99%, or equivalently, the yield loss will be 1% or less. Since this is a new piece of equipment with no prior history, we would like to show that it is capable of hitting the high yield targets. In fact, the ability of the new furnace to maintain small yield losses was part of the justification for the new equipment. A test of significance, which only involves the population mean, would be incomplete for this qualification.

> *Statements related to consistency of the output are related to the performance variation of the quality characteristic of interest, oxide thickness in this case.*

Step 3: Determine the appropriate sampling plan and collect the data

When setting up a sampling scheme for this qualification, we need to think about what our experimental unit is, the number of units that we need, our sampling scheme (that is, how will we select these units from the larger population), and the sources of variation that we want to include in this study.

Recall from the problem description that this new furnace has three zones, and each zone has its own set of process controllers to maintain the appropriate conditions. One virgin wafer is typically placed in each quartz boat in each zone of the furnace, in order to check the oxide growth and make sure that it is close to the desired target value. Therefore, the experimental unit (EU) for the furnace qualification is one virgin wafer. It should be noted, however, that we will take thickness measurements in several locations on each wafer to see how uniform the thickness layer is across the wafer. These locations within a wafer, the EU, comprise the observational units (OUs).

Decide on the level of risk

Before selecting the significance level that we are comfortable with, let's first interpret it in the context of both sets of hypothesis. A Type I error rate, α, for our hypothesis about the average thickness performance of our new oxide furnace means that we will incorrectly reject our null hypothesis in favor of the alternative hypothesis α% of the time. For example, if we select $\alpha = 0.05$ then 5% of the time, by chance alone, we will conclude that our average oxide thickness is not equal to 90 Å, when in fact it is. When honing in on an appropriate value for α we must consider the repercussions of making such an error.

What will happen if we incorrectly conclude that our furnace is not delivering our target value of an average of 90 Å of oxide? Well, we would most likely use the information that we acquired during our equipment characterization and optimization to alter some of the equipment settings in order to bring the thickness up or down, depending upon the implied direction in which it was off target, and then we would monitor the output to make sure that we tweaked the equipment correctly. Although we have a good understanding about how to make the oxide layer thicker or thinner, a good amount of time and resources will be consumed dialing the process in and we would not want to draw the wrong conclusions too often. For this hypothesis, we can live with $\alpha = 0.05$.

What about our hypothesis for the standard deviation? Once again, we need to ask ourselves what will happen if we incorrectly reject the null hypothesis in favor of the alternative. For example, if we select $\alpha = 0.05$, then 5% of the time, by chance alone, we will conclude that the standard deviation for oxide thickness is greater than 1.17 when in fact it is not. This means that we will conclude that if thickness is centered on target, then we cannot meet our yield loss goals of 1% or less when, in fact, we can. This could eventually show up in the financial reports for this process step and could also result in higher scrap rates and more tests and inspection at this step. For this hypothesis we can also live with alpha = 0.05.

Sample size calculation

The sample size calculation for the average performance is shown in Figure 4.15. For the mean, we need about 14 experimental units (virgin wafers) in order to be able to detect a shift of 1 Å in the true mean from our hypothesized value of 90 Å, given that alpha = 0.05, σ = 1.2 Å, and power = 0.8. In other words, if our true mean was higher than 91 or lower than 89 then we would want to be able to detect it 80% of the time. This choice is partially based on the stated accuracy of the controllers and the measurement error in our thickness readings. Note that the error standard deviation was obtained by rounding up to the first decimal place, our hypothesized value for σ, which was 1.165.

Figure 4.15 Sample Size and Power for Oxide Mean

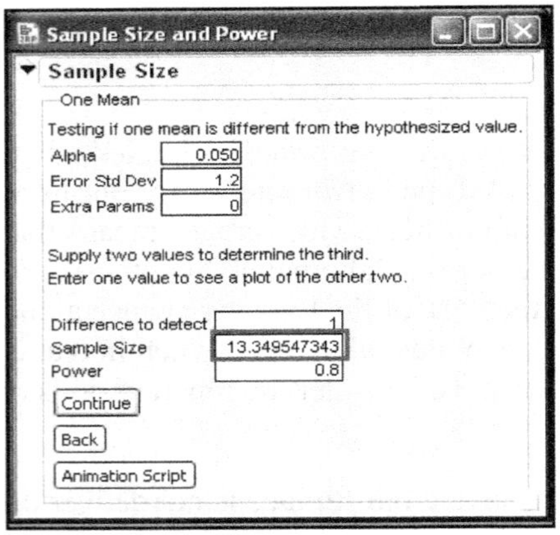

The sample size calculation for the wafer variance is provided in Figure 4.16a. JMP indicates that we need approximately 43 experimental units to detect an increase of 0.972 from our baseline variance of $(1.17)^2 = 1.37$. The value 0.972 is obtained as $(1.17 + 0.36)^2 - 1.17^2$. The first term, $(1.17 + 0.36)^2$, corresponds to an increase in wafer standard deviation of 0.36, which is equivalent to a 5% yield loss. The second term, 1.17^2, is the baseline variance. In other words, if our true standard deviation is larger than 1.53 Å, then we would like to declare it different from our standard value and reject H_0.

Figure 4.16a Sample Size and Power for Oxide Variance

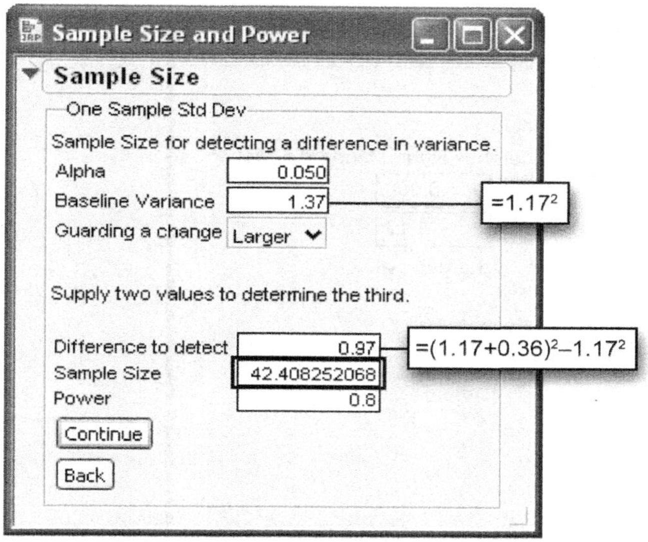

JMP 8 Note 4.1: In JMP 8 the calculations for the sample size for variance are more intuitive and easier to perform. The calculator is based on the standard deviation, and not the variance, making the **Difference to detect** calculations shown in Figure 4.16a unnecessary. We can enter the value of the baseline standard deviation, `1.17`, and the delta increase, `0.36`. Figure 4.16b shows the Sample Size and Power calculator for the standard deviation, which gives 43 wafers as the required sample size to detect a change from 1.17 Å to 1.53 Å.

Figure 4.16b JMP 8 Sample Size and Power for Oxide Standard Deviation

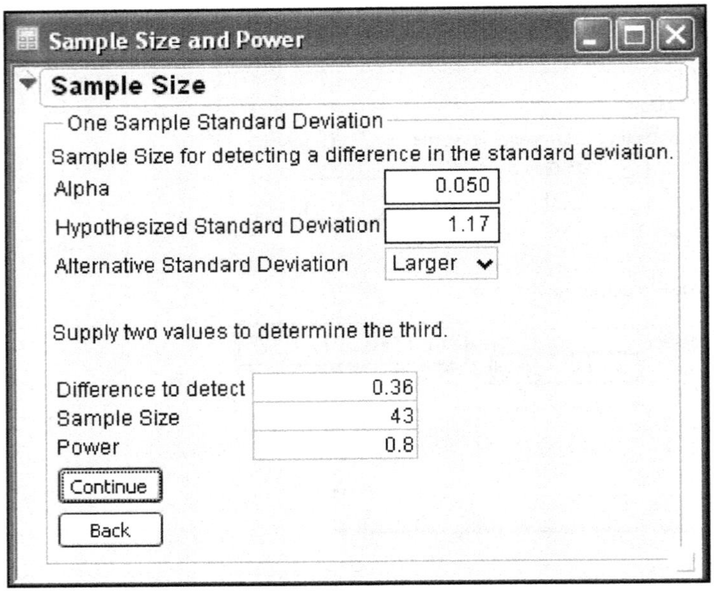

Now that we have an idea of how many experimental units we need in our study, we should think about how to sample them from our population in order to make sure that the sample is representative of the manufacturing environment. Understanding the sources of variation that are present in the furnace qualification helps us to design the appropriate sampling scheme. These sources of variation include the following: sites within each wafer, position within each quartz boat, zones within the furnace, and different runs of the furnace. As we think about our sampling scheme for this qualification we should take into account these different sources of variation. In some instances, one or more of these sources of variation can have a systematic pattern to them, and therefore potentially create different populations for our output. Historically, the furnace zones had a systematic pattern regarding the oxide growth and this new furnace, with independently controlled zones, is supposed to alleviate this problem. We would like to check for any systematic patterns among the different furnace zones, so we will use a stratified simple random sample for this study by sampling from each of the three furnace zones. To keep things simple, the sample size calculations are based only upon wafers and do not take into account all of the sources of variation described above.

The layout for the sampling plan is shown in Figure 4.17. We will make 10 runs of the furnace using a combination of dummy and product wafers in order to fill the quartz boats and get an accurate representation of manufacturing runs. We will use four virgin wafers per run by placing one virgin wafer in the top zone, two virgin wafers in the center zone (since this zone is a slightly larger than the zones on either side of it), and one virgin wafer in the bottom zone. We need a total of 40 (= 10 runs x 4 wafers per run) virgin wafers, or experimental units, for this study. The oxide thickness measurements are taken on a site (OU) within each virgin wafer (EU). Since we are also interested in site-to-site variation in this study, we will take thickness measurements in 5 locations for each virgin wafer, according to the diagram in Figure 4.17. This will result in a total of 200 = 40 (EU) x 5 (OU) measurements.

Figure 4.17 Sampling Scheme for Oxide Furnace Qualification

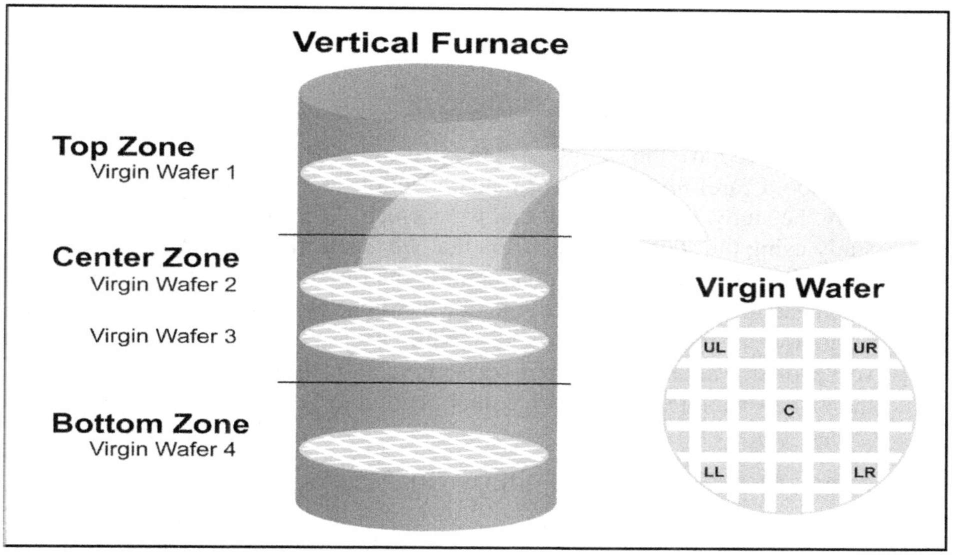

We are using 40 experimental units, which means that we can satisfy the sampling plan requirements for the average performance, which suggested 14 EUs to detect a difference of ±1 Å. However, for the performance variation we are three units short of the suggested sample size of 43 EUs. In order to figure out the impact of reducing the sample size from 43EUs to 40EUs on power, we can use the sample size calculator to solve for power instead of for sample size. This is accomplished by filling in the **Sample Size** field with 40 and leaving the **Power** field blank. Figure 4.18 shows a resulting power of 78% rather than the 80% that we originally obtained. This is close enough for our study and we should not be too concerned about this shortage.

Figure 4.18 Power Calculations for 40 Experimental Units for Oxide Furnace Qualification

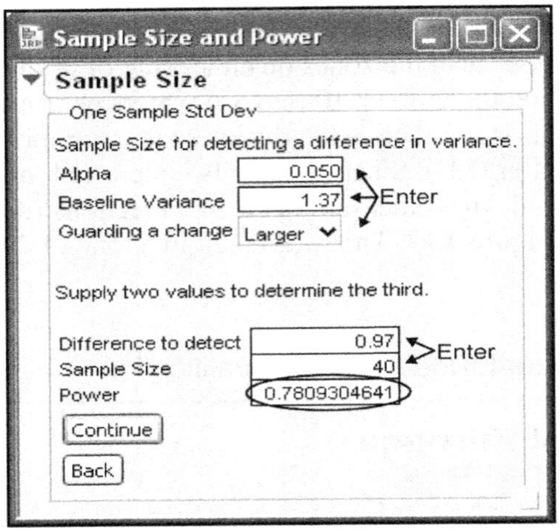

Now that we have the sampling plan (size and scheme) for the Furnace qualification, we should carefully construct a data collection sheet that can be used to gather the data from the study. It is important that we correctly identify all of the observations in this study using the sources of variation that we outlined previously. For example, the data sheet should include a column to capture the furnace location: top, center, or bottom. We also need to make sure that we include a column for comments in order to capture any excursions or problems that we have while we are running the furnace. This information will provide valuable insights if we observe any outliers, unexpected patterns, or trends in the data. A sample of the furnace qualification data is provided in Table 4.3.

Table 4.3 Sample Data Form Oxide Furnace Qualification

			Oxide Thickness Readings					
Obs. Number	**Run Number**	**Furnace Location**	**UL**	**UR**	**C**	**LL**	**LR**	**Comments**
1	1	1	88.2	91.8	90.1	90.4	92.1	
2	1	2	92.2	91.1	91.9	87.5	91.2	
3	1	3	92.3	92.4	88.1	87.4	92.6	
4	1	4	92.6	95.2	90.0	87.0	93.2	
5	2	1	89.0	92.8	90.7	89.1	92.6	
6	2	2	93.4	92.3	88.6	89.9	91.8	
7	2	3	91.3	92.0	90.2	90.0	91.6	
8	2	4	89.9	94.1	90.8	89.9	91.9	
. . .	. . .							

Step 4: Prepare the data for analysis and conduct exploratory data analysis

In order to carry out a one-sample test for the mean and the standard deviation, we need to get our data into JMP, and make sure that it is in the right format for the analysis platform that we will use.

Getting our data into JMP

The data was read into JMP from its original source, an Excel file, by clicking the **Open Data Table** icon in the JMP Starter window and making sure the **Excel Files (*.XLS)** choice is the selected file type. Once the correct filename is highlighted, we click the **Open** icon and the data is read into to a JMP data table (Figure 4.19). Note that this is a temporary table. If we want to save the data in this JMP format, then we need to select **File > Save** and enter a new name and location for the JMP table.

Is the data analysis ready?

In other words, is the data in the right format for the analysis platform we need to use? Looking at Figure 4.19 we see that each row includes information about one virgin wafer for a total of 38 rows. We should have had 40 rows of data in this JMP table, corresponding to 40 our experimental units. However, two virgin wafers from run 8 were accidentally dropped and broke while we were taking the thickness readings, thus reducing our total to 38 experimental units. The thickness readings are located in five columns, corresponding to the five locations within a wafer, and are labeled using the sampling location from within each wafer: Center, Upper Left, Lower Left, Lower Right, and Upper Right. The data types are set as **Categorical** for both RUN and FSCT (furnace section), as indicated by the red bars next to these column names in the **Columns** listing in the middle section on the left-hand side of the table. Alternatively, the data types for Center, Upper Left, Lower Left, Lower Right, and Upper Right are set as **Numeric**, as is indicated by the blue right triangle preceding the column name. The JMP table can be downloaded from the author's page for this book, http://support.sas.com/authors.

Figure 4.19 JMP Table for Oxide Furnace Qualification

Furnace Qualification

Furnace Qualificatio

Columns (8/0)
RUN
FSCT
Center
Upper Left
Lower Left
Lower Right
Upper Right
Wafer Average

Rows
All rows 38
Selected 0
Excluded 0
Hidden 0
Labelled 0

	RUN	FSCT	Center	Upper Left	Lower Left	Lower Right	Upper Right	Wafer Average
1	1	1	90.1	88.2	90.4	92.1	91.8	90.52
2	1	2	91.9	92.2	87.5	91.2	91.1	90.78
3	1	3	88.1	92.3	87.4	92.6	92.4	90.56
4	1	4	90	92.6	87	93.2	95.2	91.6
5	2	1	90.7	89	89.1	92.6	92.8	90.84
6	2	2	88.6	93.4	89.9	91.8	92.3	91.2
7	2	3	90.2	91.3	90	91.6	92	91.02
8	2	4	90.8	89.9	89.9	91.9	94.1	91.32
9	3	1	89.4	89.9	91.6	92.6	93	91.3
10	3	2	89.7	92.6	89.2	92.5	93	91.4
11	3	3	86.6	92.3	90.5	93.6	93.6	91.32
12	3	4	93.2	93	89.7	92.5	94.6	92.6
13	4	1	87.8	87.9	89.1	93.2	90.9	89.78
14	4	2	86.6	90.4	89.7	92.6	92.7	90.4
15	4	3	91.9	92.1	89	92	93.4	91.68
16	4	4	89.1	92.4	92.9	96.2	96.1	93.34
17	5	1	91.8	90	88.9	93.8	92.3	91.36
18	5	2	89.3	91.7	89.2	93.3	95.2	91.74
19	5	3	90	91.3	90	92.1	93.9	91.46
20	5	4	90.2	91.7	89.4	96.7	92.5	92.1
21	6	1	90.3	88.1	90.1	91.9	94.5	90.98
22	6	2	91.1	93.1	88.9	92.5	92.4	91.6
23	6	3	92.4	92.4	88.7	92.9	92.6	91.8
24	6	4	94.1	92.2	89.4	94.5	95.4	93.12
25	7	1	90.3	91	89.7	95	95.4	92.28
26	7	2	92.7	92.5	89.9	94.2	93.6	92.58
27	7	3	87	88.7	90.7	94.9	91.4	90.54
28	7	4	91.8	92.5	90.1	94.9	92.8	92.42
29	8	1	89	88.6	90.5	91.3	93.3	90.54
30	8	2	89.9	93.1	88.4	92.1	93.1	91.32
31	9	1	91.4	87.6	88.9	90.9	91.6	90.08
32	9	2	90.6	92.2	90.7	91.3	93.6	91.68
33	9	3	88	91.5	89.4	93.2	93.9	91.2
34	9	4	92.1	92.1	91.2	92.3	91.1	91.76
35	10	1	85.6	87	84.8	88.6	90.2	87.24
36	10	2	85.7	86.4	87	86.4	87.6	86.62
37	10	3	88.5	87.9	85.4	85.5	89.3	87.32
38	10	4	82.8	87.1	84.5	87.2	87.6	85.84

Data is adapted with permission from the Appendix to Chapter 8 of *Statistical Case Studies for Industrial Process Improvement* (Czitrom and Spagon 1997).

Although we are interested in exploring the site-to-site (OU) differences, our test of significance for both the mean and the standard deviation will be conducted using a representative thickness reading associated with each experimental unit, which is a virgin wafer. The average of the thickness readings taken at each of the five locations will be used to describe the thickness for each virgin wafer and subsequently used in the test of significance for both parameters. We need to add a new column to the table containing the wafer averages by following the steps below. Some of these steps are highlighted in Figure 4.20.

1. Double-click to the right of the last column, **Lower Right**, in the table. This adds a new column to the JMP table labeled Column 8.

2. Highlight this new column, right-click and then select **Column Info** from the drop-down list. This launches a new window containing the column information (not shown).

3. Change the **Column Name** from "Column 8" to "Wafer Average" by typing over the label already in this field.

4. Select the **Formula** option from the drop-down menu that appears under the **Column Properties** button near the bottom of the window. This action brings forward a **Formula** box to the right of the **Column Properties** drop-down menu.

5. Click the **Edit Formula** button and this launches the formula calculator in a new window.

6. From this Formula Editor window, select **Statistical** from the **Functions (grouped)** menu located next to the **OK** and **Cancel** buttons on the right-hand side of the window. This opens up another drop-down list of statistical functions that are available within JMP.

7. Select the **Mean** function from the **Statistical** functions in the drop-down list. This starts to build out the **Mean( )** function in the blank part of the formula editor. We need to add the five column names containing the five thickness readings (separated by commas) within the parenthesis of this function.

8. Select Center in the list labeled **Table Columns** in the left-hand side of the window. This automatically places this column name in the mean function, **Mean**(Center). In order to add another column to this function, we need to type a comma, and this will build out the function by adding the comma and a blank box to the right of Center. Alternatively, we can click the **^** in the calculator pad to add more arguments to this function (not shown).

9. Select Upper Left from the **Table Columns** list and then type a comma; the function should now contain **Mean**(Center, Upper Left) (not shown).

10. Repeat step 9 to add Lower Left, Lower Right and Upper Right to the **Mean** function. The final function should look like **Mean**(Center, Upper Left, Lower Left, Lower Right, Upper Right).

11. Click **Apply** and then **OK**. This populates the new column with averages of the five wafer sites and closes the Formula Editor window. Close the Column Information window, which is still open, by clicking **OK**.

Figure 4.20 Create a Column of Wafer Means

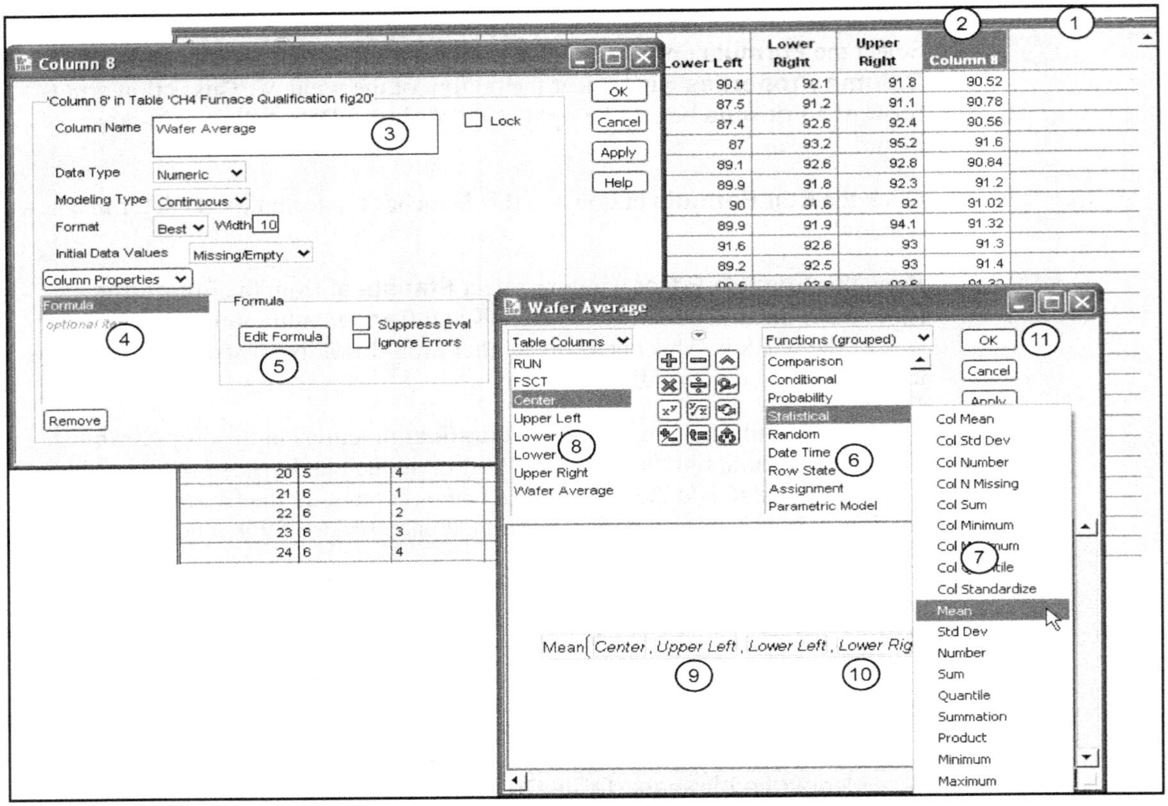

Our data is now in the correct format to proceed with our analysis. However, before we proceed with our test of significance for the average performance and performance variation in the next step, we need to check the data for any unusual observations or patterns in order to make sure that we meet the assumptions for a one-sample test. Recall that these assumptions include the following: the underlying population is normally distributed, the samples are independent from each other, and the underlying variance is constant.

In order to check our data for outliers and determine whether it is normally distributed, we will create a histogram and some summary statistics of the wafer averages. In order to generate this graph, follow the steps provided next and shown in Figure 4.21.

1. From the primary toolbar, select **Analyze > Distribution**. A dialog box appears that enables us to make the appropriate choices for our data.

2. Highlight Wafer Average from the **Select Columns** window, and then click the **Y, Columns** button. This populates the window to the left of this button with the name of the column to use for the graph.

3. Click **OK** to create the histogram.

Figure 4.21 Distribution Dialog Box for Oxide Furnace Qualification

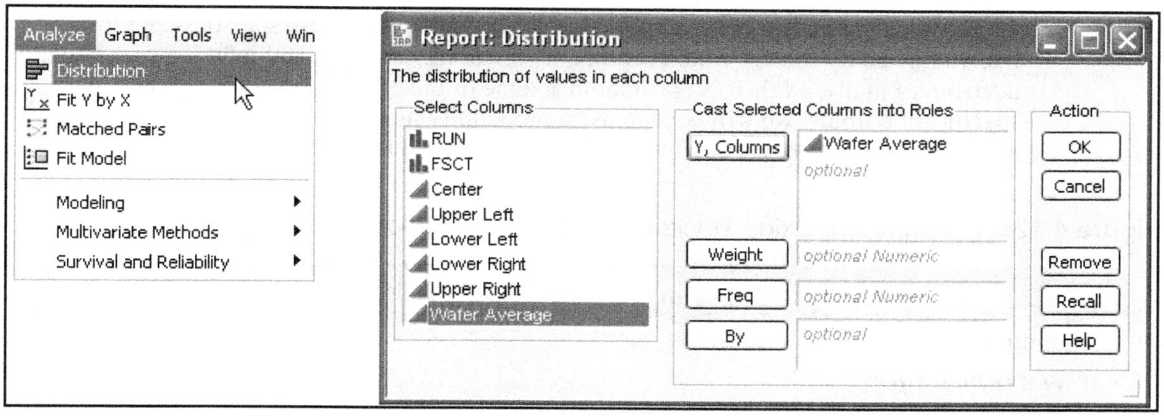

The resulting output is shown in Figure 4.22a. If the Histogram options have not been set ahead of time then the histogram is in a vertical position instead of the horizontal position shown in this figure. In order to change it to a horizontal layout, select the **Stack** option from the drop-down list that is activated by clicking the red triangle next to the word **Distributions**.

In the histogram of wafer oxide thicknesses averages we notice a group of observations to the left of 88, giving the appearance of a distribution skewed to the left. If we click these points on the histogram we can quickly find out where they are located in the JMP table because the corresponding rows are highlighted. We see that these data points are all associated with run 10. When we reviewed the notes taken during this study, we discovered that something unusual happened during this run and the data is highly suspect. In other words, there appears to be an assignable cause for the low readings associated with run 10. When we have questionable data in our study there are two things that we can do: keep it in the analysis or, *if it can be justified*, discard it. Before making a decision on which way to go, it is a good idea to carry out the analysis with and without the data in question and determine whether it changes the results of the test of significance. If the results do not change then it doesn't really matter if we keep it or discard it. However, if the results do change considerably then we must make a sound decision as to their validity.

Using all of the data, we see that the overall mean is 90.93 Å, the standard deviation is 1.65 Å, the minimum is around 85 Å, and the maximum is around 94 Å. What is this telling us about the furnace qualification? Before we can make definitive statements about the state of this qualification, we need to complete our one-sample of test of significance for the mean (= 90 Å) and the standard deviation (= 1.17 Å), and also look at the performance of the wafer averages with respect to the specification limits (LSL = 87 Å, USL = 93 Å). However, from the histogram and summary statistics we can already see that we are off center from our target value and the standard deviation looks a bit larger than desired. Another thing to consider is that both of these statistics are most likely affected by run 10, which is probably lowering our mean and inflating our standard deviation. Finally, while it is apparent that some of the 38 wafer averages exceed the specification limits, we will need to use a probability model or a tolerance interval to get a better estimate of how many wafers in our population might exceed these limits.

Figure 4.22a Histogram of Oxide Thickness Data and Suspect Readings

Furnace Qualification- Distribution of Wafer Average

Distributions

Wafer Average

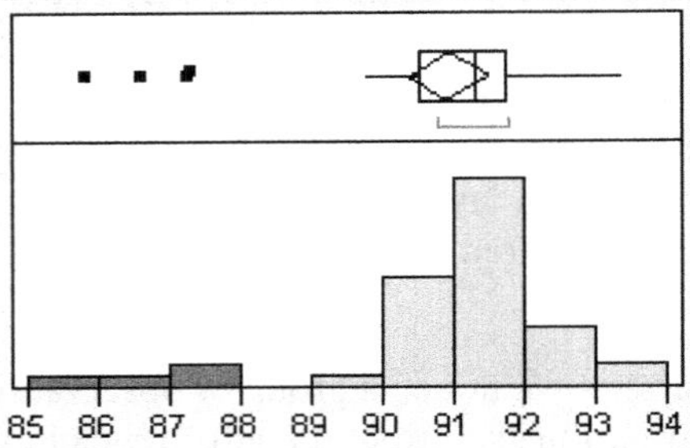

Quantiles

100.0%	maximum	93.340
99.5%		93.340
97.5%		93.340
90.0%		92.582
75.0%	quartile	91.745
50.0%	median	91.320
25.0%	quartile	90.540
10.0%		87.312
2.5%		85.840
0.5%		85.840
0.0%	minimum	85.840

Moments

Mean	90.927368
Std Dev	1.6537095
Std Err Mean	0.2682671
upper 95% Mean	91.470929
lower 95% Mean	90.383808
N	38

32	9	2	90.6	92.2	90.7	91.3	93.6	91.68
33	9	3	88	91.5	89.4	93.2	93.9	91.2
34	9	4	92.1	92.1	91.2	92.3	91.1	91.76
35	10	1	85.6	87	84.8	88.6	90.2	87.24
36	10	2	85.7	86.4	87	86.4	87.6	86.62
37	10	3	88.5	87.9	85.4	85.5	89.3	87.32
38	10	4	82.8	87.1	84.5	87.2	87.6	85.84

Are the thickness averages homogeneous?

As we discussed in Chapter 2, it is important to verify the homogeneity assumption because our statistical analysis of the data depends on this assumption (Wheeler 2005). Can we treat the thickness average data as coming from a homogeneous population? We can use an individuals and moving range chart (**Graph > Control Chart > IR**) to check the homogeneity of the thickness averages. The moving range for observation 35, Figure 4.22b, is outside the upper control limit (UCL), giving us a signal that the data may not be homogeneous. When we look at the individuals chart we notice that the last four averages, rows 35 to 38, are below the lower control limit (LCL) of 88.31 Å, which indicates that these averages are different from the rest. These four thickness averages correspond to the suspect run 10. In other words, the thickness averages are not homogenous, and the culprit seems to be run 10.

Figure 4.22b Individual Measurements and Moving Range Chart of Oxide Thickness Data

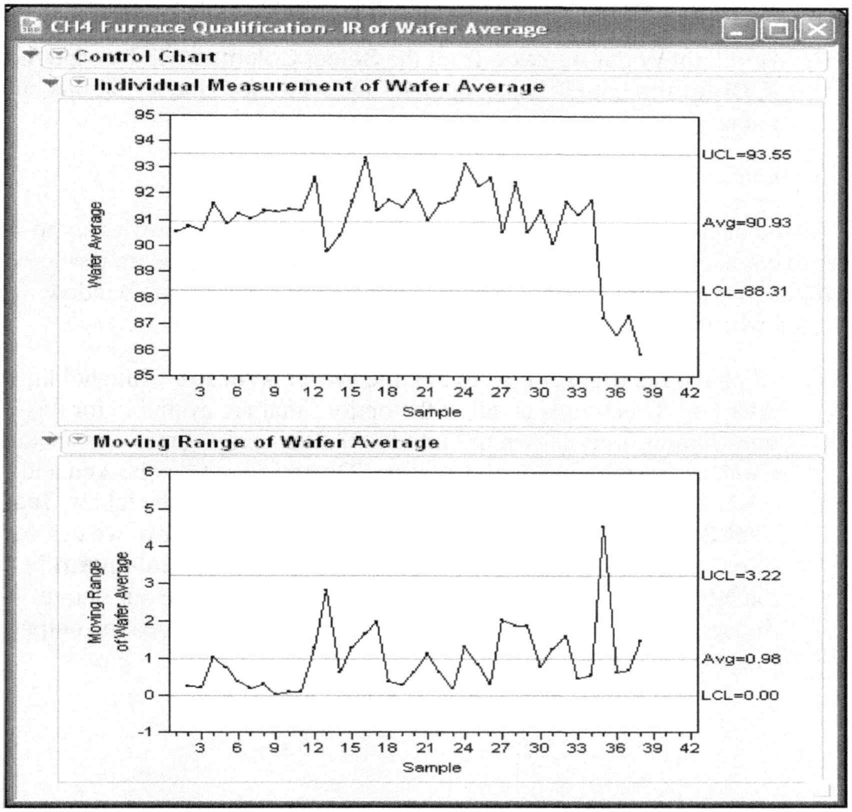

Based on our exploratory analysis of the oxide wafer averages, it looks like run 10 wafers are outliers from the rest of the wafer population, and might be influential to the outcome of the qualification. We will conduct the next step with and without run 10 and then make a decision as to its inclusion in the final qualification of the furnace.

Step 5: Perform the analysis to verify your hypotheses and to answer the questions of interest

Is the average performance of the furnace meeting the standard?

The major uncertainty in our study is to demonstrate that the new vertical furnace is capable of consistently producing wafers with a silicon dioxide layer of 90 Angstrom.

The **Distribution** Platform, within the **Analyze** menu, is used to perform tests of significance on the mean and the standard deviation. The necessary steps are provided below.

1. From the primary toolbar, select **Analyze > Distribution** and a dialog box will appear that enables us to make the appropriate choices for our data.

2. Highlight Wafer Average from the Select Columns window and then click the **Y, Columns** button. This populates the window to the left of this button with the name of the column to use for the graph.

3. Click **OK** to create the histogram.

We get the histogram produced in Figure 4.22a and discussed previously in step 4. In order to conduct our one-sample tests for the mean and the standard deviation, we need to provide our standard values for both parameters in the Options window, which is launched when we do the following.

4. Click the red triangle next to the title **Wafer Average** while holding down the ALT key. This brings up all of the options that are available for this display in one window, as is shown in Figure 4.23. We need to enter our standard values for the mean and the standard deviation. The boxes can be checked and the numbers `90` and `1.17` can be entered in the empty fields next the labels, **Test Mean** and **Test Std Dev**, respectively. While we are in this window, we can set a couple of more options. For example, **More Moments**, **Normal Quantile Plot**, and the **Normal** check boxes can all be selected to enhance our output. When all the desired check boxes are selected, click **OK** to produce the output shown in Figure 4.24.

Figure 4.23 Distribution Options for Testing the Mean and Standard Deviation

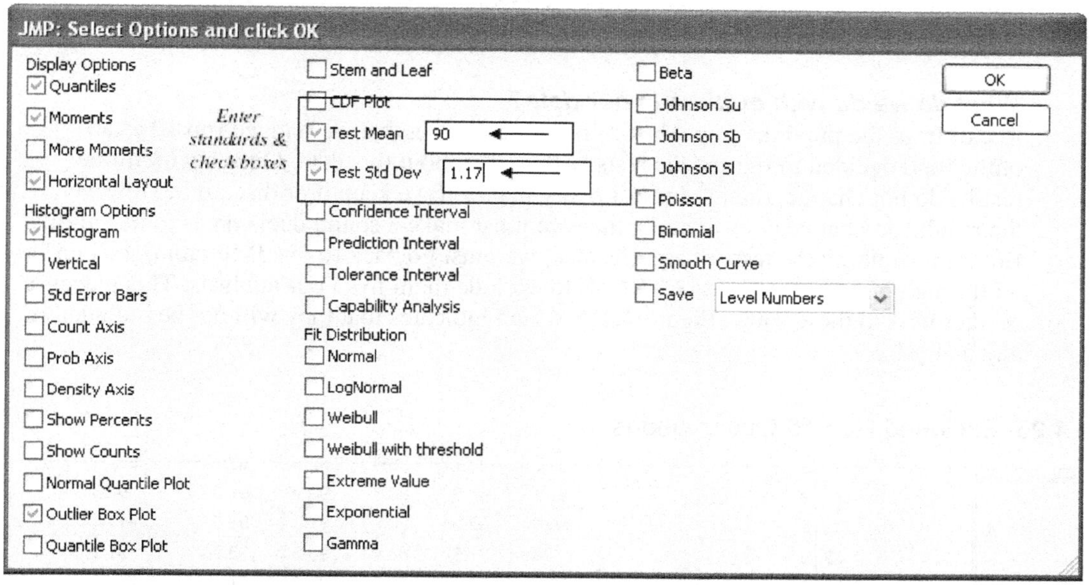

Figure 4.24 One-sample Tests for Oxide Thickness Data

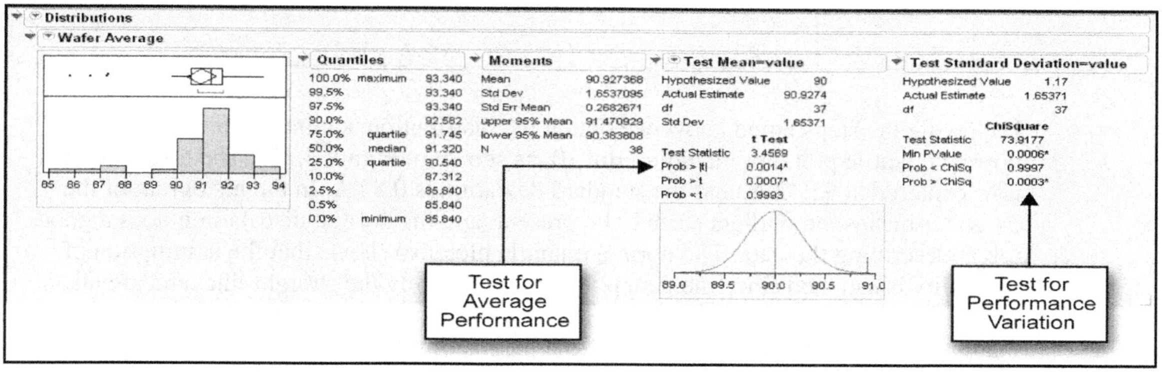

The calculated *t*-test statistic for comparing the mean to the standard value of 90, using all the data, is 3.4569 = (90.927 – 90) / (1.6537 / √38). The sign of the calculated *t*-statistic is positive, which implies that the mean is greater than our standard value of 90 Å. The *p*-value for this test statistic is 0.0014.

The calculated Chi-square test statistic for comparing the standard deviation to the standard value of 1.17, using all the data is $73.9177 = 37\times(1.6537^2 / 1.17^2)$, with a *p*-value of 0.0003. It is obvious that the standard deviation is also greater than the standard value.

What do we do with outliers in our data?

Recall from the previous step (shown in Figure 4.22b), that we flagged run 10 as an outlier and decided to run the analysis with and without this data. Once again, if the results do not change, then it doesn't really matter if we keep it or discard it. However, if the results do change considerably, then we must make a sound decision as to its validity. Before re-running the tests of significance, we must go back to our JMP table, select rows 35 through 38, and then press CTRL+E to exclude them from our analysis. This places a marker next to these rows (Figure 4.25), which indicates that they will not be included in any analysis.

Figure 4.25 Excluded Run 10 Observations

Rows	
All rows	38
Selected	4
Excluded	4
Hidden	0
Labelled	0

	31	9	1	91.4	87.6	88.9	90.9	91.6	90.08
	32	9	2	90.6	92.2	90.7	91.3	93.6	91.68
	33	9	3	88	91.5	89.4	93.2	93.9	91.2
	34	9	4	92.1	92.1	91.2	92.3	91.1	91.76
⊘	35	10	1	85.6	87	84.8	88.6	90.2	87.24
⊘	36	10	2	85.7	86.4	87	86.4	87.6	86.62
⊘	37	10	3	88.5	87.9	85.4	85.5	89.3	87.32
⊘	38	10	4	82.8	87.1	84.5	87.2	87.6	85.84

Following the steps noted above we obtain the distribution, summary statistics, and normal quantile plot, not including run 10, as shown in Figure 4.26a. The distribution is now centered at 91.42 Å, and the standard deviation is 0.82 Å. In the histogram of the 34 wafer means, no outliers seem to be present, and the normal distribution does a good job at describing the data. The normal quantile plot also shows that the assumption of normality is satisfied since the data points follow closely the straight line, and are all within the confidence (dotted) bands. The new minimum is now 88.78 Å, which is greater than the lower specification of 84 Å, while the maximum is 93.34 Å.

Figure 4.26a Distribution of Average Oxide Thickness without Run 10

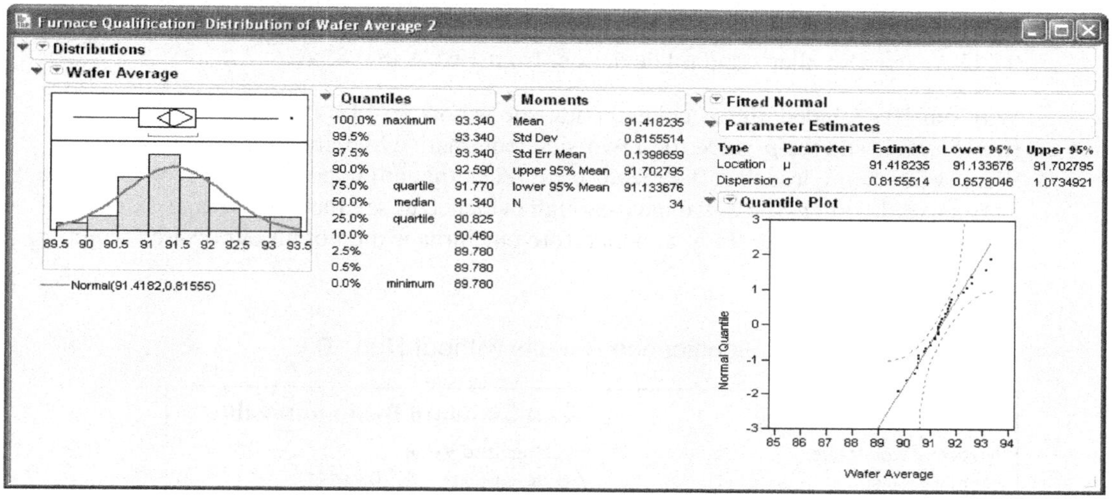

The individuals and moving range chart excluding run 10, Figure 4.26b, show all the points within the natural process variation suggesting that the data is homogeneous. However, the average is 91.42 Å, which suggests that we might not be meeting the target of 90 Å.

Figure 4.26b Individuals Chart of Average Oxide Thickness without Run 10

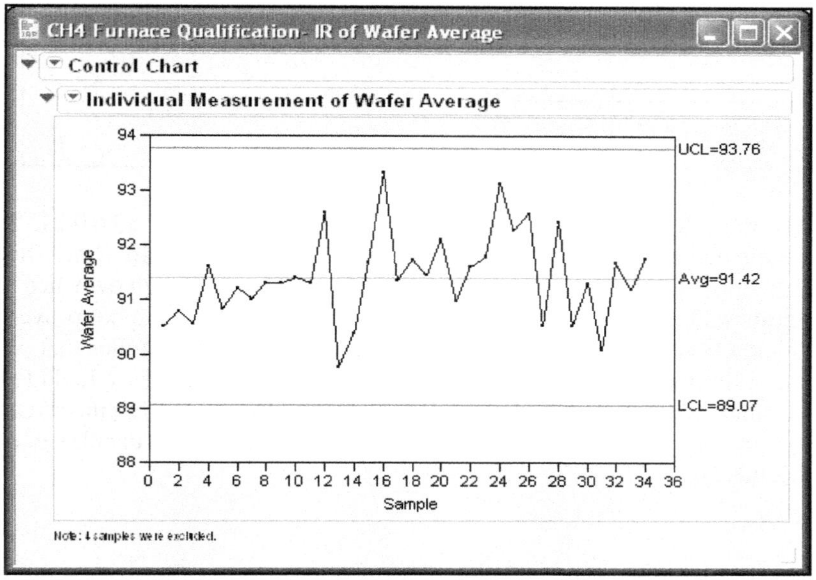

One-sample tests without run 10

Figure 4.27 shows the results of the one-sample tests without run 10. The impact of removing these four readings from run 10 resulted in a larger average, from 90.93 Å to 91.42 Å, and a smaller standard deviation, from 1.65 Å to 0.82 Å.

With run 10 excluded, the calculated *t*-statistic is now $10.14 = (91.4182 - 90) / (0.8155/\sqrt{34})$ and the *p*-value for a two-sided alternative hypothesis is now < 0.0001. The *p*-value being less than 0.05 leads us to reject the null hypothesis in favor of the alternative. In other words, we have enough evidence to say that the average oxide thickness is not equal to 90 Å, and therefore our furnace did not hit our desired target value.

Figure 4.27 Test of Significance Results without Run 10

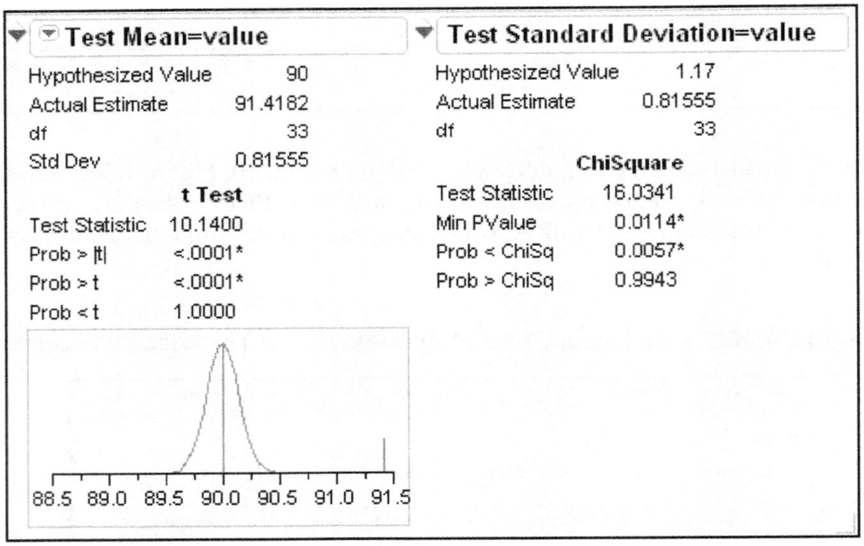

Similarly, the calculated Chi-square test statistic is now $16.034 = 33 \times (0.8155^2 / 1.17^2)$ with a one-sided hypothesis Prob > ChiSq *p*-value equal to 0.9943. Since the *p*-value is not less than our significance level of 0.05, we do not have enough evidence to reject the null hypothesis in favor of the alternative. In other words, we have not proven that the oxide thickness standard deviation is greater than 1.17 Å. This implies that we believe that we can hit our target for yield loss of 1% or less. However, if we include run 10 the results change because the *p*-value $= 0.0003 < 0.05$. Given that the run 10 readings are highly suspect and their inclusion strongly affects the results, we decided to exclude them from the analysis.

Understanding sources of variation

The test for the standard deviation indicated that, when run 10 is included, our variation is larger than expected, 1.6537 Å as compared to the target of 1.17. This was due to unusual processing conditions for run 10. Without run 10, the standard deviation is 0.8155 Å. What sources of variation contribute to this value? This is where the value of collecting additional pieces of information during our studies lies. For the furnace qualification we recorded the run number, the position within the tube (or furnace zone), and the position within the wafer. These different sources of variation, as depicted in Figure 4.28, can help us to assess where the variation is coming from, and to complete the qualification of the oxide furnace.

Figure 4.28 Sources of Variation for Oxide Furnace Qualification

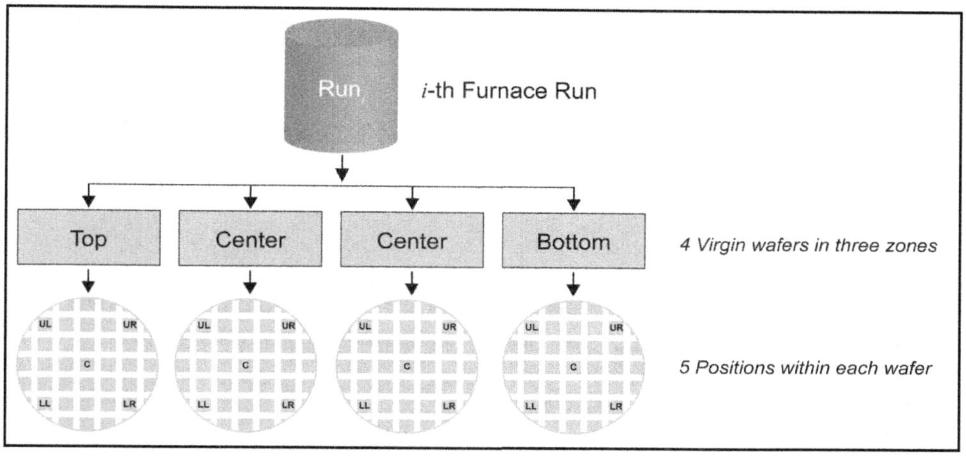

One tool that is extremely helpful for visualizing sources of variation is the **Variability Chart**. In order to use this tool we need to rearrange our data so we can have all the thickness readings in one column, plus a new column that identifies the within-wafer position of a particular thickness reading.

We need to create a new JMP table that stacks the five columns of thicknesses readings (Center, Upper Left, Lower Left, Lower Right, and Upper Right) into one column. The steps below provide the necessary details to accomplish this task and are depicted in Figures 4.29 and 4.30.

1. From the **Tables** menu in the primary toolbar, select the **Stack** option. This launches the Stack window.

2. Select Center, Upper Left, Lower Left, Upper Right, and Lower Right from the **Select Columns** menu, and then click the **Stack Columns** button. This moves the column names into the empty box on the right-hand side of the window.

3. Name the new JMP table by typing Oxide Vertical into the **Output table name** field. If this field is left blank, then JMP automatically gives the new table a generic name. Note that JMP creates a new table and does not replace the existing one that we are trying to stack.

4. Type Oxide Thickness in the **Stacked Data Column**. This column will contain all of our measured values in the new stacked JMP table.

5. Type Wafer Position in the **Source Label Column**. This column will contain the wafer position in the new stacked JMP table.

6. Click **OK** to generate the JMP table.

The stacked table should look like the one in Figure 4.31. If we want to save a permanent copy of this version of our data, then we need to select **File > Save As** in the main toolbar.

Figure 4.29 Stacking Center, Upper Left, Lower Left, Lower Right, and Upper Right

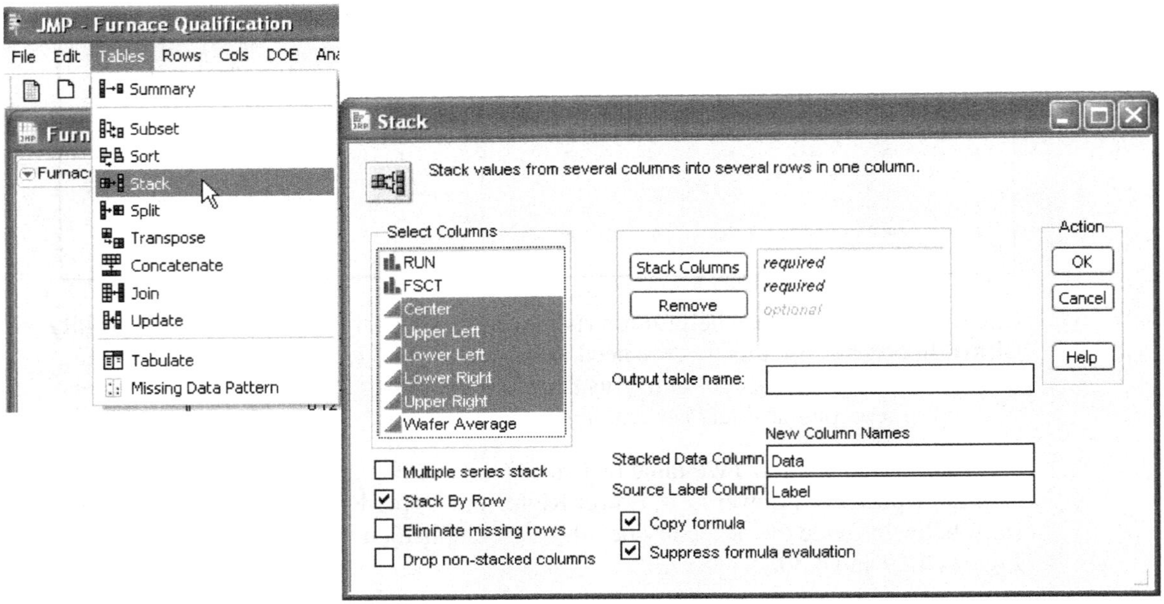

Figure 4.30 Stack Dialog Box for Oxide Thickness Table

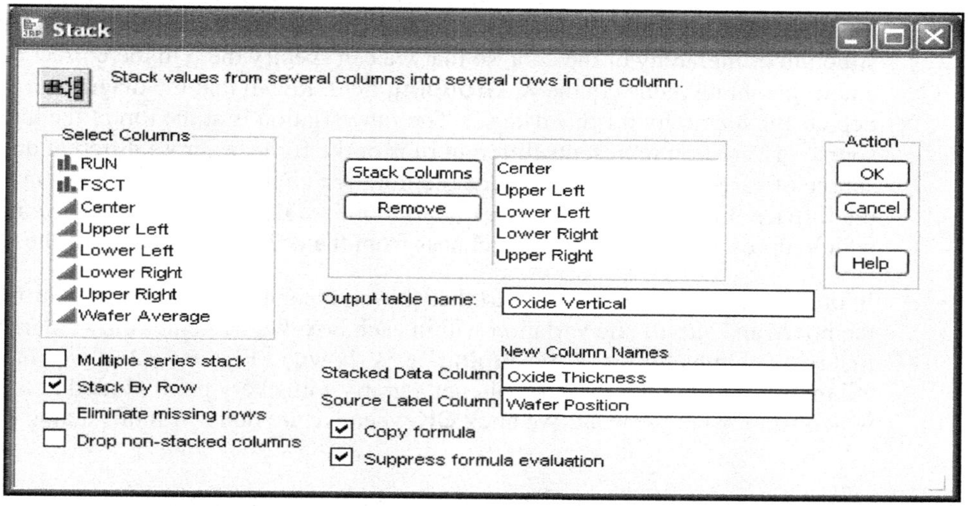

Figure 4.31 Partial JMP Table with Stacked Oxide Thickness Readings

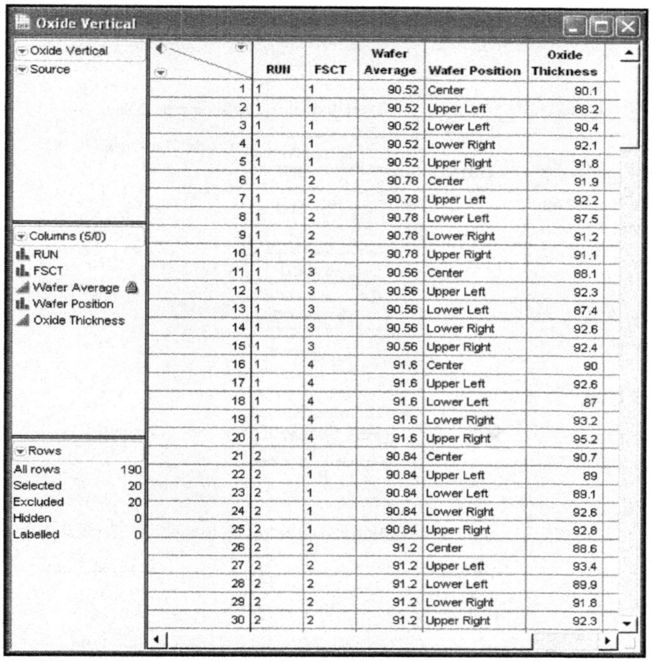

	RUN	FSCT	Wafer Average	Wafer Position	Oxide Thickness
1	1	1	90.52	Center	90.1
2	1	1	90.52	Upper Left	88.2
3	1	1	90.52	Lower Left	90.4
4	1	1	90.52	Lower Right	92.1
5	1	1	90.52	Upper Right	91.8
6	1	2	90.78	Center	91.9
7	1	2	90.78	Upper Left	92.2
8	1	2	90.78	Lower Left	87.5
9	1	2	90.78	Lower Right	91.2
10	1	2	90.78	Upper Right	91.1
11	1	3	90.56	Center	88.1
12	1	3	90.56	Upper Left	92.3
13	1	3	90.56	Lower Left	87.4
14	1	3	90.56	Lower Right	92.6
15	1	3	90.56	Upper Right	92.4
16	1	4	91.6	Center	90
17	1	4	91.6	Upper Left	92.6
18	1	4	91.6	Lower Left	87
19	1	4	91.6	Lower Right	93.2
20	1	4	91.6	Upper Right	95.2
21	2	1	90.84	Center	90.7
22	2	1	90.84	Upper Left	89
23	2	1	90.84	Lower Left	89.1
24	2	1	90.84	Lower Right	92.6
25	2	1	90.84	Upper Right	92.8
26	2	2	91.2	Center	88.6
27	2	2	91.2	Upper Left	93.4
28	2	2	91.2	Lower Left	89.9
29	2	2	91.2	Lower Right	91.8
30	2	2	91.2	Upper Right	92.3

Now we can create a plot to look at the different sources of variation using the **Variability/Gage Chart** in the **Graph** menu. The dialog box for this graph is shown in Figure 4.32. It is important to know how the sources of variation are organized, and the structure or hierarchy of the data, so that we can specify them in the correct order when entering column names in the **X, Grouping** field. Recall that the diagram in Figure 4.28 depicts the hierarchy for this data set. The run variation is at the top of the hierarchy; this source of variation represents different runs of the furnace across different days. The next source of variation is FSCT (furnace section), which represents the different zones within the furnace. Finally, the last source of variation is wafer position (or site-to-site variation), which represents differences in thickness from the center to the edges of the wafer.

In order to create a variability chart that shows run and furnace zone variation between the boxes and site-to-site variation within each box, we will enter only the names of the first two columns for our **X, Grouping**, as is shown in Figure 4.32. If we entered all three columns in this field then we would get a graph with every point plotted in a trend plot, which is not what we want. We click **OK** when we are done with this dialog box.

Figure 4.32 Variability Chart Dialog Boxes for Oxide Thickness Data

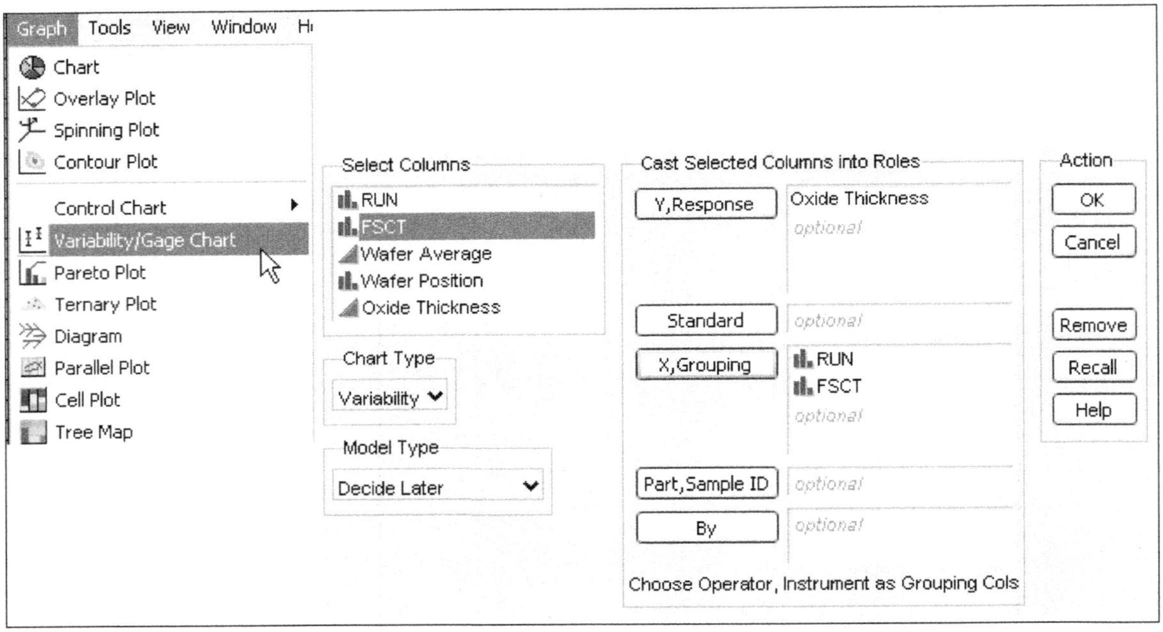

A variability chart is produced using the sources of variation described above. However, we will enhance the graph before showing the output by holding the ALT key and clicking the small red triangle next to the words Variability Gage. Select the options shown in Figure 4.33 and click **OK** to produce the plot shown in Figure 4.34, which does not include run 10.

Figure 4.33 Variability Chart Options

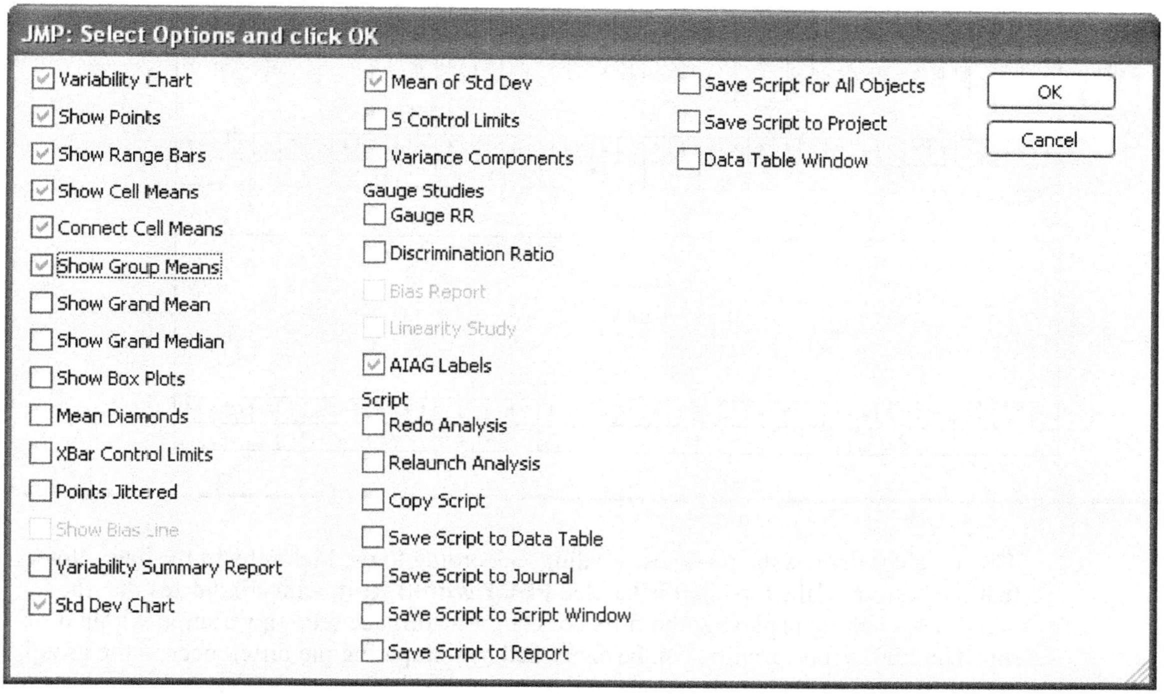

Figure 4.34 Variability for Oxide Thickness Data

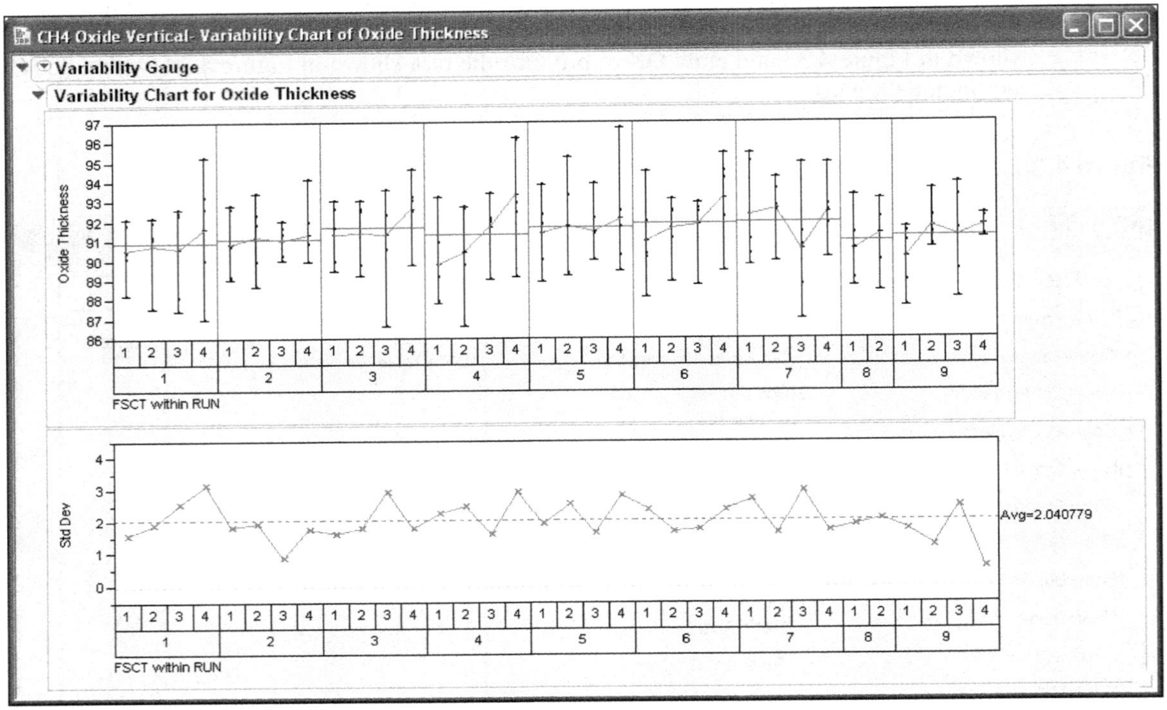

The top chart depicts the thickness readings according to the hierarchy in the data. Note that the bottom of the top chart is labeled **FSCT within Run**, which indicates that the variability chart is displaying the furnace section-to-furnace section variation within a run. The run-to-run variation can be examined by comparing the differences in the group means, shown by the (purple) horizontal lines, across the nine runs. The variation in the four furnace zones (FSCT) can be examined by looking at the difference in the (blue) lines within each of the nine runs. For example, run 4 shows the wafer oxide thickness to increase from the top of the furnace (position 1) to the bottom of the furnace (position 4). Finally, the variation among different locations within each wafer is given by the height of each of the vertical bars, for each furnace zone and each run. The bottom of the furnace (position 4) in run 1 shows a larger within-wafer variation, taller line, than the other three positions.

The bottom chart shows the within-wafer, location-to-location, standard deviation for each furnace position within a given run. The average within-wafer standard deviation is 2.0408 Å. We note, as before, that for run 1 the bottom of the furnace (position 4) has a larger within-wafer standard deviation than the other 3 positions. Actually, for run 1, it looks like the variation is increasing from top to bottom.

Which are our largest sources of variation in oxide thickness measurements? The run-to-run variation appears to be the smallest source of variation in this data set. It is also obvious that all of the runs have average thicknesses that are above our target value of 90 Å because all of the (purple) lines are above 90 Å. The variation among the furnace zones seems to be the next largest source of variation in our data set. In fact, runs 1, 3, 4, and 6 have the thickest oxide layers at the bottom of the furnace (position 4). Finally, the largest source of variation seems to be the within-wafer variation, with a standard deviation of about 2 Å.

Step 6: Summarize the results with key graphs and summary statistics

We should select key graphs and summary tables from the analyses performed in step 5 to help us answer our uncertainties, give us insight into the performance of the new oxide furnace, enable us to make claims about the future performance of the furnace, and quantify the uncertainty in our results.

A great way to summarize the results of one-sample significance tests is by means of confidence intervals. Recall from Chapter 2 that a statistical interval helps us to quantify the uncertainty surrounding the mean and standard deviation. Confidence intervals can be used to demonstrate whether the average performance (or performance variation) is meeting the standard. For displaying average performance against specification limits we use a histogram overlay with specification limits, along with a capability analysis.

Comparing to a standard using confidence intervals

This is accomplished in JMP, from within the **Distribution** platform, by selecting **Confidence Interval** in the drop-down menu that is displayed by clicking the red triangle next to **Wafer Average**. This brings up the window shown in Figure 4.35a, where one can select the type of interval and the confidence (default 95%).

Figure 4.35a Selection of Confidence Intervals

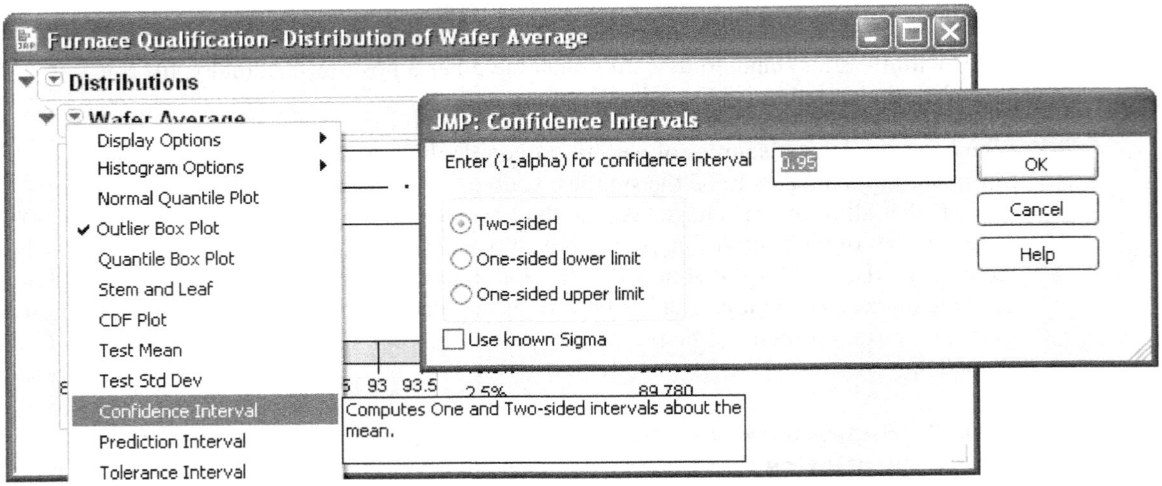

Once you click **OK**, JMP generates two-sided 95% confidence intervals for the mean and standard deviation as is shown in Figure 4.35b. To perform the significance test we check to see whether the standard value is within the given confidence intervals. Figure 4.35b is then a great way to show if we have enough evidence for our performance claims.

For the average performance, the standard value of 90 Å is not within the 95% confidence interval, [91.13368 Å; 91.70279 Å]; and it is lower than the lower bound of 91.13 Å, indicating that the average performance is actually greater than the standard.

Figure 4.35b Confidence Interval Output Excluding Run 10

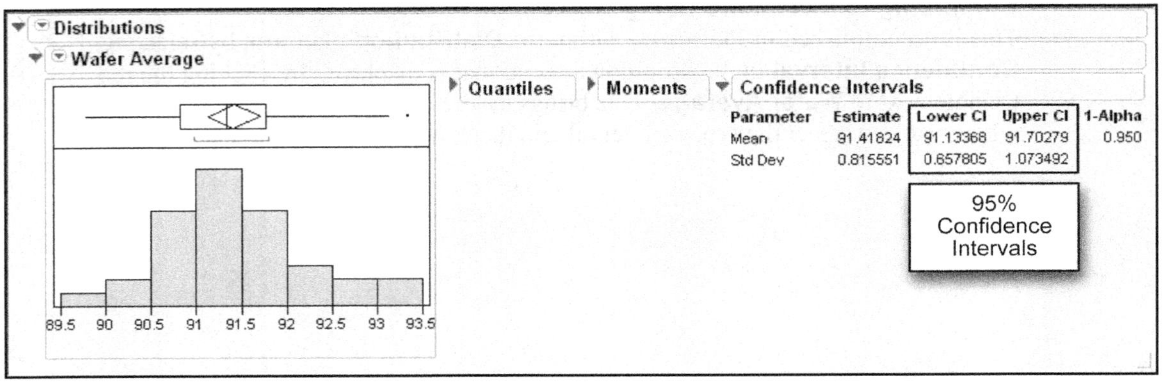

For the performance variation, the 95% confidence interval, [0.657805 Å; 1.073492 Å], does not contain our standard of 1.17 Å, which is larger than the upper bound of 1.073492 Å, indicating that the performance variation is actually less than the standard.

Comparing the thickness averages to the specification limits

The specification limits for the virgin wafer averages are LSL = 87 Å and USL = 93 Å. We can add our specification limits to the histogram displaying the average wafer thickness measurements in order to get an idea of our yield loss with the new Oxide Furnace. We can do this by selecting the **Capability Analysis** in the menu that is displayed by clicking the red triangle next to the response name, **Wafer Average**. This brings up a dialog box that contains fields for our upper and lower specification limits and our target value. Enter in the appropriate values and make sure that **Long Term Sigma** check box is selected, as is shown in Figure 4.36. This implies that process capability indices will be calculated using the standard deviation of all of the wafer averages, and not from a short-term estimate obtained from a control chart. Finally, click **OK** when this is done.

Figure 4.36 Process Capability Dialog Box

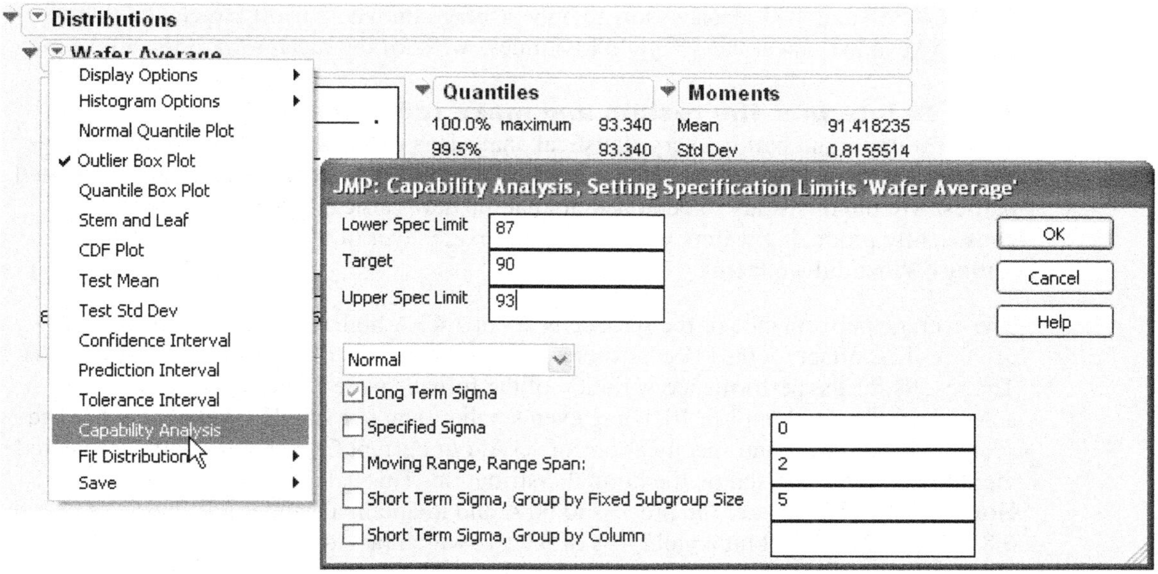

The capability analysis is shown in the Figure 4.37. It is clear from the histogram that our process mean is shifted to the right by about 1.5 Å, which moves the entire distribution toward the upper specification limit. The process capability index C_{pk} = 0.647 gives a projected yield loss of 2.62% based on the normal distribution. This value is higher than the current yield loss of the furnace that is being replaced, which is about 1%. The process capability index C_p = 1.23. This reflects the potential C_{pk} if we center the process and, since we are off target, $C_p \neq C_{pk}$. The actual yield loss is shown in the column **%Actual**, which shows an overall yield loss, **Total Outside**, of 5.88%.

Figure 4.37 Process Capability Output for Oxide Thickness Excluding Run 10

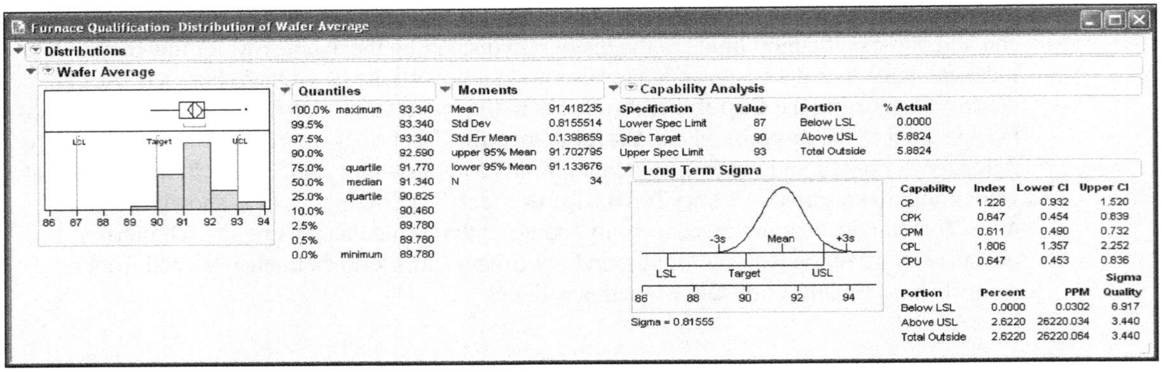

Figures 4.35b and 4.37 clearly show that the average thickness is off target, and that we will not be able to meet the 1% yield loss unless we shift the mean toward the 90 Å target.

Step 7: Interpret the results and make recommendations

The most important part in every statistical analysis is to translate our statistical findings according to the context of the situation at hand. What do we report back to the interested parties? We ran this study to demonstrate that the new vertical furnace is capable of consistently producing wafers with a silicon dioxide layer of 90 Å, using the SOP settings. What did we learn?

The average performance of the furnace is about 1.42 Å higher than the specified target of 90 Å. This offset seems to occur more often at the bottom of the furnace (Figure 4.34). The results for the performance variation of the furnace were a nice surprise, with the actual variation, without run 10, being even smaller than our standard value of 1.17. The problem is that given the specifications of ±3 Å, our current C_{pk} is 0.647, with a projected yield loss of 2.62% based on the normal distribution. This is larger than the goal of 1%. However, if we can center the process to 90 Å and maintain a standard deviation of 0.8155 Å, then we will hit a yield loss of 0.02%, which far exceeds our goal of 1%.

Based on the results of this study, we recommend the installation of the new furnace after centering the process to the target value of 90 Å, which can be accomplished by adjusting the oxidation time. Further studies can be conducted to reduce the variation within the wafer and improve upon the thickness profile from zone to zone by adjusting the zone temperatures and gas flows.

4.5 Testing Equivalence to a Standard

A test of significance helps to demonstrate a significant discrepancy between the hypothesis that the average performance is equal to the standard (that we assume to be true) and our data. If the *p*-value is smaller than the pre-specified significance level, then we can say that the average performance is not meeting the standard. However, if the *p*-value is larger than the pre-specified significance level, all we can say is that we "believe" the average performance to be close to the standard, but we do not know this for a fact. A test of equivalence, on the other hand, demonstrates that the average performance is equivalent, up to a range of acceptable values, to the standard. Below we show how steps 2 and 5 from the 7-step method need to be modified to conduct an equivalence test.

Step 2: Specify the hypotheses of interest

The null hypothesis states that the absolute value of difference between the average performance and the standard value is greater than a pre-specified bound, δ. The alternative states that absolute value of difference between the average performance and the standard value is less than a pre-specified bound, δ.

H_0: $|\text{Average Performance} - 90\text{ Å}| > \delta$ (Assume)

H_1: $|\text{Average Performance} - 90\text{ Å}| \leq \delta$ (Prove)

Choosing the equivalence bound

The pre-specified bound, δ, is chosen to define a range of values for which the difference between the average performance and the standard is irrelevant from the practical point of view. Let's say that for our example as long as the average thickness is between ±1 Å of the target of 90 Å, we consider the average performance to be on target, or more accurately, to be *equivalent* to the target. The null and alternative hypothesis then becomes as follows:

H_0: $|\text{Average Performance} - 90\text{ Å}| > 1\text{ Å}$ (Assume)

H_1: $|\text{Average Performance} - 90\text{ Å}| \leq 1\text{ Å}$ (Prove)

Step 5: Perform the analysis to verify your hypotheses

Performing a one-sample test of equivalence can be done using the script, **Equivalence Test with Standard.JSL**, that we have written for this purpose. Once you open the script you can run it by holding down CTRL+R, or by selecting the running person icon in the top menu. Figure 4.38 shows the dialog box where one indicates the variable containing the data, Wafer Average, the target value, 90 Å, and the equivalence window, ±1 Å.

Figure 4.38 One-Sample Equivalence Test

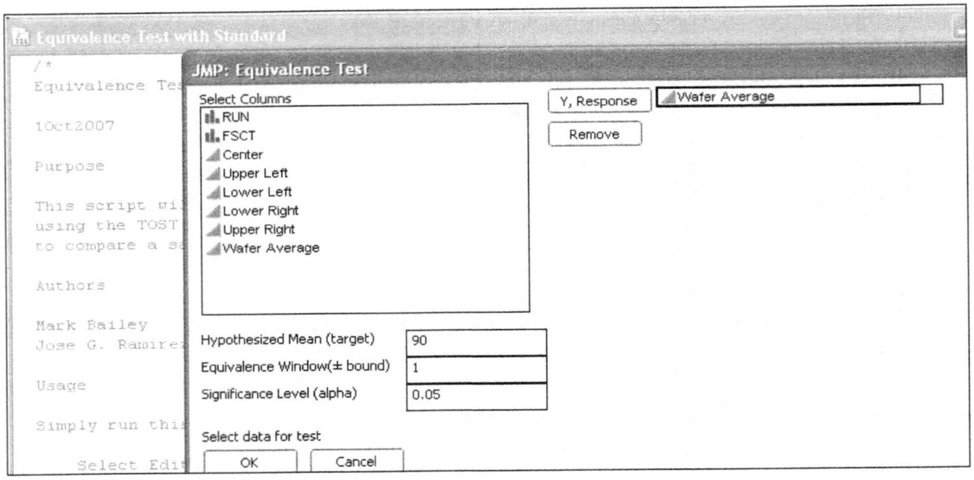

The output of the test is shown in Figure 4.39. The results show the output of two one-sided *t*-tests and the corresponding *p*-values under the **Equivalence Test for mu=90±1** heading. For an equivalence test to be significant both *p*-values need to be less than the significance level, 0.05, in this case. Since one of the *p*-values, 0.997, is greater than 0.05, we fail to reject the null hypothesis |Average Performance − 90 Å | > 1 Å. In other words, we cannot say that our average performance is equivalent, in the ±1 Å window, to 90 Å. In fact, the average performance is 91.42 Å, which is greater than the upper equivalence bound of 91 Å.

Figure 4.39 One-Sample Equivalence Test Results

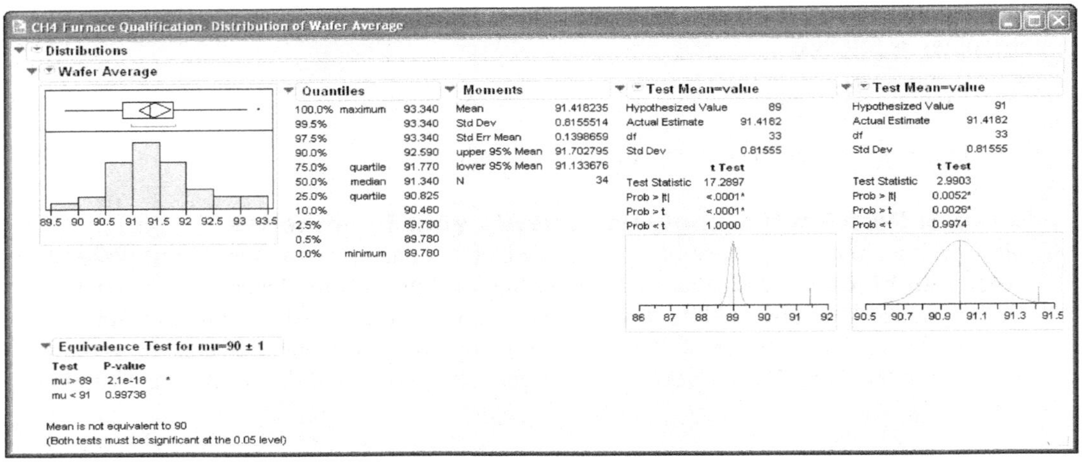

4.6 Summary

In this chapter we learned how to use a one-sample test of significance to provide insight into the average performance, and the performance variation, of a product, process, or material. This type of analysis is useful when we have only one population that we want to investigate and we are not deliberately trying to induce a signal in the system. We showed how to set up a one-sample study and provided step-by-step instructions for how to use JMP to help us conduct sound statistical analysis and make data-driven decisions. Some of the key concepts and practices that we should remember from this chapter are shown below.

- Include multiple sources of variation in these types of studies.
- Don't forget to clearly define the population of interest and make sure that the sample is representative of the population.
- Include tests for both the mean and the variation.
- Use the scientific method to make sure that the calculated statistics are helping us support or refute our hypothesis.
- Use probability models in place of sampling statistics to get better estimates of the performance of the population (that is, estimating yield loss).
- Always translate the statistical results back in the context of the original question.

4.7 References

Czitrom, V., and P.D. Spagon. 1997. *Statistical Case Studies for Industrial Process Improvement (ASA-SIAM Series on Statistics and Applied Probability).* Philadelphia, PA: Society for Industrial Mathematics.

Natrella, M.G. 1963. *Experimental Statistics, NBS Handbook 91*. Washington, D.C.: U.S. Department of Commerce, National Bureau of Standards. Reprinted 1966.

Wheeler, D.J. 2005. *The Six Sigma Practitioner's Guide to Data Analysis.* Knoxville, TN: SPC Press, Inc.

Chapter 5

Comparing the Measured Performance of Two Materials, Processes, or Products

Chapter Goals

- Become familiar with the terminology and concepts needed to compare average performance and performance variation of two materials, products, or processes.
- Use JMP to compare average performance and performance variation of two populations.
- Understand when to analyze your data with the **Fit Y by X** platform versus the **Matched Pairs** platform.
- Translate JMP statistical output in the context of a problem statement.
- Present the findings with the appropriate graphs and summary statistics.

5.1 Problem Description

What do we know about it?

You are a senior chemist working for an analytical laboratory that services several industries, including chemical, electronics, and plastics. One of your key customers in the chemical industry is using silver as a catalyst in an oxidation reaction, which requires that the atomic weight of silver be measured with very low uncertainty. They have contracted your lab to measure the atomic weight of silver (or more precisely, the average atomic mass of silver) on a regular basis.

Note that most elements occur in nature as mixtures of isotopes (forms of the same atom). For example, silver occurs in ^{107}Ag (atom percent = 51.839170) and ^{109}Ag (atom percent = 48.16083) in a ratio $^{107}Ag/^{109}Ag$ of around 1.07. The average atomic mass or atomic weight of silver is calculated using the atomic masses of the isotopes and their amount in a sample (see, for example, Brown, LeMay, and Bursten, 1994, p. 76). In other words, the atomic weight is a weighted average of the isotope masses and their relative abundance.

What is the question or uncertainty to be answered?

Your lab has one high precision solid sample mass spectrometer that is designed to determine the isotopic ratio composition of a variety of elements, including silver, with high precision. Due to the increase in demand for this type of analysis, you recently purchased a second mass spectrometer of the same make, but a more recent model of the one that you currently have in the lab. The current spectrometer is held to a stringent calibration protocol and is frequently checked for the quality of its output.

Because your lab will be making the silver atomic mass determination for this customer on a regular basis, you would like to determine whether the new isotope ratio mass spectrometer is performing to the same level of performance as the current one. This way you can make sure that your customer has confidence in the results generated from either of the mass spectrometers.

How do we get started?

The primary objective of this study is to demonstrate that two mass spectrometers are both precise and accurate to a traceable standard, and that their outputs do not differ "much" from each other. Because of the equal atom composition in the mixes, the linearity is not deemed to be a problem. To get started, we will need to do the following:

1. Describe our requirements in terms of a statistical framework that enables us to demonstrate, with a certain level of confidence, that the output of the two mass spectrometers is not significantly different from each other.
2. Translate the statistical findings in a way that clearly conveys the message without getting lost in the statistical jargon.

What is an appropriate statistical technique to use?

Two mass spectrometers are used to measure the isotopic ratio of silver from a reference sample, which, in turn, are used to determine the atomic weight of silver. We will use *two-sample significance tests* to compare the average performance and performance variation of the two instruments. In Section 5.3 of this chapter we review the key concepts that are relevant to two-sample significance tests, and in Section 5.4 we introduce a step-by-step approach for performing two-sample significance tests using JMP.

5.2 Key Questions, Concepts, and Tools

Several statistical concepts and tools, introduced in Chapter 2, are relevant when comparing the measured performance of two materials, processes, or products. Table 5.1 outlines these concepts and tools, and the "Problem Applicability" column helps to relate these to the comparison of the two mass spectrometers.

Table 5.1 Key Questions, Concepts, and Tools

Key Questions	Key Concepts	Problem Applicability	Key Tools
Is our sample meaningful?	Inference: from a random and representative sample to a population	Our customer is demanding accurate and precise data for the atomic weight of silver. This implies that the instruments must be accurate to a representative traceable standard. The need for precise or repeatable data suggests that the results should be consistent across different instruments, different times, and different technicians (a traditional gauge R&R or EMP (Evaluating the Measurement Process) study). How do we ensure that each instrument receives "identical" samples, so that any detected difference is attributable to the two mass spectrometers and not differences in the samples measured on each?	Sampling schemes and Sample size calculator
Are we using the right techniques?	Probability model using the Normal distribution	There is no reason to suspect that the atomic weight of silver is not symmetrical about an average value. There aren't any boundary conditions in the measurement technique that would truncate the response values and skew the distribution. Therefore, the normal distribution should provide an adequate description of the data.	Normal quantile plots to check distributional assumptions
What risk can we live with?	Decision theory	This measurement technique requires a high degree of precision and we want to make sure that our customer will have confidence in our results. Therefore, we would like to take a conservative approach and design this study with low risk so that we are not going to make the wrong decisions.	Type I and Type II errors. Statistical power and confidence

(*continued*)

Table 5.1 (*continued*)

Key Questions	Key Concepts	Problem Applicability	Key Tools
Are we answering the right question?	Two-sample significance tests	We need to demonstrate that the two mass spectrometers provide "nearly identical" results. First of all, we need to demonstrate that the average output from both is similar and close to a standard value. This can be accomplished by using a two-sample *t*-test for the means or a paired *t*-test. We also need to demonstrate that the instruments have similar repeatability and reproducibility A test comparing the two variances would assist with this item.	Paired or *t*-test to compare the means and *F*-test to compare the variances
How do we best communicate the results?	Visualizing the results	We need to demonstrate accurate and precise performance of both instruments. This can be visualized very effectively with a comparative histogram that is overlaid with the reference value and descriptive statistics. In addition, we want to examine the data for trends with a trend plot or process behavior chart.	Comparative histograms, box plots, confidence intervals

5.3 Overview of Two-Sample Significance Test

5.3.1 Description and Some Applications

A two-sample significance test derives its name from the comparison of the average performance of two entities or conditions (such as two pieces of equipment, two raw material suppliers, or two equipment settings) in order to determine whether they are similar or not. The two items that we want to compare can be combined into one factor using a name that represents the intent of the comparison. The factor can be described as "supplier" for the comparison of two suppliers and as "equipment" for the comparison of two pieces of equipment. In other words, the factor is stated in a way that represents the primary comparison, and the levels in a way that represents the two items that are being directly compared. As we described in Chapter 2, this type of study involves one factor with two levels.

> *In Six Sigma parlance this is a study for a response, Y, and a factor, X, with two levels, X_1 and X_2.*

For these types of situations, we can either intentionally alter the levels of the causal factor, hoping to see a signal in the performance metric, or we can passively collect data under the two conditions, or levels of our factor. This results in two samples of measurements taken from the two different populations of interest that are used to compare their average performance and performance variation.

Some examples from science and engineering where a two-sample significance test would be appropriate include the following:

- Determining whether the application of a new coating to a painted surface of a metal object decreases flaking, chipping, and loss of luster under harsh weather conditions as compared to the standard coating.
- Quantifying the impact of two different processing temperatures for the deposition of metal on a wafer, which enhances its thickness uniformity.
- Selecting a second source supplier for a solvent whose product performs as well as the current supplier based on product specifications.
- Ensuring that the performance of a product, which is manufactured in two different countries, is indistinguishable to the customer.
- Deciding if a humidity-controlled warehouse is needed to prevent performance degradation of stored inventory for extended periods of time as compared to storage in a regular warehouse.
- Understanding if two operators get the same results when using calipers to measure the thickness of a gasket.

What are the criteria by which we declare two items different? Similar to the one-sample significance test, the two main criteria revolve around understanding and comparing the average performance, as well as the performance variation. These two situations will give rise to a two-sample significance test for the means and an additional test for the variances.

A special situation arises, for example, when we take the measurements on the same experimental units before and after an experimental treatment or condition. There are a couple of applications described previously that fall into this situation. For example, the comparison of the performance of two operators using calipers to measure gasket thicknesses might require that the operators measure the *same* set of gaskets using the *same* calipers. The *pairing* occurs because, for a given caliper, both operators measure the *same* gasket, and this is repeated for all the gaskets in the study. This works to our advantage because we have only *one population* of gaskets to deal with. In other words, both operators will measure each experimental unit, namely, a gasket, using the same caliper. We refer to this as a "paired" study and we must take special care in designing and executing the study and analyzing the results using a paired *t*-test.

In Sections 5.3.2 and 5.3.4 we discuss how two-sample significance tests can be used to compare measured performance of two entities with respect to average performance and performance variation. In Section 5.3.3, we describe when a paired-*t*-test is preferable and how it can be analyzed. Finally, in Section 5.4, we show how to carry out this type of comparison for the two mass spectrometers using JMP.

5.3.2 Comparing Average Performance of Two Materials, Processes, or Products

In the previous chapter, we showed how we could draw conclusions about the mean (or the central tendency) of the measured performance for one material, process, or product with respect to a standard. In this situation, we compared the mean from our sample with our stated standard value, *k*, and used a simple *t*-test to determine whether the sample mean was significantly far enough away from *k* to be declared different. In this chapter we extend these concepts to situations when we want to compare the means of two populations to each other, and determine whether they are different enough to be declared significant with respect to the noise in the populations. In statistics, tests for comparing two population means (or the average performance) are referred to as two-sample significance tests for the mean, and the most common one is the two-sample *t*-test.

When we carry out a two-sample significance test for the means we are interested in drawing a conclusion about the means (or the central tendencies) of the measured performance of two materials, processes, or products. For example, we might want to compare the average performance of a raw material attribute for a part supplied by two different suppliers and determine whether they are comparable. Figure 5.1 illustrates the statistical concepts behind a two-sample *t*-test for the means. The population representing the raw material attribute from Supplier A can be described by a normal distribution with a certain mean (central tendency) that we refer to as μ_A. Similarly, the population representing the same raw material attribute from Supplier B can be described by a normal distribution with a certain mean (central tendency) that we refer to as μ_B. We take random and representative parts from Supplier A and Supplier B, measure the performance attribute on all of the parts, and then carry out a two-sample *t*-test to determine whether the two means, μ_A and μ_B, are significantly different from each other.

Figure 5.1 Comparing Two Populations, A and B

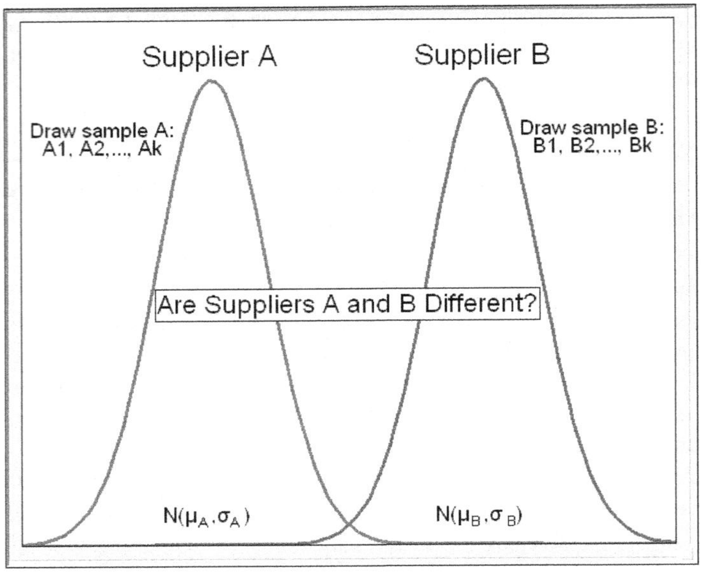

What do we mean when we say that the two supplier means are significantly different from each other? Similar to the one-sample *t*-test, there are three ways that we can compare the two means. We can check if the mean from Supplier A is greater than, less than, or simply not equal to the mean of Supplier B. The statistical jargon for these three types of comparisons (equal, greater than, or less than) involves using a null (H_0) and an alternative (H_1) statistical hypothesis that contain statements about the average performance of the two population means, μ_A and μ_B. The three ways that we can write the null (H_0) and alternative (H_1) hypothesis are shown below.

1. *The Average performance of population 1 is different from population 2*

   H_0: Average Performance (μ_1) = Average Performance (μ_2) (Assume)
   H_1: Average Performance (μ_1) $\neq$ Average Performance (μ_2) (Prove)

This is a two-sided alternative since we do not care about the average performance of the first population being less or greater than the average performance of the second population, just that it is different from it. The null hypothesis (H_0) assumes that the average performance of both population means, μ_1 and μ_2, is the same. On the other hand, the alternative hypothesis (H_1) states that the average performances of the two populations are not equal to each other. If our significance test favors the alternative then we have *proven*, in the sense of having enough evidence, that the average performance of population 1 is different from the average performance of population 2, without making any statements regarding the direction of departure in the two means. This is the most popular choice for many applications.

> *A two-sided alternative hypothesis does not make any claims about the direction of departure of the mean from our standard.*

Sometimes, we want to prove with statistical rigor that the two population means are similar and not just assume it to be true. For example, many performance claims are stated in way to suggest that their performance is just as good as or no worse than a competitor's performance. This is equivalent to flipping the two hypotheses around and putting the equality statement in the alternative statement, H_1: Average Performance (μ_1) = Average Performance (μ_2). This is referred to as an *Equivalence Test* and is discussed in Section 5.5.

> *We assume that the average performances of the two populations are equal since we can never really prove it.*

2. *The Average performance of population 1 is greater than population 2*

H_0: Average Performance (μ_1) $\leq$ Average Performance (μ_2) (Assume)
H_1: Average Performance (μ_1) $>$ Average Performance (μ_2) (Prove)

In this one-sided alternative hypothesis, H_1, we are interested in knowing whether the average performance of population 1 is greater than the average performance of population 2. Here, a rejection of the null hypothesis, H_0, will show, with statistical rigor, that the average performance of population 1 is larger than the average performance of population 2.

> *Proving the alternative hypothesis means that we have enough evidence (proof) to demonstrate, or establish, the validity of claims in the alternative hypothesis.*

The decision to write the hypothesis in this manner is application specific. For example, if we want to prove that a second entity (represented by μ_1) is superior in performance to the current (represented by μ_2) before we can use it, then we would want to set up an appropriate one-sided alternative hypothesis. For example, we might want to change suppliers for a liquid adhesive that we use in our products but only want to consider suppliers that have adhesives with a higher adhesion bond. This would lend itself to the following one-sided alternative hypothesis: H_1: Avg. Bond Strength New Supplier > Avg. Bond Strength Existing Supplier.

3. *The Average performance of population 1 is lower than population 2*

   H_0: Average Performance (μ_1) $\geq$ Average Performance (μ_2) (Assume)
   H_1: Average Performance (μ_1) $<$ Average Performance (μ_2) (Prove)

Finally, for this one-sided alternative we are interested in knowing whether the average performance of population 1 is less than the average performance of population 2. If we do not have enough evidence to reject H_0 then we *assume* that the mean of population 1 is larger or equal to the mean of population 2. However, if we reject H_0 in favor of H_1 then we have shown with statistical rigor that the average performance of population 1 is smaller than the average performance of population 2.

Once again, the specific application will help us decide whether we want to write the hypothesis in this manner. If we want to prove that a second entity (represented by μ_1) is inferior to the current (represented by μ_2), we can use this alternative hypothesis. For example, a recent process problem at a chemical plant is thought to be the result of switching to a new supplier. We might want to show that a chemical impurity (for example, nickel) in the polymer provided by the new supplier is greater than what it was in the polymer provided by the previous supplier. This can be written as: H_1: Avg. Chemical Impurity Current Supplier < Avg. Chemical Impurity New Supplier.

The Two-sample *t*-Statistic

In the previous chapters we introduced the concept of a test statistic representing the signal-to-noise ratio found in the data. The larger the signal relative to the noise, the more likely the difference will be declared statistically significant. In the two-sample *t*-test, for the average performance of two populations, we need to calculate a signal-to-noise ratio and determine whether it is large. The general form of a two-sample *t*-test can be stated as follows:

$$t = \frac{\text{Signal}}{\text{Noise}} = \frac{(\text{Estimated Avg. Population 1} - \text{Estimated Avg. Population 2})}{\text{Combined Performance Noise}}$$

The numerator, or signal, in this test statistic is represented by the difference in the two population means. If the two means are far apart then this difference (signal) will be large. On the other hand, if the two means are similar, then their difference should be close to 0. The signal then gets divided by the noise, which is the pooled noise from both populations, unless the populations' variations are different in which case a special version of the *t*-test is used.

When we have a priori knowledge of the two population variances, then the signal-to-noise ratio is a *z*-statistic. The two population means must be estimated using the sample means and the two variances are "known" based on prior knowledge. In the formula which follows, n_1 and n_2 are the number of samples that were taken from each population.

$$Z_0 = \frac{Signal}{Noise} = \frac{(\bar{X}_1 - \bar{X}_2)}{\sqrt{\frac{\sigma_1^2}{n_1} + \frac{\sigma_2^2}{n_2}}}$$

This signal-to-noise ratio is a Z-score that follows a standard normal distribution under the assumption that the null hypothesis is true. Similar to the one-sample *t*-test described in the previous chapter, the magnitude of the *z*-statistic for a two-sample test on the average performances gives us a sense of how big the signal is with respect to the noise (Figure 5.2) and if we have enough evidence to reject the null hypothesis. For example, if the calculated *z*-statistic is ≤ 1, then it is highly unlikely that we will reject the null hypothesis and conclude that the average performances are different from each other. However, if the calculated *z*-statistic is ≥ 2.5, then there is a good chance that we will reject the null hypothesis and conclude that the average performances are different from each other, since the signal is two-and-a-half times the noise.

Figure 5.2 Areas under the Standard Normal Distribution

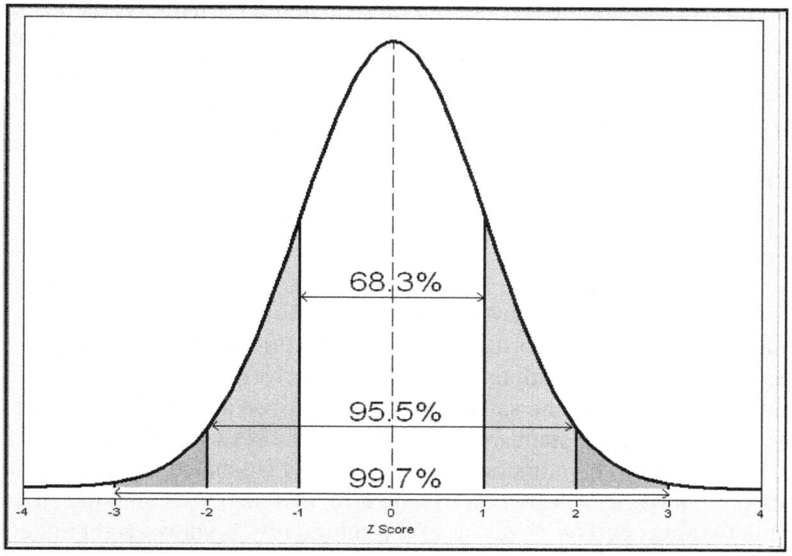

In addition to its magnitude, the sign of the *z*-statistic gives us insight regarding where our average performance for population 1 is relative to population 2. This value is positive if the average performance of population 1 is greater than population 2 and negative if the average of population 1 is less than population 2. For the null hypothesis of equality of the two population means, if our calculated *z*-statistic is large and negative then we have proven that the population means are different and that the mean from population 1 is smaller than population 2 but, if it is large and positive, then we have proven that the two population means are different with the mean from population 1 being greater than the mean from population 2. If our test statistic is small, regardless of its sign, then we assume, since we do not have enough evidence to refute it, that the two population means are not significantly different from each other.

Since we rarely know the true value of the performance variations, we have little use for the *z*-statistic, Z_0, and need to use one that accounts for the estimation of the population standard deviations from the data. The resulting signal-to-noise ratio for this scenario, assuming that the two population variances are not significantly different, is provided below.

$$t = \frac{Signal}{Noise} = \frac{(\bar{X}_1 - \bar{X}_2)}{\sqrt{s_p^2\left(\frac{1}{n_1} + \frac{1}{n_2}\right)}} = \frac{(\bar{X}_1 - \bar{X}_2)}{\sqrt{\left(\frac{(n_1 - 1)s_1^2 + (n_2 - 1)s_2^2}{n_1 + n_2 - 2}\right)\left(\frac{1}{n_1} + \frac{1}{n_2}\right)}}$$

Here n_1 and n_2 are the sample sizes of the random and representative samples taken from each population. Although preferable, it is not a requirement for these two samples to be equal in size. In the formula above the pooled variance s_p^2 is calculated as a "weighted average" of the two populations' variances s_1^2 and s_2^2, as shown below.

$$s_p^2 = \left(\frac{(n_1 - 1)s_1^2 + (n_2 - 1)s_2^2}{n_1 + n_2 - 2}\right)$$

As an illustration Figure 5.3a shows the partial JMP output for a two-sample *t*-test that compares the average performance of the output from, for example, two pieces of equipment using thickness readings. The average performances for the New and Old equipment, calculated from the sampled data, are 54.488 and 50.537, respectively. Similarly, the performance standard deviations for the New and Old equipment are 1.87 and 1.77, respectively. The numerator, or signal, for *t* is labeled **Difference** in the JMP output and is -3.9508. It is important to note how JMP is performing the subtraction between the two means. The direction of the subtraction is shown right under the **t Test** banner and occurs in reverse alphanumeric order as **Old–New**. The negative sign for *t* implies that the new equipment produces thicker output than the old equipment.

Figure 5.3a Partial JMP Output for a Two-sample t-Test

Means and Std Deviations

Level	Number	Mean	Std Dev	Std Err Mean	Lower 95%	Upper 95%
New	20	54.4882	1.87033	0.41822	53.613	55.364
Old	20	50.5374	1.77002	0.39579	49.709	51.366

t Test

Old-New ←

Assuming equal variances

Difference	-3.9508	t Ratio	-6.86131
Std Err Dif	0.5758	DF	38
Upper CL Dif	-2.7851	Prob > |t|	<.0001*
Lower CL Dif	-5.1165	Prob > t	1.0000
Confidence	0.95	Prob < t	<.0001*

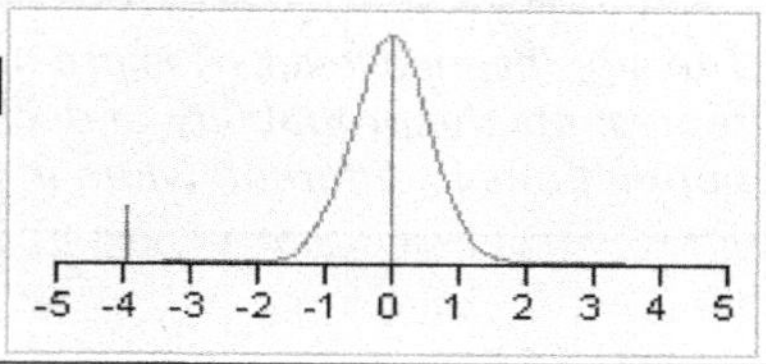

The denominator, or noise, for t is labeled **Std Err Dif** in the JMP output and is 0.5758. When the two-sample sizes are equal to n, the formula for the denominator reduces to the square root of $(s^2_1 + s^2_2)/n = \sqrt{(1.87033^2 + 1.77002^2)/20} = 0.5758$. Finally, we can complete the calculation for the t-statistic by taking the ratio of the signal to the noise or, -3.9508 / 0.5758 = -6.86131; i.e., the signal is about 6.86 times larger than the noise. This value is labeled **t Ratio** in the JMP output.

For this example, if we use the standard normal distribution as a guideline, then the signal-to-noise ratio of -6.86 is very large and most likely statistically significant. However, just as we did for a one-sample t-test for the average performance, we should obtain the p-value for this test statistic and use it to determine whether we should reject our null hypothesis. When we have to estimate the standard deviations of the two populations, which is usually the case, we must use the Student's t-distribution, with the appropriate degrees of freedom, to find the p-value associated with the t-statistic.

For a two-sample test for two means, the degrees of freedom are given by a formula $(n_1 + n_2 - 2)$, which reduces to $2(n - 1)$ when the sample sizes are equal. Recall from the last chapter that the p-value is dependent on the type of comparison that we want to test. For example, if our alternative hypothesis is that the average performance of the old is greater than the new (H_1: $\mu_{old} > \mu_{new}$), then the p-value is the area under the Student's t-distribution to the right of the calculated t-statistic, since the result of **Old–New** is positive. Similarly, if our alternative hypothesis is that the average performance of the old is less than the new (H_1: $\mu_{old} < \mu_{new}$), then the p-value is the area to the left of the calculated t-statistic, since the result of **Old–New** is negative. If the alternative hypothesis is that the two average performances are different, without any direction implied, then the p-value is the area to the right of the calculated positive t-statistic plus the area to the left of the calculated negative t-statistic.

In Chapter 2 we saw that when performing significance tests we can never be certain. In other words, when testing hypothesis we allow ourselves to a "make a mistake" a certain percent of the time. That is, a certain percent of the time we are going to say that the two average performances are not equal when, in fact, they are. This error rate (significance level), α, which is pre-specified *before* we perform the test, serves as the cutoff point for our significance test decisions. If the *p*-value $< \alpha$, then we have enough evidence to prove our alternative hypothesis, but if the *p*-value $\geq \alpha$, we do not have enough evidence to say that our average performances are significantly different from each other. The value of α is typically selected to be 0.05.

> *The significance level, α, is pre-specified **before** we perform the test, and represents the probability of declaring that the average performance of the two populations is different, when in fact it is the same (Type 1 error).*

Figure 5.3b Areas under the t-Distribution

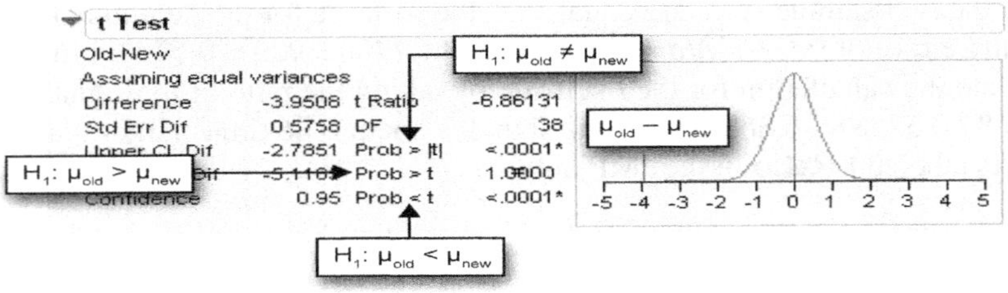

Figure 5.3b shows the JMP output with the *p*-values corresponding to the three different types of hypothesis. The symbols following the "Prob" (> |t|, > t, < t) correspond to the three ways in which we can specify the alternative hypothesis. The correct alternative depends upon our problem statement and scientific hypothesis. JMP puts an * next to those *p*-values that are < 0.05, the default significance level cutoff value. If the *p*-value is less than 0.0001, then JMP labels it as "<0.0001" instead of printing the exact value.

1. H_0: Average Performance Old = Average Performance New
   H_1: Average Performance Old ≠ Average Performance New
   p-value < 0.0001* Note: exact *p*-value = 3.79064 x 10-8

2. H_0: Average Performance Old ≤ Average Performance New
   H_1: Average Performance Old > Average Performance New
   p-value = 1.0000 Note: exact *p*-value = 0.99999998

3. H_0: Average Performance Old ≥ Average Performance New
   H_1: Average Performance Old < Average Performance New
   p-value < 0.0001* Note: exact *p*-value = 1.89532×10^{-8}

For this illustrative example we did not discuss the problem statement or scientific hypothesis and, therefore, did not specify the alternative hypothesis of interest. However, we can examine the three *p*-values and determine what we have *proven*, if anything, about this data. The *p*-value for hypothesis (1) is < 0.0001, which indicates that we have enough evidence to say that the average thicknesses produced by the two pieces of equipment are different from each other. The *p*-value for hypothesis (2) is 1, which indicates that we do not have enough evidence to say that the average thickness produced by the old machine is greater than the average thickness produced by the new machine. Finally, the *p*-value for hypothesis (3) is < 0.0001, and we can say that the average thickness produced by the old machine, 50.537, is less than the average thickness produced by the new machine, 54.488.

As before, the *p*-values for all three alternative hypotheses are related to one another. As long as we know the *p*-value for one of the hypothesis statements, we can calculate the *p*-values for the other two alternative hypothesis statements. The following equations, using the labels in the JMP output, can be used to understand how these three *p*-values are related:

1. Prob < t + Prob > t = 1
2. 2 x min{ Prob < t, Prob > t } = Prob > |t|.

We can verify these equations for the output in Figure 5.3b using the exact *p*-values:

1. Prob < t + Prob > t = $1.89532 \times 10^{-8} + 0.99999998 = 1$
2. 2 x min{ Prob < t, Prob > t } = $2 \times 1.89532 \times 10^{-8} = 3.79064 \times 10^{-8}$.

The validity of the two-sample *t*-test rests upon the following assumptions:

1. The distribution of the data generated by the underlying populations can be approximated by a normal distribution.
2. The two population variances must be similar (homogeneous).
3. The experimental units must be independent from each other within a sample and across the two samples.

JMP makes it easy to check for assumptions 1 and 2, and in the case of unequal variances it provides adjustments to the calculated test statistic. Since this requires us to conduct a significance test on the population variances, this is covered in a subsequent section. Assumption 3 is dependent on the way the study is conducted and cannot be checked easily. The best defense is to set up the study appropriately and use randomization.

Two-sample t-test assumptions:
1. approximate normality
2. homogeneous variances
3. independent observations

5.3.3 What to Do When We Have Matched Pairs

Sometimes the assumption that the experimental units are independent across the two samples is violated because the two samples share the same experimental units. At first, this might seem to be a problem, but in fact it can work to our advantage. For example, if we are able to measure the response of interest on the *same* experimental units under the two conditions that we are studying, then we have a more direct comparison of the impact of the two conditions that is not influenced by the population-to-population variation, because we have only one population of experimental units to deal with. Recall the caliper example given in Section 5.3.1, where two operators measure the same sample of gaskets using a given caliper. If we use two different samples of gaskets, one for each operator, we have to deal not only with the gasket-to-gasket variation within a sample of gaskets, but also the variation between gaskets from the two different samples. By using just one sample of gaskets we eliminate this last source of variation, thereby reducing the noise in our signal-to-noise ratio. More examples of this type of data collection scheme are shown in Table 5.2.

Table 5.2 Examples of Matched Pairs Studies

Question	Experimental Unit	First Measurement	Second Measurement
1. Does a low-fat diet help reduce cholesterol?	1 Person	Cholesterol taken on each individual before diet change	Cholesterol taken on each individual six months after diet change
2. Is there a bias in thickness readings of a steel part measured with two different calipers?	1 Steel Part	Thickness measured on each part with the first caliper	Thickness measured on each part with the second caliper
3. Does the addition of a cleaning step after soldering increase the number of shorts on a circuit board?	1 Circuit Board	Number of shorts counted before cleaning step on each circuit board	Number of shorts counted after cleaning step on each circuit board
4. Will an unassembled dollhouse shrink if it sits in inventory for too long in humid conditions?	1 Doll House	Dimensional measurements for critical components in each doll house before packaging and storage	Dimensional measurements for critical components after two weeks of storage in humid conditions

For each of these examples we notice that each experimental unit is measured twice. The first measurement is taken either *before* the application of the experimental treatment (as in questions 1, 3, and 4), or with the first level of the experimental factor (as in question 2). The second measurement is taken either *after* the application of the experimental treatment, or with the second level of the experimental factor. Why do we need to collect data on the same sample of experimental units? Why not collect data on different samples of experimental units, one for each treatment or condition? As we mentioned before, we do this in order to eliminate the effect of the sample-to-sample variation on the overall noise, which increases our chances of detecting a signal. Consider example 2. We can measure 10 steel parts with the first caliper and 10 different steel parts with the second caliper. However, if we want to make a claim about the bias between the two calipers, we want to make sure that the bias is due to caliper differences and not due to differences between the two samples of steel parts. By using a single set of steel parts with both calipers we eliminate the sample-to-sample variation, and reduce the overall variation.

The analysis of data coming from a matched pair study is somewhat different from the analysis of the two-sample *t*-test described in Section 5.3.2. The easiest way to analyze this type of data is to reduce it to a one-sample *t*-test for the mean by working with the difference of the two responses, for example, $Y_{diff} = Y_{Before} - Y_{After}$. The analysis becomes a one-sample *t*-test in which the average difference is compared to zero; i.e., there is no difference between the before and after measurements.

Let's look at an example of a paired *t*-test using the data shown in Figure 5.4 for a study comparing the thickness of a part before and after it goes through a drying oven. There are 20 parts, which have been randomly selected for this study. The thickness readings before the drying operation are recorded in the column labeled Before, while the thickness readings of the cooled down parts after the drying operation are recorded in the column labeled After. Finally, a third column has been added to the table that is calculated using the *Difference = Before – After* measurements. This is easily calculated using the **Formula Editor**.

Figure 5.4 Data for Matched Pairs Oven Study

Part Number	Before	After	Difference
1	49.5592294	49.5057271	0.05350225
2	46.1876191	47.0910696	-0.9034505
3	50.657628	54.1759149	-3.5182869
4	56.6665101	49.960186	6.70632418
5	44.5461434	47.9412698	-3.3951264
6	48.7792797	49.6692651	-0.8899854
7	51.5007731	48.5154044	2.98536866
8	54.1819861	49.1006974	5.08128877
9	54.2330787	48.0234804	6.2095983
10	47.5156514	48.0541635	-0.5385121
11	51.1309402	44.707539	6.42340114
12	50.1864018	52.0462595	-1.8598577
13	51.2778255	55.3440887	-4.0662633
14	48.621866	55.6343786	-7.0125126
15	51.4979148	50.2331063	1.26480856
16	57.0156529	53.3422514	3.67340152
17	51.9937232	43.8934637	8.10025953
18	56.2554939	49.1623141	7.09317982
19	50.2599325	51.2409045	-0.980972
20	56.8741092	47.9059595	8.96814973

If there is no difference in thickness (no shrinkage) before and after the oven, then we expect the differences to be distributed around zero. If the mean of the distribution of differences is above or below zero, we expect either shrinkage or growth respectively.

We will use a one-sample *t*-test for the mean to carry out our significance test on the impact of the oven on the parts thickness. This is easily accomplished by selecting **Analyze > Distribution** and then selecting the Difference column as the **Y, Columns** in the dialog box. We then select the red triangle to the left of the word Difference above the histogram and select the **Test Mean** option. The hypothesis that we are testing is that the mean of the differences is zero. The null hypothesis assumes that the differences are 0, i.e., the oven does not impact the average thickness of the parts. The alternative hypothesis states that the oven does have an effect on the average thickness of the parts.

H_0: $\mu_{differences} = 0$ (Assume—Oven does not affect average part thickness)

H_1: $\mu_{differences} \neq 0$ (Prove—Oven does affect average part thickness)

The output for the one-sample *t*-test is shown in Figure 5.5, with a calculated *t*-statistic of 1.6052. This signal-to-noise ratio is in the gray zone, and we will need to rely upon the *p*-value to determine the outcomes of our significance test. If we use a two-sided alternative with $\alpha = 0.05$ then the *p*-value of 0.1249 leads us to conclude that there is not enough evidence to reject H_0, and we assume that the oven does not affect the thickness of the parts. Although there are 40 measurements in this example, the degrees of freedom for a paired *t*-test are equal to the number of pairs, the number of samples, minus one, or 20 pairs minus 1 = 19 degrees of freedom. The mean of the differences (Before–After) is 1.66972, with a 95% confidence interval (-0.5074; 3.8468), which does include 0, and indicating that the actual value of the mean of the distribution of differences could be zero.

Figure 5.5 Paired t–test Analysis for Oven Data Using One-Sample t-Test

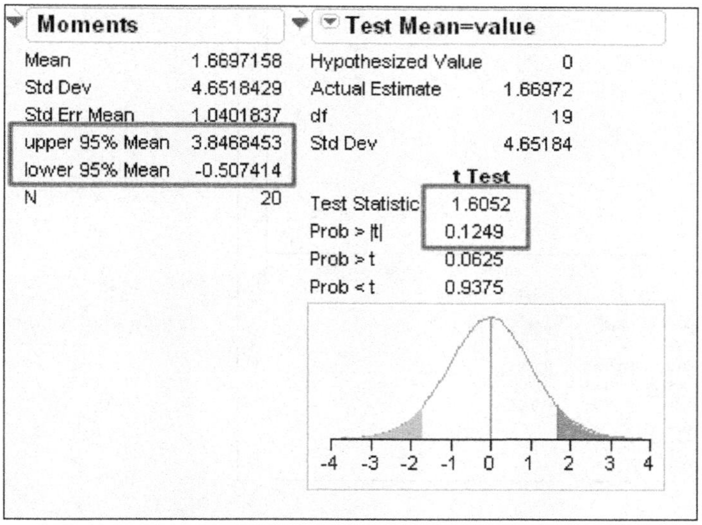

We can also conduct a paired *t*-test using the **Analyze > Matched Pairs** platform in JMP. To use this platform the data must be configured the same way as is shown in Figure 5.4, with two columns for the Before and After measurements. Figure 5.6 shows the output of the matched pairs platform that includes a **Difference vs. Mean** plot, the Before (51.4471) and After (49.7774) means, the **Mean Difference** (1.6697), and a 95% confidence interval for the **Mean Difference** (-0.5074; 3.8468). The calculated *t*-statistic, t Ratio = 1.6052, the degrees of freedom (19), and the corresponding *p*-values (the two-sided *p*-value = 0.1249). These results are identical to those in Figure 5.5 generated using the **Distribution** platform. Unique to this platform is the **Difference vs. Mean** plot, which is equivalent to a 45° rotation of a plot for After versus Before. One can quickly see that all the differences are between -10 and 10, and that the mean difference, represented by the red solid line is greater than 0. The dotted lines represent the 95%

confidence interval, and we see that the lower 95% bound is below 0, which indicates that 0 (no difference) is a likely value, and, therefore, we do not have enough evidence to say that the drying operation is changing the thickness of the parts. The **Difference vs. Mean** plot is a great way to summarize the results of a matched-pair analysis.

Figure 5.6 Output from Matched Pairs for the Oven Data

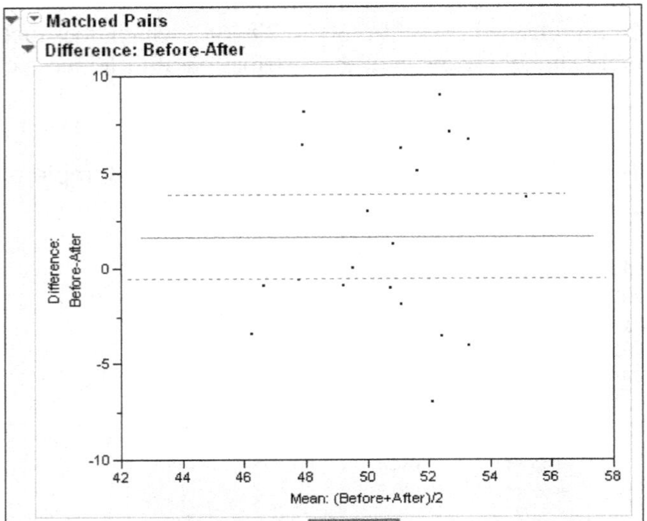

<table>
<tr><td>Before</td><td>51.4471</td><td>t-Ratio</td><td>1.605212</td></tr>
<tr><td>After</td><td>49.7774</td><td>DF</td><td>19</td></tr>
<tr><td>Mean Difference</td><td>1.66972</td><td>Prob > |t|</td><td>0.1249</td></tr>
<tr><td>Std Error</td><td>1.04018</td><td>Prob > t</td><td>0.0625</td></tr>
<tr><td>Upper95%</td><td>3.84685</td><td>Prob < t</td><td>0.9375</td></tr>
<tr><td>Lower95%</td><td>-0.5074</td><td></td><td></td></tr>
<tr><td>N</td><td>20</td><td></td><td></td></tr>
<tr><td>Correlation</td><td>0.03824</td><td></td><td></td></tr>
</table>

5.3.4 Comparing the Performance Variation of Two Materials, Processes, or Products

Recall that there are three assumptions (normality, independent observations, similar variances) that must be met in order for us to be able to perform a one-sample or two-sample Student's *t*-test, or as we will see in Chapter 6, an analysis of variance. The two-sample Student's *t*-test formula in Section 5.3.2 shows how the two-sample *t*-test is a function of the pooled variance s_p^2. In order for us to be able to pool the variances of the two populations we need to check if the noises in the two populations are similar in magnitude, i.e., that the variances are homogeneous. Apart from testing the homogeneity of variances we need to quantify and understand the variation in our data. By focusing on the performance variation, we can look for ways to reduce variation in our products, processes, and materials to achieve higher process capability, i.e., increase the likelihood that product characteristics meet customer expectations.

As it was the case for testing the average performance, we begin by specifying the null and alternative hypothesis. There are three ways that we can write the null (H_0) and alternative (H_1) hypotheses for comparing the standard deviations of two populations:

1. *The performance variations of the two populations are different from each other*

   H_0: Performance Variation (σ_1) = Performance Variation (σ_2) (Assume)
   H_1: Performance Variation (σ_1) $\neq$ Performance Variation (σ_2) (Prove)

   This is a *two-sided* alternative since we do not care about one performance variation being less than or greater than the other, just that they are different from each other. The null hypothesis (H_0) *assumes* that the population standard deviations, σ_1 and σ_2, are equal to each other. On the other hand, the alternative hypothesis (H_1) states that our performance variations are not equal to each other. If our significance test favors the alternative hypothesis then we have proven with statistical rigor that the populations' standard deviations are different, but have made no statements regarding their direction of departure from each other.

> *We assume that that the two performance variations are equal to each other since we can never really prove it*

2. *The performance variation of the first population is greater than the second*

   H_0: Performance Variation (σ_1) $\leq$ Performance Variation (σ_2) (Assume)
   H_1: Performance Variation (σ_1) > Performance Variation (σ_2) (Prove)

   In this one-sided alternative hypothesis, we are interested in knowing whether the performance variation of the first population is greater than the performance variation of the second population. Here, a rejection of the null hypothesis will show with statistical rigor that the first population has a larger standard deviation than the second population, i.e., that the first standard deviation is inferior to the second one. For example, we might not want to qualify a new supplier if we demonstrate that the performance variation of a key attribute is larger than that for the current supplier. For this situation, the null hypothesis assumes that the performance variation of the first product, process, or material is equal to or less than that of the second product, process, or material unless proven otherwise.

3. *The performance variation of the first population is smaller than the second*

   H_0: Performance Variation (σ_1) $\geq$ Performance Variation (σ_2) (Assume)
   H_1: Performance Variation (σ_1) < Performance Variation (σ_2) (Prove)

Finally, for this one-sided alternative we are interested in knowing whether the performance variation of the first population is less than the performance variation of the second population. If we do not have enough evidence to reject the null hypothesis, then we assume that the performance variation of the first product, process, or material is equal to or greater than that of the second product, process, or material unless proven otherwise. However, if we reject the null hypothesis in favor of the alternative, then we have shown, with statistical rigor, that the first population's standard deviation is less than the second population's standard deviation. This form of the hypotheses is again useful if we want to demonstrate that the first standard deviation is superior to the second one. For example, in order to switch to a new test method for measuring the viscosity of a solution, we might want to demonstrate that the new method has less variation than the existing method.

The F-Statistic

For comparing the performance variation of two processes, products, or materials we use a test statistic that is the ratio of the sample variances of the two populations:

$$F = \frac{s_1^2}{s_2^2}$$

> *The F test is sensitive to departures from normality, which can result in inaccurate p-values.*

If the sample standard deviations, s_1 and s_2, are similar to each other then the ratio of their squared values will be close to 1. If the first sample standard deviation, s_1, is much larger than the second sample standard deviation, s_2, then their ratio squared will be greater than 1. If the first sample standard deviation, s_1, is much smaller than the second sample standard deviation, s_2, then their ratio squared will be less than 1. Note that the *F*-test statistic cannot be negative because is the ratio of two squared terms.

As before, we need a yardstick to decide whether our calculated test statistic is "large"; in other words, to decide whether the signal is larger than the noise. The *F*-distribution, written $F_{v1, v2}$, is skewed to the right, with a lower bound of 0, and is described using two sets of degrees of freedom, ν_1 and ν_2, corresponding to the degrees of freedom of the numerator and the denominator, respectively. The *p*-value, which is the area under the *F*-distribution beyond the calculated *F*-statistic, is used to assess the significance of the test. As before, if the *p*-value is less than the pre-specified critical value, α, then we reject the null hypothesis in favor of the alternative hypothesis.

Let's conduct a test for the thickness variation of the two pieces of equipment in Section 5.3.2. Figure 5.7 shows the mean and standard deviations for the new and old pieces of equipment.

Figure 5.7 Comparing the Performance Variation of Two Pieces of Equipment

Means and Std Deviations

Level	Number	Mean	Std Dev	Std Err Mean	Lower 95%	Upper 95%
New	20	54.4882	1.87033	0.41822	53.613	55.364
Old	20	50.5374	1.77002	0.39579	49.709	51.366

The *F*-statistic is easily calculated by squaring the ratio of the two-sample standard deviations. The form of our alternative hypothesis determines which standard deviation goes in the numerator and which in the denominator. For example, if we are testing H_1: $\sigma_{Old} > \sigma_{new}$ then $F = \sigma^2_{old} / \sigma^2_{new} = 1.77^2 / 1.87^2 = 0.896$, while if we are testing $\sigma_{new} > \sigma_{Old}$ then $F = \sigma^2_{new} / \sigma^2_{old} = 1.87^2/1.77^2 = 1.116$. To determine whether we have rejected any of the null hypothesis, we compare the *p*-value from the test to our pre-specified significance level, $\alpha = 0.05$.

The JMP output, shown in Figure 5.8, shows a graphical representation of the two-sample standard deviations for the new and old equipment, along with descriptive statistics and five different tests for testing the two variances. In this chapter we will focus on the **F-test 2-sided** and cover the other tests in Chapter 6.

Statistics Note 5.1: The *F*-test for comparing two variances assumes that they come from independent and normally distributed populations. We should make sure that we check the normality assumption since the *F*-test is sensitive to departures from normality, which can result in inaccurate *p*-values.

Figure 5.8 JMP Output for Testing Two Population Variances

Tests that the Variances are Equal

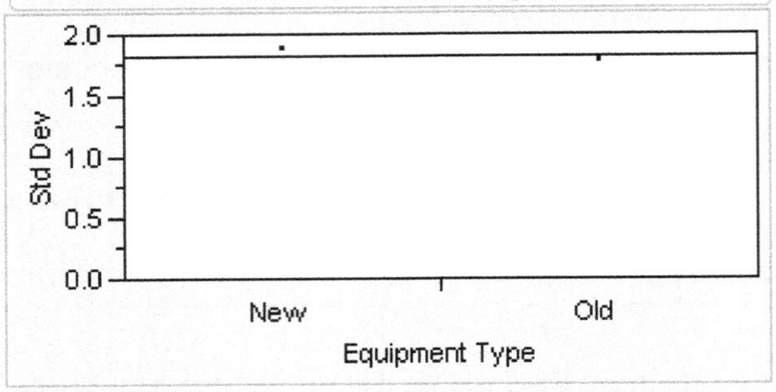

Level	Count	Std Dev	MeanAbsDif to Mean	MeanAbsDif to Median
New	20	1.870332	1.474210	1.466812
Old	20	1.770020	1.432632	1.403630

Test	F Ratio	DFNum	DFDen	p-Value
O'Brien[.5]	0.0773	1	38	0.7825
Brown-Forsythe	0.0330	1	38	0.8569
Levene	0.0158	1	38	0.9005
Bartlett	0.0562	1	.	0.8126
F Test 2-sided	1.1166	19	19	0.8126

As the label implies, JMP carries out a test for a two-sided alternative hypothesis, H_1: $\sigma_{old} \neq \sigma_{new}$, and calculates the test statistic, **F Ratio**, as the ratio of the largest to the smallest variance: $\sigma^2_{max} / \sigma^2_{min} = \sigma^2_{new} / \sigma^2_{old} = 1.87^2/1.77^2 = 1.1166$. This *F*-statistic is distributed as an *F*-distribution with 19 degrees of freedom (number of samples minus 1) for both the numerator and the denominator, $F_{19, 19}$.

> **JMP Note 5.1**: When displaying the descriptive statistics, the levels for the factor are always listed in alphabetical order; however, the F Ratio will always be the ratio of the larger variance to the smaller variance.

The *p*-value for H_1: $\sigma_{old} \neq \sigma_{new}$ is 0.8126 > 0.05 (see also Figure 5.9), so we don't have enough evidence to say that the two population standard deviations are different from each other. Therefore, we assume that the homogeneous variances for the two-sample *t*-test have not been violated.

What about the One-Sided Hypothesis?

The output in Figure 5.8 shows only the *p*-value for the two-sided hypothesis. We can use the *p*-values equations to help calculate the following *p*-values for the two one-sided hypotheses.

1. Prob < F + Prob > F = 1
2. 2 × min{ Prob < F, Prob > F } = Prob > F 2-Sided

The Prob > F 2-Sided = 0.8126 (H_1: $\sigma_{old} \neq \sigma_{new}$), which is two times the minimum *p*-value for the two one-sided alternative hypotheses H_1: $\sigma_{old} > \sigma_{new}$ and H_1: $\sigma_{old} > \sigma_{new}$. The minimum *p*-value is then 0.8126 / 2 = 0.4063. Since the two *p*-values for the one-sided alternative hypothesis must add to 1, the last *p*-value must be 1 – 0.4063 = 0.5937. We now need to identify which *p*-value corresponds to each alternative hypothesis. JMP calculates the *F*-test statistic with σ_1 representing the larger sample standard deviation and σ_2 representing the smaller standard deviation. For our example the new equipment has a larger standard deviation (1.870332) than the old equipment (1.77002). The one-sided *p*-value for testing H_1: $\sigma_{max} > \sigma_{min}$ is half the *p*-value that is output by JMP for the two-sided alternative hypothesis; i.e., the *p*-value = 0.8126 / 2 = 0.4063 corresponds to H_1: $\sigma_{new} > \sigma_{old}$. The *p*-value corresponding to H_1: $\sigma_{new} < \sigma_{old}$ is 1 – 0.4063 = 0.5937. The correct assignments of the three *p*-values are shown below:

1. H_0: Performance Variation (σ_{new}) = Performance Variation (σ_{old})
   H_1: Performance Variation (σ_{new}) ≠ Performance Variation (σ_{old}),
   p-value = 0.8126
2. H_0: Performance Variation (σ_{new}) ≤ Performance Variation (σ_{old})
   H_1: Performance Variation (σ_{new}) > Performance Variation (σ_{old}),
   p-value = 0.4063
3. H_0: Performance Variation (σ_{new}) ≥ Performance Variation (σ_{old})
   H_1: Performance Variation (σ_{new}) < Performance Variation (σ_{old}),
   p-value = 0.5937.

All these *p*-values lead us to conclude that the performance variation of the old and new equipment are similar, and that we have not violated the *homogeneous variance* assumption for the two-sample *t*-test for the two means.

> **JMP Note 5.2:** Unlike the case for comparing two means, we do not get a graphical representation of the *F*-distribution to help us visualize the results. An *F*-distribution based on 19 degrees of freedom, both for the numerator and the denominator, is shown in Figure 5.9. The area to the right of the calculated test statistic, 1.1166, is 0.4063. This *p*-value means that there is a 40.63% probability of obtaining a test statistic as large or larger than the one observed (1.1166). The area to the left of the calculated test statistic, 1.1166, is 0.5937 implying that there is a 59.37% probability of obtaining a test statistic as small or smaller than the one obtained using this sampled data.

Figure 5.9 *p*-Values for the F-Test

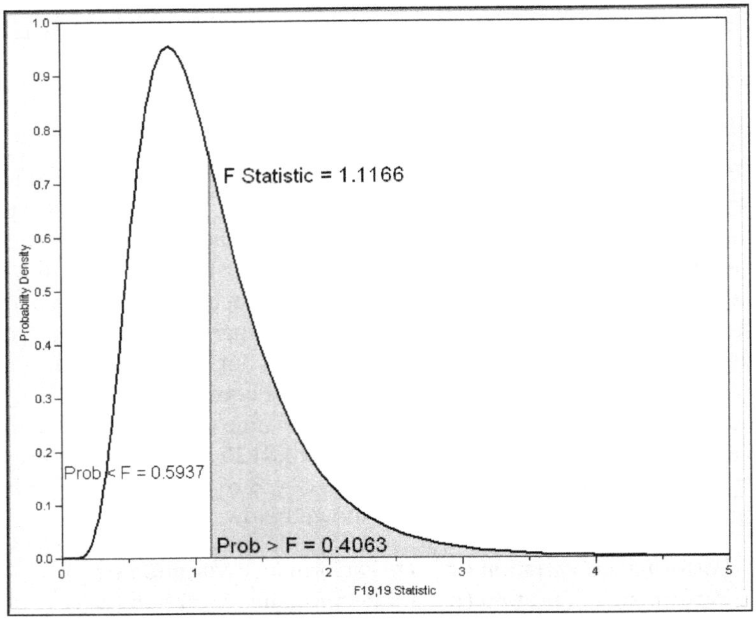

5.3.5 Sample Size Calculations

As was pointed out in the last chapter, for any meaningful analysis, we need to understand how the data is going to be collected, what they represent, and how many we need. Let's revisit the important concepts of experimental and observational unit. The *experimental unit (EU)* is the smallest unit that is affected by the "treatment" or condition we want to investigate. On the other hand, the *observational unit (OU)* is the smallest unit on which a measurement is taken. In many situations the experimental unit is the same as the observational unit in the sense that only one measurement is taken per EU. But in other cases multiple measurements, OUs, are taken on a single experimental unit. This difference is important because the output from the sample size calculations controls how many experimental units we need in our study, not how many times an experimental unit is measured, or how many different measurements we need per experimental unit. Remember, the number of observational units per EU increases the number of measurements but not the sample size. For example, if we need 10 samples (EUs) in our study and decide to measure each experimental unit three times, then we will end up with 30 measurements but only 10 samples. Observational units do not count as replications.

> *Observational units do not count as replications.*

The *sampling plan* for our hypothesis test consists of the *sample size*, which answers the question of how many experimental units are needed, and the *sampling scheme*, which answers the question of how to select the experimental units from the larger population. In Chapter 2 we saw that the two most common sampling schemes are the "simple random sample" and "stratified simple random sample. Our knowledge of the *sources of variation*, which are present in our products, processes, and materials, is also relevant in how we sample our population. Sometimes we choose to include many of these sources of variation in our sampling plans, such as when qualifying a new piece of equipment. Other times we want to minimize the sources of variation that are present, such as if we want to isolate the impact of a process knob on a piece of equipment under "ideal" conditions.

So how many samples do we need to be relatively certain about the outcome?

The sample size calculator for testing two-population means can be found under **DOE > Sample Size and Power** (Figure 5.10). There are several inputs that are required for these calculations, and they are described in detail next. As of JMP 7 the calculator has functionality only for determining sample sizes for **Two-sample Means** (Figure 5.10), but not for two variances.

Figure 5.10 Accessing Sample Size and Power Calculations in JMP

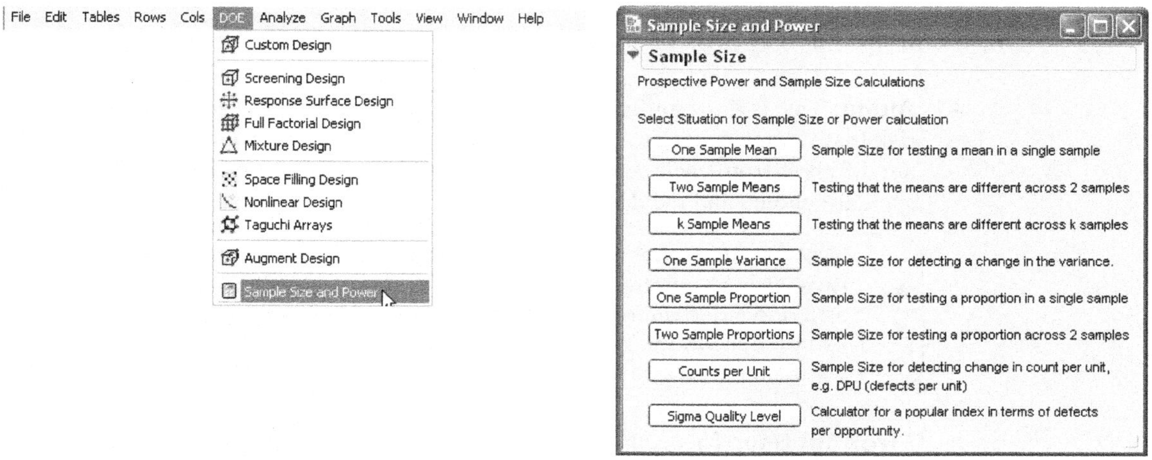

Two-sample Means (Average Performance of Two Populations)

JMP provides an easy to use sample size and power calculator (Figure 5.11a). What follows is a discussion of the inputs needed to use the calculator for calculating the number of samples or experimental units needed to compare two population means. These are the same inputs that were required for the one-sample test for the mean described in Chapter 4.

Figure 5.11a Sample Size Calculator for Two Sample Means

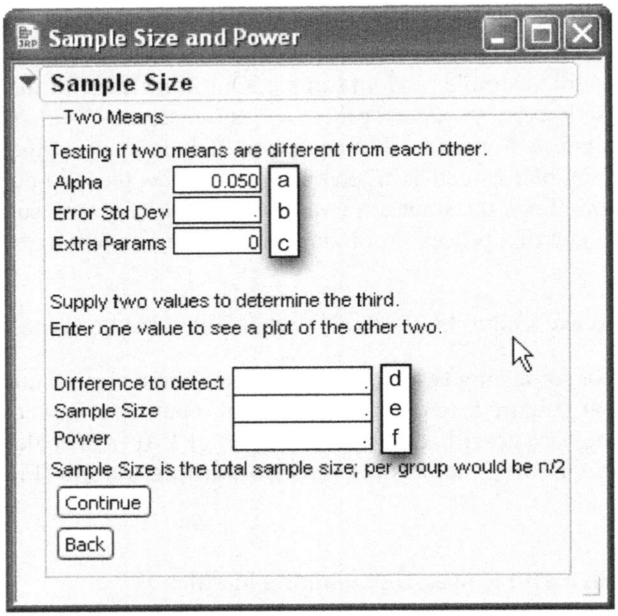

Inputs to the Sample Size Calculator

a. **Alpha:** This is the significance level of the test, and this defines the amount of risk that we are willing to live with for incorrectly stating that the two means are different from each other, when in fact they are not. An alpha of 0.05 is the standard choice and the default in JMP.

> **JMP Note 5.3:** JMP performs the sample size calculation assuming a two-sided alternative hypothesis. If we set up our alternative hypothesis as a one-sided alternative then we should use 2α in this field. For example, if we are willing to live with $\alpha = 0.05$ for our one-sided significance test on our mean then we should enter `0.10` in this field.

b. **Error Std Dev:** This is the noise in our data, after removing the effect of any differences due to changing the factor levels (if a model has been fit to the data, then this is the Root Mean Square Error or RMSE). It is usually described by the standard deviation, σ, from one of the populations. Historical data is helpful for providing estimates of both standard deviations. Once again, the two-sample *t*-test assumes these to be equal. If you suspect that they are not equal, then use the value of the larger one. If this is not possible because we have no historical data (or we are dealing with a new process or material, for example), we can enter a value of 1 in this field. If we do this, then we need to make sure that we specify the "difference to detect" in terms of multiples of this unknown standard deviation.

c. **Extra Params:** This field is used for studies involving more than one factor (multi-factor experiments), which is a topic not covered in this book. We will leave this blank for the two-sample *t*-test for the means.

d. **Difference to detect:** This is the smallest practical difference in the two-population means that we would like to detect. This value must be positive and larger than 0. For example, a difference to detect of 2 implies the following:

1. H_0: Average Performance μ_1 = Average Performance μ_2 versus
   H_1: Average Performance $\mu_1 \neq$ Average Performance μ_2,

We want to be able to detect a difference ($|\mu_1 - \mu_2| > 2$) a large percent of the time.

2. H_0: Average Performance $\mu_1 \leq$ Average Performance μ_2 versus
   H_1: Average Performance $\mu_1 >$ Average Performance μ_2,

We want to be able to detect a difference ($\mu_1 > 2\ \mu_2$) a large percent of the time.

3. H_0: Average Performance $\mu_1 \geq$ Average Performance μ_2 versus
   H_1: Average Performance $\mu_1 <$ Average Performance μ_2 ,

We want to be able to detect a difference ($\mu_1 < 2\ \mu_2$) a large percent of the time.

Note that if we specify our **Error Std Dev** as 1 then the **Difference to detect** must be interpreted in multiples of standard deviations. For example, if we set **Error Std Dev** = 1 and **Difference to detect** = 0.5 this implies that we would like to detect a difference between the two means of 0.5σ units. This is a nice trick because it solves the problem of not knowing the noise in the system. However, once we get an estimate of noise from the data, the difference that we are able to detect will be a function of this estimated noise, and in some cases, this noise (i.e., variation) can be too large to detect the difference in the two population means that we are after.

e. **Sample Size:** This is the minimum sample size that is recommended for the study. We need to collect measurements on these many experimental units in order to be able to detect our practical difference, **Difference to detect**, with a specified **Power**. For a two-sample *t*-test for the means, this number represents the total sample size for the study. We divide this number by 2 in order to get the number of samples (experimental units) for each population.

f. **Power:** Recall from our discussion in Chapter 2 that the power of a test is the probability of detecting a difference between the average performances of the two populations, when in fact there is a difference. Being a probability, the **Power** takes on values between 0 and 1 and can be converted to a percentage by multiplying it by 100. Generally speaking, we would like to have a significance test with a high probability of detecting an actual difference in the population means if the difference exists. We recommend starting with a power of at least 0.8 and adjusting it up or down as needed. Don't be too greedy. The more power you want, the more samples you will need!

There are several ways that we can use the JMP sample size and power calculator. We can input values for **Alpha, Error Std Dev, Difference to detect,** and **Power** and let JMP calculate the sample size. For example, let's say we need to figure out how many samples we need to detect, with high power, and Average Performances difference of 1.5 standard deviations. We enter **Alpha** = `0.05`, **Error Std. Dev.** = `1`, **Difference to Detect** = `1.5`, and **Power** = `0.9`. In other words, we want to be able to detect a difference between the two population means equal to one and one half the standard deviation (1.5), 90% of the time, when in fact the difference is 1.5 times the noise or larger.

Figure 5.11b Inputs for Sample Size Calculator for Two Sample Means

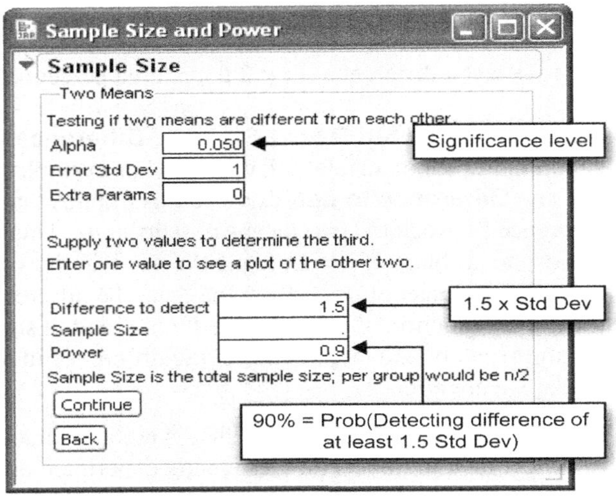

When we click **Continue** using the parameters outlined previously (Figure 5.11b), we get a total sample size of 20.8 (Figure 5.12). Yes, it is normal to have fractional sample sizes. We recommend that you round up since this always increases the power, if only by a little bit. For this example, we will round up to a number that is divisible by 2 for a total of 22 experimental units. Since we usually try to have an equal number of samples per population, we require 11 experimental units for each of the two populations.

Figure 5.12 Calculated Sample Size for Two Sample Means

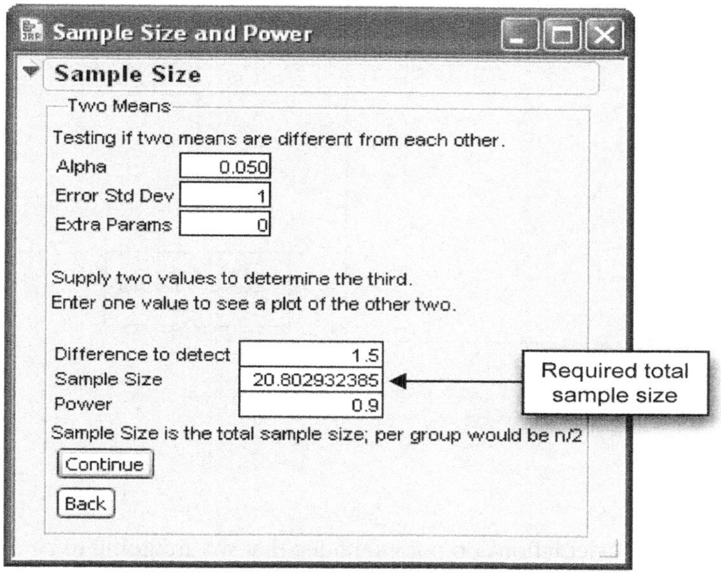

Just as we did with the sample size calculations for one-sample tests on the mean, we can also use this interface to explore different scenarios by generating sample size curves as a function of **Power**, or the **Difference to detect**. This is very, very useful when we want to explore the gains of investing in additional experimental units or in selecting the minimum sample size needed. For the example shown in Figure 5.11b, what if we want to see what differences we can detect as sample size goes up, holding our power at 90%? In the sample size calculator we omit **Difference to Detect** and click **Continue**, and this will generate a plot of sample size versus difference to detect. Using the same inputs, **Alpha** = `0.05`, **Error Std. Dev.** = `1`, and **Power** = `0.9` we get the curve shown in Figure 5.13.

To get a better view of the curve we needed to expand the y-axis to reach 200 and the x-axis to reach 2. Now the plot clearly shows the point of diminishing returns for adding experimental units in our study. There is a high price to pay in order to detect smaller differences in the two population means. For example, a difference of 0.5σ or less requires at least 170 experimental units in total, 85 per population. However, if we want to increase our ability to detect smaller differences then, if we double our sample size from 22 to 44 units, we can detect a 1σ difference between the two population means, compared with 1.5σ.

Figure 5.13 Sample Size Curve for Testing Two Means

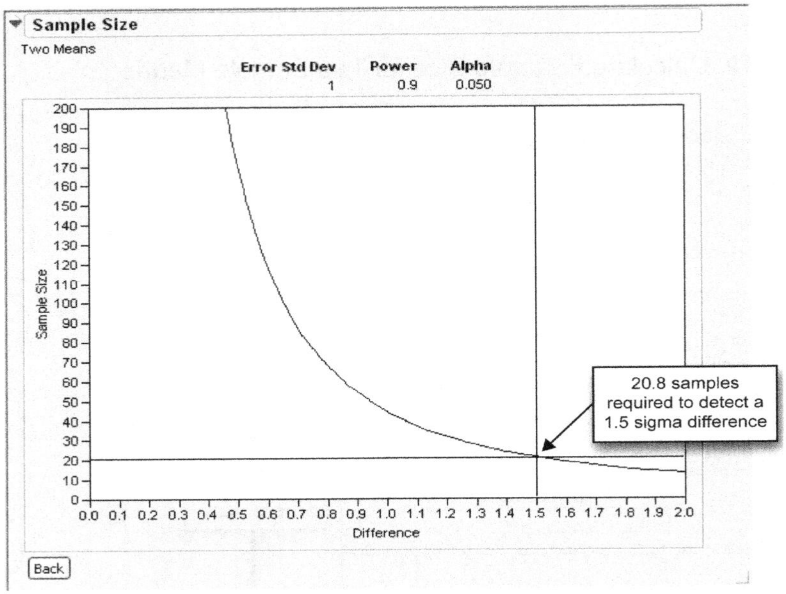

Caution: Sample size calculations do not guarantee that we are going to be able to exactly detect a specified difference between two materials, two products, or two processes that are indeed dissimilar. The values that we input into the various fields of the JMP **Sample Size and Power** calculator helps us to a get an idea of the number of units that we need in order to answer our uncertainties with some degree of confidence.

5.4 Step-by-Step JMP Analysis Instructions

We will illustrate how to conduct a two-sample *t*-test to compare the new isotope ratio mass spectrometer to the current one following the 7 steps shown in Table 5.3. The first two steps should be stated before any data is collected or analyzed, and they do not necessarily require the use of JMP. However, Steps 3 through 6 are conducted using JMP and the output from these steps helps us to complete Step 7.

Table 5.3 Step-by-Step JMP Analysis for Mass Spectrometer Comparison

Step	Objectives	JMP Platform
1. Clearly state the question or uncertainty.	Make sure that we are attempting to answer the right question with the right data.	Not applicable.
2. Specify the hypotheses of interest.	Is there evidence to claim that the performance of the mass spectrometers is different? Are the spectrometers unbiased?	Not applicable.
3. Determine the appropriate sampling plan and collect the data.	Identify how many samples (experimental units and observational units) are needed. State the significance level, α, that is, the risk we are willing to take.	• **DOE > Sample Size and Power**
4. Prepare the data for analysis and conduct exploratory data analysis.	Visually compare the distributions of the two spectrometers. Look for outliers and possible data entry mistakes. Do a preliminary check of assumptions (normality and similar variances).	• **Analyze > Fit Y by X** • **Graph > Control Chart > IR Chart**
5. Perform the analysis to verify your hypotheses, and to answer the questions of interest.	Are the observed differences (the discrepancies between the data and our hypothesis) real or due to chance? We want to be able to detect a difference between the spectrometers, and estimate its size. Check the analysis assumptions to make sure your results are valid.	• **Analyze > Fit Y by X > Unequal Variances, Means / ANOVA / Pooled t**
6. Summarize the results with key graphs and summary statistics.	Find the best graphical representation of the results that give us insight into our uncertainty. Select key output for reports and presentations.	• **Analyze > Fit Y by X** • **Graph > Control Chart > IR Chart**
7. Interpret the results and make recommendations.	Translate the statistical jargon into the context of the problem. Assess the practical significance of your findings.	Not applicable.

Step 1: Clearly state the question or uncertainty

A new mass spectrometer was added to the lab in order to determine the isotopic ratio composition of various elements, including silver. Because our customers require a high degree of accuracy and precision in the results, it is imperative that the measurement performance of the new and old mass spectrometers is nearly identical. Therefore, the main objective of this study is to determine whether there is a bias between the instruments and with respect to a Standard Reference Material (SRM), and if the precision of both instruments is equivalent and acceptable for the applications for which they are used.

The atomic weight of silver, calculated from the isotopic ratio determined by the mass spectrometers, will be used to evaluate the quality of the readings taken with these two analytical instruments. Calibration mixes prepared by blending weighed portions of chemically pure ^{107}Ag and ^{109}Ag isotopes are used to compare the two mass spectrometers following documented procedures.

Note: This example is based on a study conducted by the National Institute of Standards and Technology (NIST) to measure the atomic weight of a reference sample of silver using two nearly identical mass spectrometers. The results from the study were reported in "The Absolute Isotopic Abundance and Atomic Weight of a Reference Sample of Silver" (Powell 1982). The atomic weight of silver data used in this chapter is available from the data sets archives of the Statistical Reference Data Sets of NIST (http://www.itl.nist.gov/div898/strd/anova/AtmWtAg.dat).

Step 2: Specify the hypotheses of interest

Performance comparisons are done based on the average performance and the performance variation. The problem description from Step 1 suggests that the two mass spectrometers, the existing and the new, must provide similar results giving an equality null hypothesis:

$$H_0\text{: } \mu_{\text{Mass Spectrometer New}} = \mu_{\text{Mass Spectrometer Old}} \quad \text{(Assume)}$$
$$H_1\text{: } \mu_{\text{Mass Spectrometer New}} \neq \mu_{\text{Mass Spectrometer Old}} \quad \text{(Prove)}$$

This hypothesis statement assumes that the average performance (the atomic weight of silver determined from the isotopic ratio obtained from the two mass spectrometers) is statistically similar, unless there is enough evidence to prove they are statistically different. The alternative hypothesis, H_1, is a two-sided alternative, because there is nothing in the problem description that suggests a desire to prove that one mass spectrometer produces larger (or smaller) readings, on average, than the other one. These hypotheses deal with the relative bias between the two instruments.

Statistics Note 5.2: We "prove" the alternative hypothesis in the sense of establishing by evidence a fact or the truth of a statement (The Compact Oxford English Dictionary).

In order to detect a bias with respect to a given standard we need to compare the average performance of each instrument to a standard reference material of silver, with a known atomic weight. Below are the two one-sample hypothesis statements for comparing the average performance of each mass spectrometer to the reference silver standard of 107.86815 atomic mass units (amu).

a) H_0: $\mu_{\text{Mass Spectrometer New}} = 107.86815$
H_1: $\mu_{\text{Mass Spectrometer New}} \neq 107.86815$

b) H_0: $\mu_{\text{Mass Spectrometer Old}} = 107.86815$ (Assume)
H_1: $\mu_{\text{Mass Spectrometer Old}} \neq 107.86815$ (Prove)

We also need to determine whether the performance variation is similar for the two mass spectrometers. Since homogeneity of variance is one of the required assumptions of the *t*-test this test is always done first. The corresponding two-sided hypotheses are as follows:

H_0: $\sigma_{\text{Mass Spectrometer New}} = \sigma_{\text{Mass Spectrometer Old}}$ (Assume)
H_1: $\sigma_{\text{Mass Spectrometer New}} \neq \sigma_{\text{Mass Spectrometer Old}}$ (Prove)

While the homogeneity of variance assumption is important, it is not the only reason for writing a hypothesis for comparing the variances of the two mass spectrometers. Understanding the precision, or repeatability, of the analytical instruments is just as important as quantifying any bias to a known standard, and is often the focus for traditional gauge R&R or EMP (Evaluating the Measurement Process) studies. Since our customers are confident with the output from the older mass spectrometer, we might consider testing if the variation in the new mass spectrometer is larger than the variation in the existing one, by using the following hypothesis statement:

H_0: $\sigma_{\text{Mass Spectrometer New}} \leq \sigma_{\text{Mass Spectrometer Old}}$ (Assume)
H_1: $\sigma_{\text{Mass Spectrometer New}} > \sigma_{\text{Mass Spectrometer Old}}$ (Prove)

We will assume that the new instrument has equal or better repeatability than the older one, and will be concerned if we prove, with statistical rigor, that the variation is actually larger than the current one.

Step 3: Determine the appropriate sampling plan and collect the data

In this section we will determine how many experimental units are needed to compare the two mass spectrometers used to determine the average atomic weight of silver. What is the experimental unit (EU) for our study? The definition for the experimental and observational unit is specific to the context of the problem we are trying to solve. In order to define the experimental unit used for comparing the average atomic weights produced by the two mass spectrometers, we must understand the mechanics of how they are used to obtain the atomic weight of a sample of silver. The specific details of this procedure are given in Powell et al. (1982). Each atomic weight reading requires a calibration mix, which requires a certain amount of the reference sample of silver. Therefore, our experimental unit consists of one calibration mix. Since there is only one reading for each calibration mix measured on a mass spectrometer, our observational unit is also one calibration mix.

Decide on the level of risk

How do we make an appropriate selection for the significance level, α, of the test for the average and variation hypotheses statements provided in Step 2? Let us consider their specific interpretation as applied to the reasons for this study. Recall that a Type I error is the error of incorrectly rejecting the null hypothesis, and that the significance level, α, is the probability of making a Type I error. For example, if we select $\alpha = 0.05$ for the average performance hypothesis, then 5% of the time, by chance alone, we will incorrectly conclude that the two mass spectrometers produce different values for the average atomic weight of silver. What are the implications of committing this type of error in our study? Should we be more conservative or more liberal in our choice of α? Since we are concerned with reporting highly accurate values we want to be a bit more conservative and select $\alpha = 0.01$.

Similarly, for the two-sided performance variation hypothesis, if we select $\alpha = 0.05$, then 5% of the time, by chance alone, we will conclude that the standard deviations for the two instruments are different, when in fact they are not different. This will cause us to use a different test statistic for conducting our hypothesis tests for the means. For the one-sided alternative hypothesis, if we select $\alpha = 0.05$, then 5% of the time, by chance alone, we will conclude that the standard deviation of the atomic weight of silver readings for the new mass spectrometer is larger than the standard deviation for the older one, when in fact it is not larger. For these hypotheses, we will also be a bit more conservative and select $\alpha = 0.01$.

Sample size calculation

We need to select **DOE > Sample Size and Power** in order to get access to the sample size calculator for testing **Two-sample Means**. Because we are not sure about what value to enter for our **Error Std Dev**, we enter `1` in this field to represent a general value of our standard deviation. Our choices for the other fields are shown in Figure 5.14, with α = `0.01` and a **Difference to detect** = `1` standard deviation. Since we do not want our customers to be able to notice a difference in the readings taken from the two mass spectrometers, we want to be able to detect a small difference, 1 sigma, between the two-sample means with high power.

Figure 5.14 Inputs for Sample Size Graph for Mass Spectrometer Study

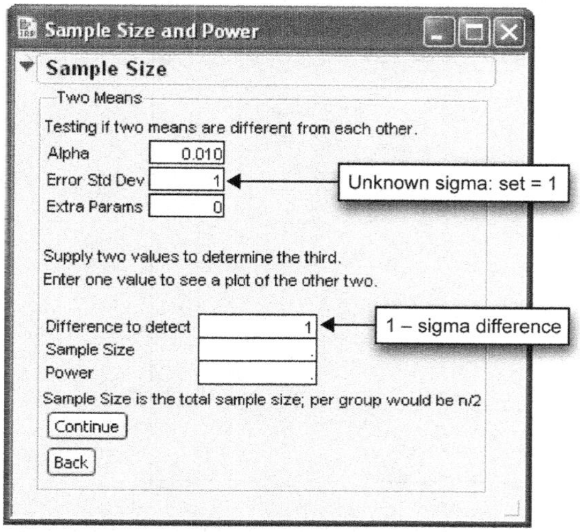

In order to investigate our choices of sample size, we will generate a curve of the power as a function of our total sample size, by leaving these two fields blank in the dialog box and clicking **Continue**. The curve of power versus total sample size is shown in Figure 5.15. It should be obvious that total sample size required for our study increases as the power increases. Recall that the power is the probability of detecting a difference in our two means with the *t*-test, of 1 standard deviation or less, when it truly exists. In order to have a power of about 80% (77% to be exact), we will need a total of 48 experimental units, or 24 per group. This graph is a very useful negotiating tool because we can ask the people involved how many experimental units they are willing to provide in order to be able to detect a difference of a given size. Trade-offs can be made in terms of sample size and power. In this situation 48 EUs is manageable and, since we want to detect a small difference in the means of the two instruments, 48 becomes the combined sample size.

*A **Power vs. Sample Size** graph is a useful negotiating tool.*

Figure 5.15 Sample Size vs. Power Graph for Mass Spectrometer Study

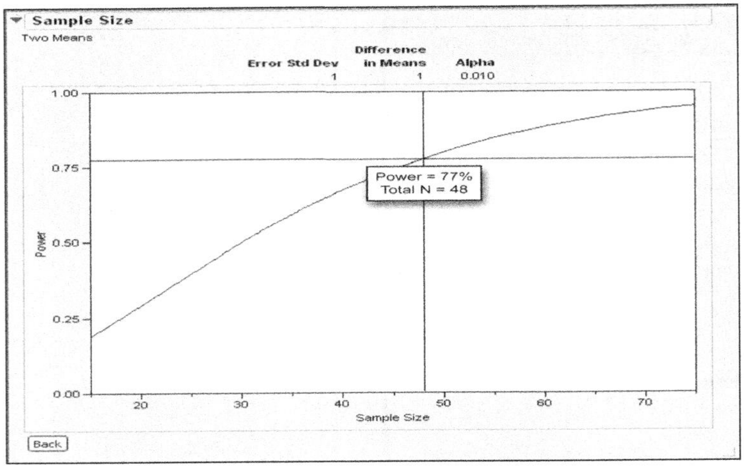

The experimental unit for this study is one calibration mix. How do we obtain the 48 experimental units from the total population of experimental units? We need to prepare 48 calibration mixes by blending weighed portions of ^{107}Ag and ^{109}Ag solutions. We then randomly assign 24 mixes to the old mass spectrometer, and 24 to the new mass spectrometer.

Now that we have determined the sampling plan for the comparison of the two mass spectrometers, we should carefully construct a data collection sheet that can be used to gather the data from the study. The random assignment of mixes to spectrometers is accomplished by using a random number generated using the **Random Uniform** generator in the **Random** function category in the **Formula Editor**. It is also important that we capture all relevant information such as the operator who took the measurements, and always include a column for comments in order to capture any excursions or problems that we have while conducting the study. This information will provide valuable insights if we observe any outliers, unexpected patterns, or trends in the data. A portion of the mass spectrometer data is provided in Table 5.4.

Table 5.4 Sample Data for Comparing Two Mass Spectrometers

Calibration Mix	Randomization	Mass Spectrometer	Operator	Atomic Weight	Comments
29	0.0016095	Old	Ann	107.8681604	
22	0.02507942	New	Joe	107.8681333	
25	0.0556518	Old	Ann	107.8681079	
18	0.06289465	New	Joe	107.8681518	
48	0.0925643	Old	Ann	107.8681368	
27	0.09377345	Old	Ann	107.8681513	
2	0.1013341	New	Joe	107.8681465	
43	0.24608794	Old	Ann	107.8681469	

Step 4: Prepare the data for analysis and conduct exploratory data analysis

The first thing that we need to do is get our data into JMP and make sure that it is in the right format for the analysis platform used to carry out a two-sample *t*-test for the means, and a test for the equality of variances. The data was read into JMP from its original source, an Excel file, by selecting the **Open Data Table** icon in the **JMP Starter** window and selecting **Excel Files (*.XLS)** for file types. Once the correct filename is selected, we click the **Open** icon and the data is converted to a JMP data table. Note that if we want to save the data in this JMP format we need to select **File > Save** and enter a new name and location for the JMP table.

The JMP table for the mass spectrometer study is shown in Figure 5.16, and can be downloaded from the author pages for this book at support.sas.com/authors. Each row includes information about one calibration mix and its atomic weight of silver for a total of 48 rows. It should be noted that all of the atomic weight measurements for both instruments are in one column, labeled Ag Weight. The information about which mass spectrometer the weight comes from is contained in the column labeled Mass Spectrometer, and which operator used the spectrometer is listed in the column Operator. There is also a column, Randomization, that contains the order in which the calibration samples were measured. We should also note that the data type and modeling type for the atomic weight is numeric and continuous, while for the mass spectrometers it is character and nominal.

Figure 5.16 JMP Table for Mass Spectrometer Study

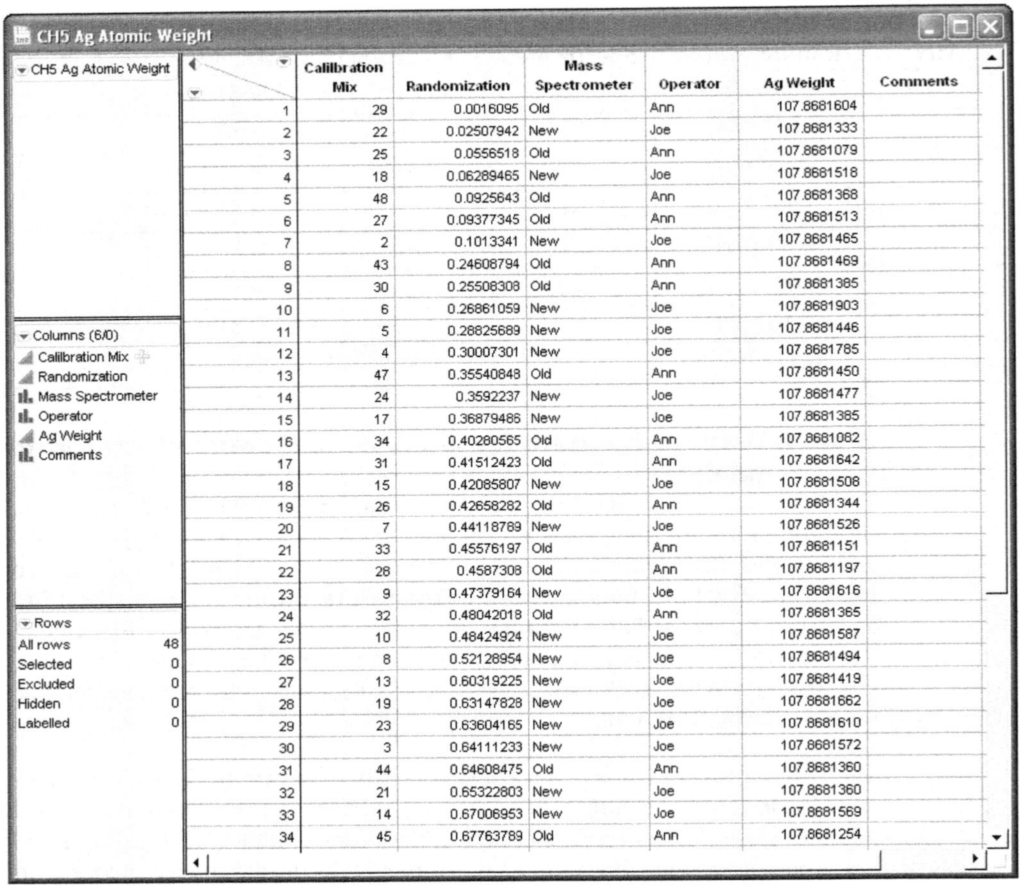

	Calilbration Mix	Randomization	Mass Spectrometer	Operator	Ag Weight	Comments
1	29	0.0016095	Old	Ann	107.8681604	
2	22	0.02507942	New	Joe	107.8681333	
3	25	0.0556518	Old	Ann	107.8681079	
4	18	0.06289465	New	Joe	107.8681518	
5	48	0.0925643	Old	Ann	107.8681368	
6	27	0.09377345	Old	Ann	107.8681513	
7	2	0.1013341	New	Joe	107.8681465	
8	43	0.24608794	Old	Ann	107.8681469	
9	30	0.25508308	Old	Ann	107.8681385	
10	6	0.26861059	New	Joe	107.8681903	
11	5	0.28825689	New	Joe	107.8681446	
12	4	0.30007301	New	Joe	107.8681785	
13	47	0.35540848	Old	Ann	107.8681450	
14	24	0.3592237	New	Joe	107.8681477	
15	17	0.36879486	New	Joe	107.8681385	
16	34	0.40280565	Old	Ann	107.8681082	
17	31	0.41512423	Old	Ann	107.8681642	
18	15	0.42085807	New	Joe	107.8681508	
19	26	0.42658282	Old	Ann	107.8681344	
20	7	0.44118789	New	Joe	107.8681526	
21	33	0.45576197	Old	Ann	107.8681151	
22	28	0.4587308	Old	Ann	107.8681197	
23	9	0.47379164	New	Joe	107.8681616	
24	32	0.48042018	Old	Ann	107.8681365	
25	10	0.48424924	New	Joe	107.8681587	
26	8	0.52128954	New	Joe	107.8681494	
27	13	0.60319225	New	Joe	107.8681419	
28	19	0.63147828	New	Joe	107.8681662	
29	23	0.63604165	New	Joe	107.8681610	
30	3	0.64111233	New	Joe	107.8681572	
31	44	0.64608475	Old	Ann	107.8681360	
32	21	0.65322803	New	Joe	107.8681360	
33	14	0.67006953	New	Joe	107.8681569	
34	45	0.67763789	Old	Ann	107.8681254	

Our data is "analysis ready" so we can proceed with our significance test for the parameters of interest. However, before we proceed with our significance test for the average performances and performance variations, it is always a good idea to take a quick look at the data for outliers, typos, or any other unusual results. We also need to check the assumptions for the two-sample *t*-test for the means. Namely, we need to check that the underlying populations can be described by a distribution that closely follows the normal distribution, that the samples are independent from each other, and that the variances of the two mass spectrometers are equivalent.

In order to check our data for outliers, check the normality of the samples, and determine whether the variances are homogeneous, we need to launch into the **Fit Y by X** platform, see Figure 5.17.

Steps for Launching the Fit Y by X Platform

1. From the primary toolbar, select **Analyze > Fit Y by X**. A dialog box will appear that enables us to make the appropriate choices for our data.
2. Select Ag Weight from the **Select Columns** window and then click **Y**, **Response**. This populates the window to the right of this button with the name of our response.
3. Select Mass Spectrometer from the **Select Columns** window and then click **X**, **Factor**. This will populate the window to the right of this button with the name of the factor under investigation.
4. Click **OK**.

Figure 5.17 Fit Y by X Dialog Box for Mass Spectrometer Comparison

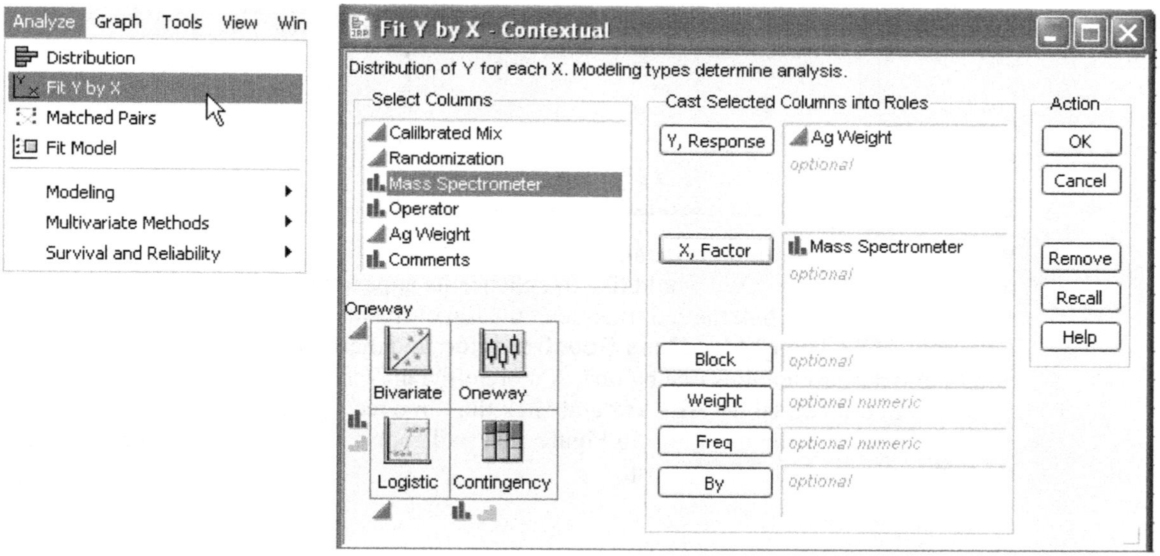

The default JMP output is shown in Figure 5.18, with the atomic weight of silver plotted on the y-axis and the values for the two mass spectrometers plotted on the x-axis.

Figure 5.18 Graph from Fit Y by X Platform Comparing Mass Spectrometers

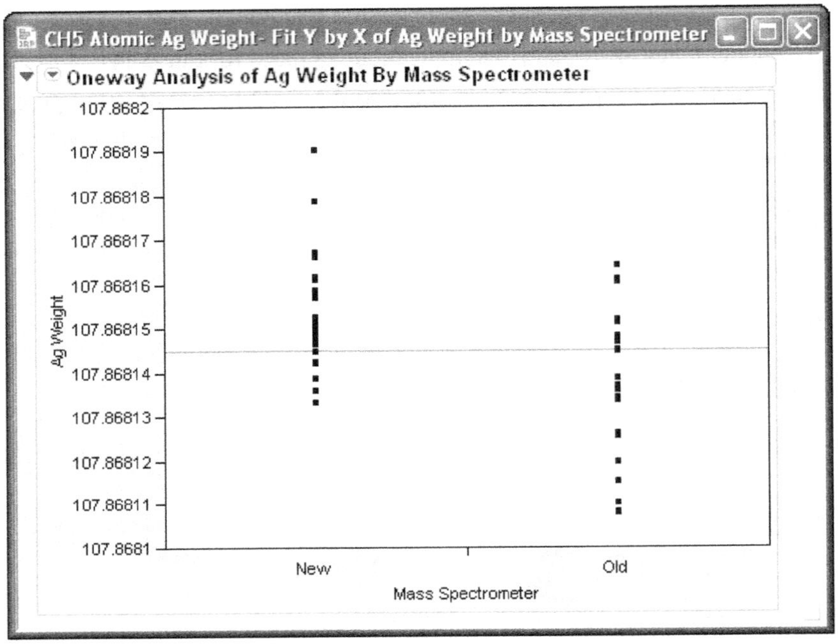

While this plot is useful in its default form to look for any outliers and get a general sense of the two distributions, we would like to enhance its appearance by turning on some additional features. Click the red triangle at the top of the plot, next to the label **Oneway Analysis of Ag Weight by Mass Spectrometer**, and then select **Display Options** to turn on the desired features one by one. A more efficient way to turn on multiple display options is to hold down the ALT key and click the same red triangle at the top of the plot. This will launch the menu shown in Figure 5.19 and enable us to select multiple features by clicking in the boxes of interest.

Figure 5.19 Options for Changing Plot Features in Fit Y by X

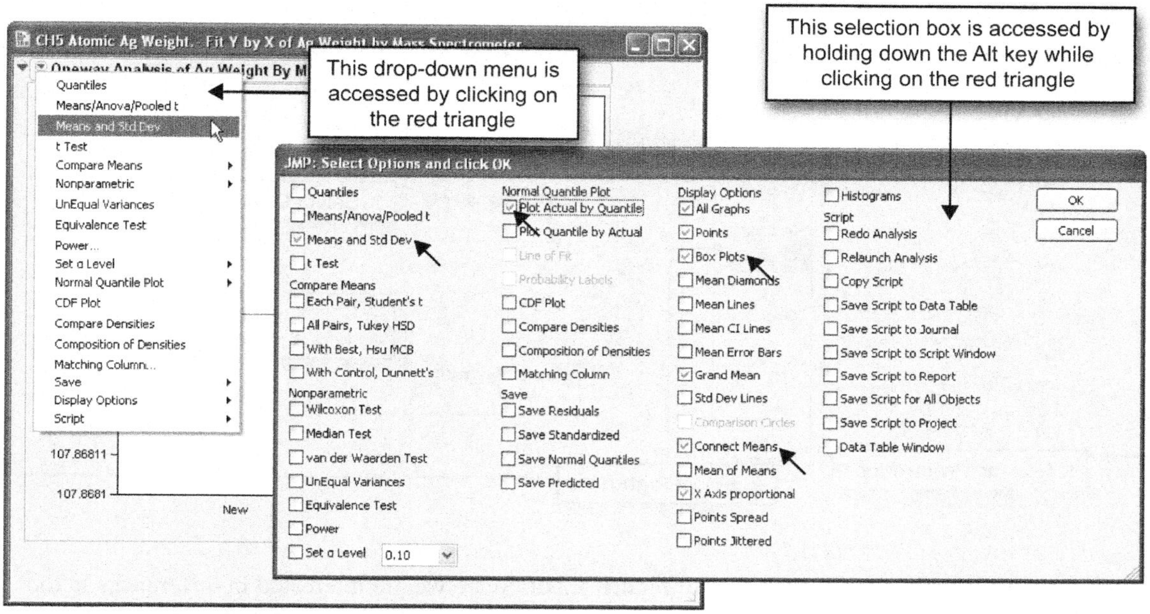

Finally, if we want these preferences to apply to all analysis we conduct in this platform, then we can permanently set them from the **JMP Starter > Preferences** icon, Figure 5.20.

Figure 5.20 JMP Menu for Setting System Preferences for the Oneway Platform

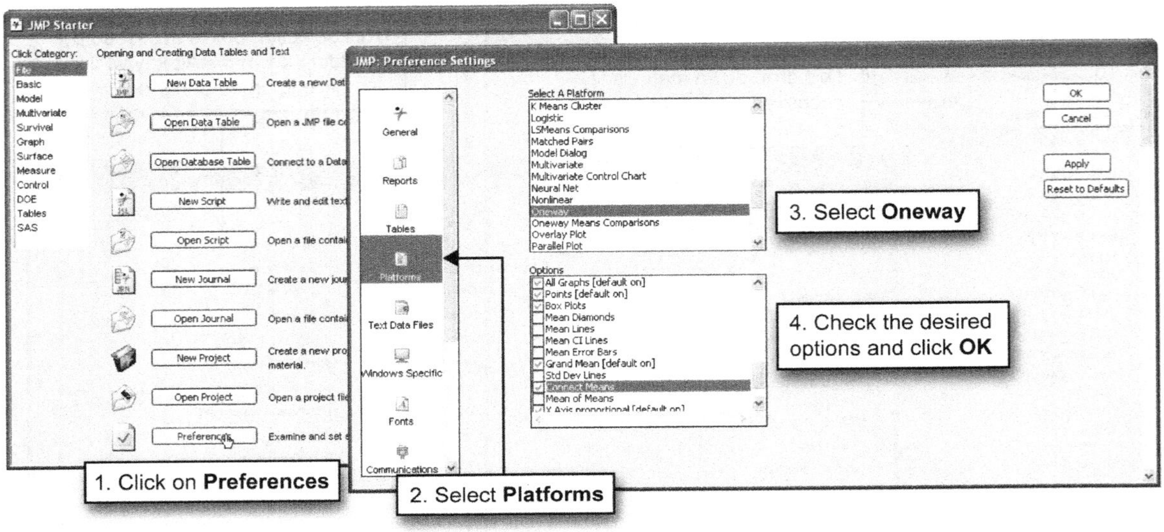

The enhanced plot is shown in Figure 5.21a. Since we are interested in differences in the readings out to the 5th or 6th decimal place, we need to expand the number of significant digits so we can see the differences. This can be accomplished by double-clicking on the **Mean** values and then specifying that the number of decimal places = 7 and field width = `15` and clicking **OK**. We can also expand the other fields in this table in a similar manner.

Figure 5.21a Enhanced Plot for Comparison of Mass Spectrometers

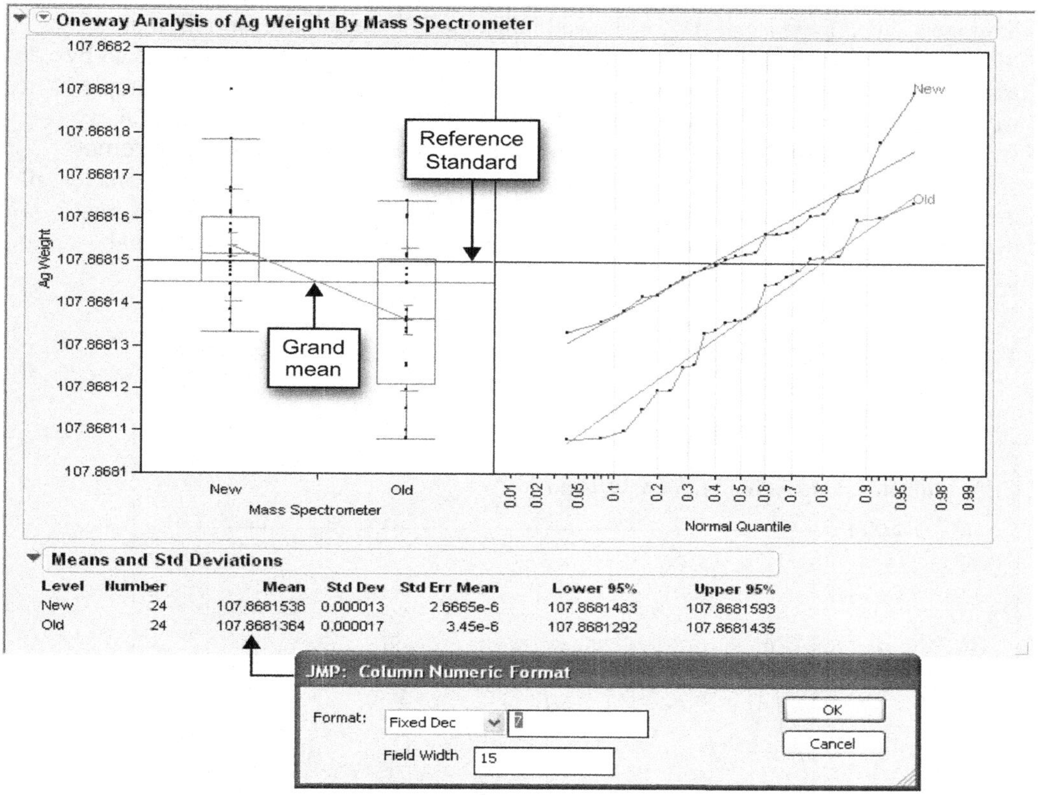

Level	Number	Mean	Std Dev	Std Err Mean	Lower 95%	Upper 95%
New	24	107.8681538	0.000013	2.6665e-6	107.8681483	107.8681593
Old	24	107.8681364	0.000017	3.45e-6	107.8681292	107.8681435

We should examine the means and the standard deviations in conjunction with the box plot to get an initial sense for the differences in the means and variances of the two mass spectrometers, and also to look for outliers. There appears to be one atomic weight measurement for the new mass spectrometer that is beyond the upper fence of the box plot. If you hover over the point with your mouse, you can see that it is observation 10 in the JMP table. When we go back to check the logs and the comments, we do not uncover anything unusual about this point. We can also take a quick look at the normality of the two populations by examining the normal quantile plots that correspond to the two mass spectrometers. With the exception of perhaps one outlier (observation 10 that was previously mentioned), the data points in the two plots seem to follow the lines pretty well and do not suggest a significant departure from normality. You can also look at the distribution of the data by turning on the **Histograms** option within the **Display Options** (not shown). We will also show how to check this assumption using the residuals from the model in a subsequent step.

Are the data homogeneous?

Can we treat the data generated by each of the spectrometers as coming from homogeneous populations? As we showed in Chapter 2, we can use an Individual Measurement process behavior chart to answer this question. We first sort the data by Mass Spectrometer and Randomization (the order in which the calibration samples were taken). We now select **Graph > Control Chart > IR** and in the Control Chart window we select Ag Weight and click **Process**, and then select Mass Spectrometer and click **Phase**. The resulting chart is shown in Figure 5.21b. For each spectrometer, all the points are within the control (including the highest value for the new spectrometer—the one we thought it was a possible outlier), and they do not display any unusual patterns.

Figure 5.21b Individual Measurements Chart of Ag Weight to Check Homogeneity

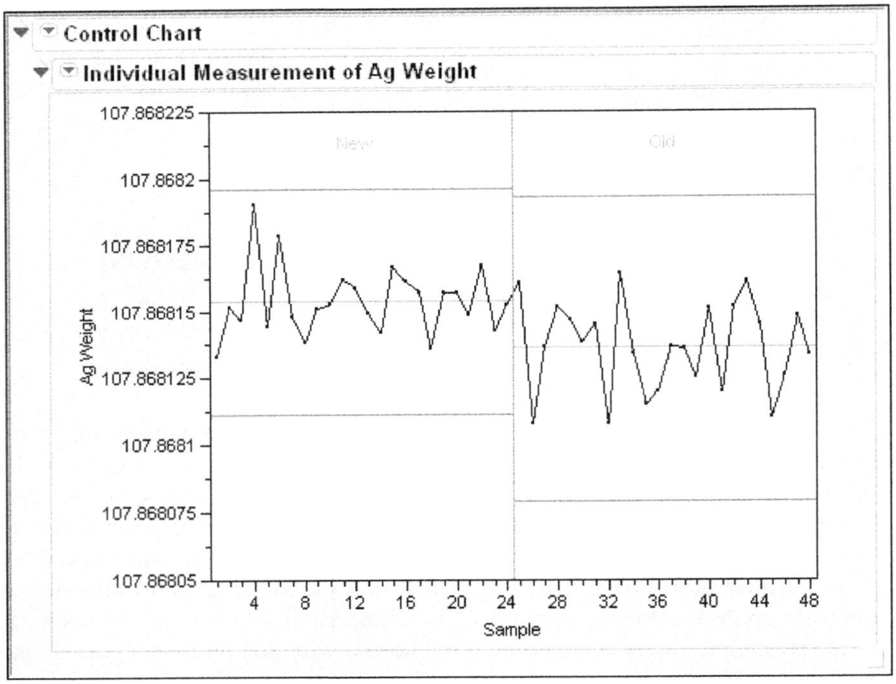

Figure 5.21b gives us a sense for the possible outcomes for the hypothesis tests for comparing the means and standard deviations for the new and old mass spectrometers. By comparing the center lines we see that the average atomic weight of silver is higher for the new mass spectrometer, about 1.74E-5 units, than for the old mass spectrometer, 107.8681538 amu versus 107.8681364 amu, respectively. In comparing the two means to

the standard reference value of 107.86815 amu, it appears that the new mass spectrometer comes closer, on average, to this value. In addition, by looking at the control limits we see that the ones for the new mass spectrometer are narrower, and therefore have less variation, than the ones for the old mass spectrometer. From Figure 5.21a the standard deviations for the new and old are 0.000013 amu versus 0.000017 amu, respectively. Is this difference large enough to cause us to violate the assumption of equal variances between the two populations required for a two-sample *t*-test? We will check this assumption in Step 5.

Based on our exploratory analysis of the data we would speculate that the two means might be different and that the mean of the old mass spectrometer is statistically different from our reference standard. It is not evident whether the difference in the two standard deviations will be statistically significant.

Step 5: Perform the analysis to verify your hypotheses

The major uncertainty in our study is how well the performances of the two mass spectrometers agree with each other. We want to compare the average performance and the performance variation of the two mass spectrometers. We also would like to check for possible bias in each mass spectrometer as compared to a standard.

Average performance

1. H_0: $\mu_{\text{Mass Spectrometer New}} = \mu_{\text{Mass Spectrometer Old}}$ (Assume)
   H_1: $\mu_{\text{Mass Spectrometer New}} \neq \mu_{\text{Mass Spectrometer Old}}$ (Prove)

Performance variation

2. H_0: $\sigma_{\text{Mass Spectrometer New}} = \sigma_{\text{Mass Spectrometer Old}}$ (Assume)
   H_1: $\sigma_{\text{Mass Spectrometer New}} \neq \sigma_{\text{Mass Spectrometer Old}}$ (Prove).

These hypotheses can be tested using the **Fit Y by X** platform. We should first test the second hypothesis since it is relevant for verifying the assumption of variance homogeneity.

Are the variances the same for the two mass spectrometers?

This question is easily answered by clicking on the small red triangle above the box plot and selecting **UnEqual Variances** from the drop-down menu, as is shown in Figure 5.22. Notice the tooltip that is automatically displayed and describes the purpose of the selection.

Figure 5.22 Unequal Variance Selection from Fit Y by X

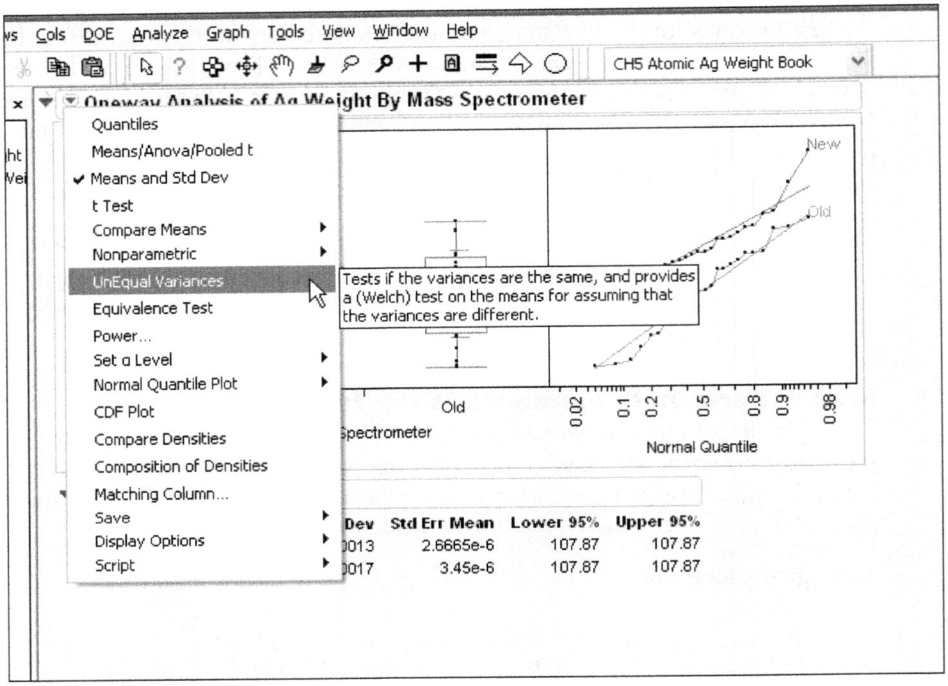

This selection adds output right below the table for the means and standard deviations for the two mass spectrometers. The output shown in Figure 5.23 shows a plot of the two standard deviations for the new and old mass spectrometers. The y-axis reference line in this plot is an estimate for the combined standard deviations, assuming that they are equal. We have already noted that the new mass spectrometer has slightly smaller variance, but is it statistically significant? JMP provides several test statistics to test the hypothesis of homogeneous variances. We will revisit some of these in Chapter 6. As was described in Section 5.3.4 of this chapter, we will use the F-test to test the homogeneity of variances hypothesis since the normality assumption holds (Figure 5.21a).

Figure 5.23 Output for Unequal Variances Test

Tests that the Variances are Equal

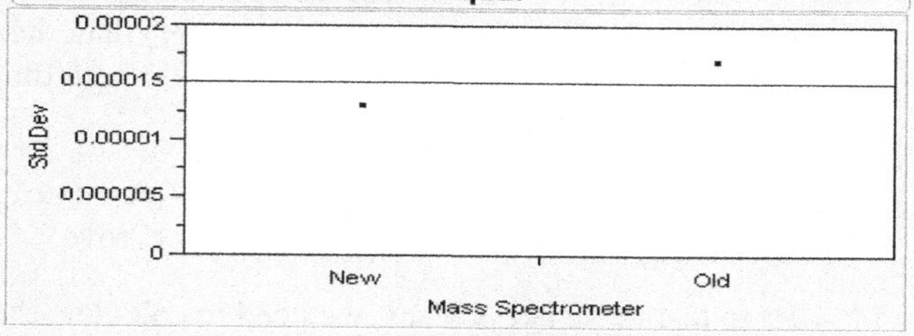

Level	Count	Std Dev	MeanAbsDif to Mean	MeanAbsDif to Median
New	24	0.0000131	0.0000097	0.0000095
Old	24	0.0000169	0.0000136	0.0000136

Test	F Ratio	DFNum	DFDen	p-Value
O'Brien[.5]	1.6905	1	46	0.2000
Brown-Forsythe	2.4108	1	46	0.1274
Levene	2.2626	1	46	0.1394
Bartlett	1.4778	1	.	0.2241
F Test 2-sided	1.6740	23	23	0.2241

$F = \sigma^2_{old} \div \sigma^2_{new}$

$F = 0.0000136^2 \div 0.0000131^2$

$F = 1.6740$

$H_0: \sigma_{new} = \sigma_{old}$

$H_1: \sigma_{new} \neq \sigma_{old}$

Welch Anova testing Means Equal, allowing Std Devs Not Equal

F Ratio	DFNum	DFDen	Prob > F
15.9467	1	43.252	0.0002*

t Test
3.9933

The test statistic for the *F*-test is 1.6740, which is the ratio of the two variance estimates for the new and old mass spectrometers. Recall from the discussion in Section 5.3.4 that JMP always calculates this ratio by taking the larger variance divided by the smaller variance, and therefore, it will always be larger than 1. We interpret this test statistic as follows: the variance for the current mass spectrometer is 1.67 times as large as the variance for the new mass spectrometer. We also need to obtain the *p*-value in order to determine whether the F statistic is "large." This *p*-value will come from an *F*-distribution with 23 degrees of freedom for the numerator and the denominator. Don't forget, the *numerator degrees of freedom* corresponds to the sample size for the old mass spectrometer, while the *denominator degrees of freedom* corresponds to the sample size for the new mass spectrometer. The corresponding *p*-value for $F_{23,23} = 1.6740$ is 0.2241. The area to the right of 1.6740, 22.41%, is the probability of obtaining, by chance, an F statistic as large as or larger than the one we obtained in this study, under the assumption that the null hypothesis is true. Since the *p*-value is greater than our significance level, 0.01, this indicates that the data does not provide enough evidence to conclude that the two variances are different.

Statistics Note 5.3: The Brown-Forsythe test is more robust against departures from normality, and is our test of choice for comparing variances (Chapter 6). In this case the *p*-value for the Brown-Forsythe test is 0.1274 (Figure 5.23), which indicates that the two population variances are homogeneous.

Is the noise in the new spectrometer the same or better than the noise in the old spectrometer?

We also noted that in a gauge R&R or EMP study, we might be interested in testing if the variance of one instrument is larger (or smaller) than the variance of the other instrument. In Step 2 we specified the following hypotheses regarding the variances of the old and new mass spectrometers:

H_0: $\sigma_{\text{Mass Spectrometer New}} \leq \sigma_{\text{Mass Spectrometer Old}}$ (Assume)
H_1: $\sigma_{\text{Mass Spectrometer New}} > \sigma_{\text{Mass Spectrometer Old}}$ (Prove)

In order to test these hypotheses, we need to calculate the *p*-value by hand using the formulas shown in Section 5.3.4. The *F*-statistic = $\sigma^2_{\text{New}} / \sigma^2_{\text{Old}} = 0.0000131^2 / 0.0000136^2 = (1.6740)^{-1} = 0.5974$. The *p*-value can either be obtained using the **F Distribution** option from the **Probability Function** in the formula editor, or by calculating it from the *p*-value (0.2241) given in Figure 5.23. Since we know that the *p*-value for the two-sided alternative is twice the *p*-value for the one-sided alternative with the smallest *p*-value, we start by dividing 0.2241 in half to obtain $p\text{-value}_1 = 0.11208$. The *p*-value for the other one-sided alternative hypothesis is $p\text{-value}_2 = 1 - 0.11208 = 0.8879$. Now that we have the *p*-values for both one-sided alternatives, we need to figure out which one belongs to the alternative hypothesis we want to explore, H_1: $\sigma_{\text{Mass Spectrometer New}} > \sigma_{\text{Mass Spectrometer Old}}$.

Because of the way that JMP computes the test statistic for the two-sided alternative hypothesis, the *p*-value derived by dividing the two-sided *p*-value in half, $p\text{-value}_1 = 0.11208$ corresponds to H_1: $\sigma_{\text{Mass Spectrometer Old}} > \sigma_{\text{Mass Spectrometer New}}$. Therefore, the $p\text{-value}_2 = 0.8879$ is for the alternative hypothesis of interest, H_1: $\sigma_{\text{Mass Spectrometer New}} > \sigma_{\text{Mass Spectrometer Old}}$ which indicates that we do not have enough evidence to say that the variation in the new spectrometer is larger than the variation in the old spectrometer. Therefore, we can assume that the new mass spectrometer is as precise as the old.

Is the average performance of the two mass spectrometers similar?

Here are the two hypotheses:

H_0: $\mu_{\text{Mass Spectrometer New}} = \mu_{\text{Mass Spectrometer Old}}$
H_1: $\mu_{\text{Mass Spectrometer New}} \neq \mu_{\text{Mass Spectrometer Old}}$

As was noted earlier, the test statistic and p-value for testing the hypothesis will depend upon the outcome for the test of the two variances. If the two variances are shown to be different, we use Welch's ANOVA to test the two means. This output can be found right below the JMP output for testing the two variances, and is shown in Figure 5.24. As we noted above, since the p-value = 0.2241 for the two variances *F*-test is greater than 0.01, we do not need to use Welch's ANOVA.

In order to obtain the two-sample *t*-test, click the small red triangle at the top of the window, next to the **Oneway Analysis** and select **Means/Anova/Pooled t**, as shown in Figure 5.25. This is the correct choice for when the two population variances are assumed equal and can be pooled into one error term. Just for completeness, we also show a second choice, **t Test**, for computing the test statistic for the means when the variances are not equal (Welch's ANOVA).

Figure 5.24 JMP Output for Welch's ANOVA

Tests that the Variances are Equal

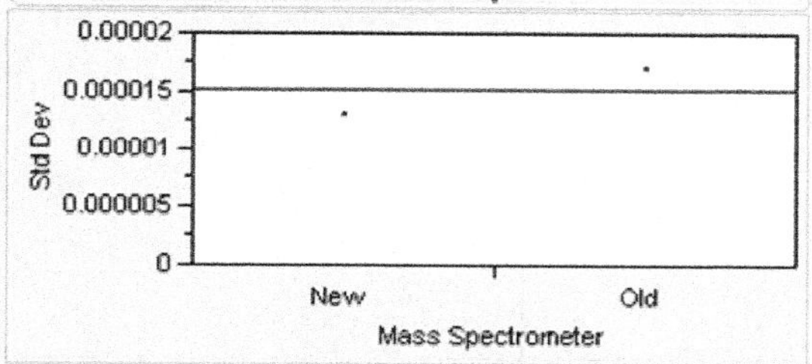

Level	Count	Std Dev	MeanAbsDif to Mean	MeanAbsDif to Median
New	24	0.0000131	0.0000097	0.0000095
Old	24	0.0000169	0.0000136	0.0000136

Test	F Ratio	DFNum	DFDen	p-Value
O'Brien[.5]	1.6905	1	46	0.2000
Brown-Forsythe	2.4108	1	46	0.1274
Levene	2.2626	1	46	0.1394
Bartlett	1.4778	1	.	0.2241
F Test 2-sided	1.6740	23	23	0.2241

Welch Anova testing Means Equal, allowing Std Devs Not Equal

F Ratio	DFNum	DFDen	Prob > F
15.9467	1	43.252	0.0002*

t Test
3.9933

$H_0: \mu_{new} = \mu_{old}$
$H_1: \mu_{new} \neq \mu_{old}$

Welch's ANOVA is used to compare the means if the standard deviations are statistically different.

Figure 5.25 Using a t-Test to Compare Two Means in JMP

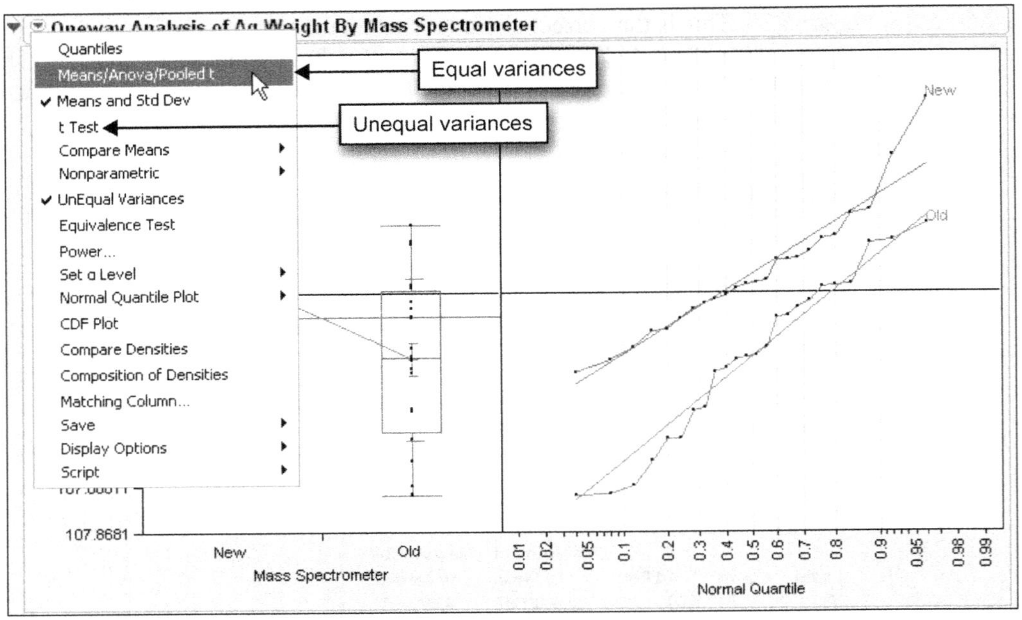

The output for our selection of **Means/Anova/Pooled t** is shown in Figure 5.26. By default JMP display **Mean Diamonds** along with **Boxplots**. The Mean Diamonds show the group mean, the centerline, and a 95% confidence interval, the height of the diamond. The lines above and below the centerline represent overlap marks that can be used to compare the groups when the sample sizes of the two groups are equal.

Figure 5.26 t-Test Output for Comparing Mass Spectrometers Means

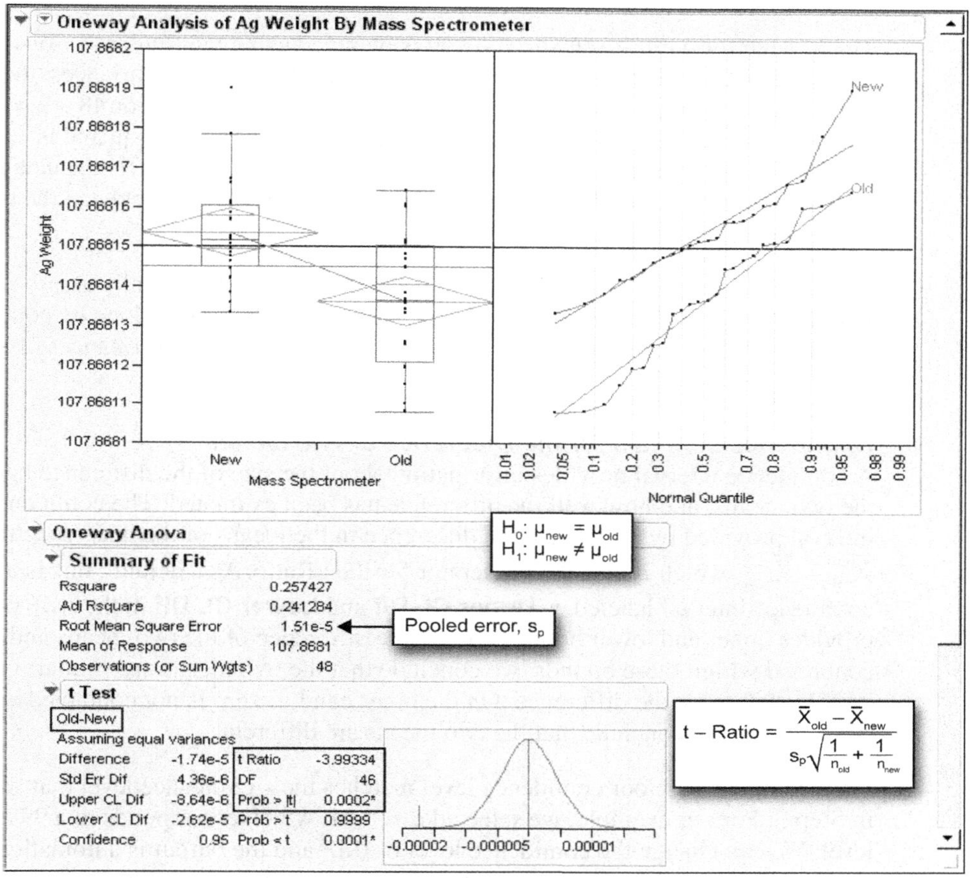

The test statistic that we need can be found next to the label **t Ratio** and is equal to −3.99. Note that the equation for calculating this test statistic is shown in Figure 5.26 and is easily derived from the JMP output by taking the ratio of the statistic next to the label **Difference** and the **Std Err Dif**, which is equal to -1.74E-5 ÷ 4.36E-6 = −3.99.

Remember that the t Ratio is a signal-to-noise ratio so we can say that our signal is close to four times as large as our noise. What is our signal in this context? It is the difference between the two means, $\text{mean}_{\text{Old}} - \text{mean}_{\text{New}}$. If the two means were very similar their difference would be close to zero—no signal. The sign of the *t*-statistic tells which mean is larger (smaller) than the other. Unless you use value ordering, JMP always takes the difference of the means arranged in descending alphabetical order, **Old–New** in our case. (See Figure 5.26.) The negative sign implies that the sample mean for the new mass spectrometer is larger than the sample mean for the old mass spectrometer, by about

1.74e-5 amu. This was previously observed in the box plots and descriptive statistics shown in this section.

The *p*-value for our test statistic = –3.99 is obtained using the Student's *t*-distribution with 46 degrees of freedom. Since we were able to pool the two variances, the degrees of freedom are simply the total sample size minus 2 (two means), or 48 – 2 = 46. For our two-sided alternative hypothesis, the *p*-value is labeled Prob > |t| and is equal to 0.0002. Because this *p*-value is less than our significance level, 0.01, the data provides *enough* evidence to say that the average atomic weight of silver between the two mass spectrometers is different.

Since a statistical difference has been detected, we can now officially assess it by estimating the bias between the mass spectrometers. This can be done by computing a confidence interval for the difference between the average performances of the two spectrometers.

Confidence Intervals for the Difference of Two Means

A confidence interval provides information about the size of the difference between the two means, and how well the difference has been estimated. The confidence interval provided by JMP is for the difference in the means of the mass spectrometers, $\overline{X}_{Old} - \overline{X}_{New}$, which is also the numerator for the t Ratio. As a default, this is a 95% confidence interval labeled as **Upper CL Dif** and **Lower CL Dif** in the JMP output. This provides upper and lower bounds around the difference of the two means and, if zero is contained within these bounds, we conclude that the two means are similar; i.e., zero is a possible value for the difference. On the other hand, if zero is not contained within these bounds, then we conclude that the two means are different.

We must make sure our confidence level matches the significance level that we specified in Step 3. For our example, we selected $\alpha = 0.01$, which corresponds to a 99% confidence level. We can change the confidence level in JMP and the output is automatically updated to reflect our selection. We simply click the small red triangle in the banner of the window and select **Set α Level** and choose .01 from the drop-down menu, as is shown in Figure 5.27.

Figure 5.27 Changing α Levels to Achieve Desired Confidence Level

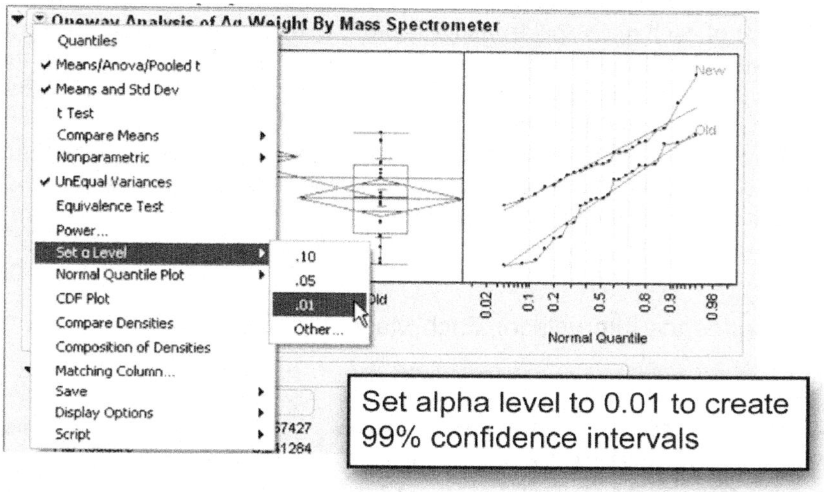

The JMP output, shown in Figure 5.28, is updated to reflect our new choice for alpha. The confidence interval for the difference of -1.74e-5 is (-0.00003; -5.7e-6), which does not contain zero. In other words, a bias of 1.74E-5 amu exists between the old and new mass spectrometers, with the new mass spectrometer giving higher readings. Since the two groups have equal sample sizes, we can also use the overlap marks of the **Mean Diamonds** to compare the groups. Since the overlap marks for the old and new diamonds do not overlap (Figure 5.26) we can say, with 99% confidence, that the two means are different.

Figure 5.28 t-Test Confidence Interval for the Difference

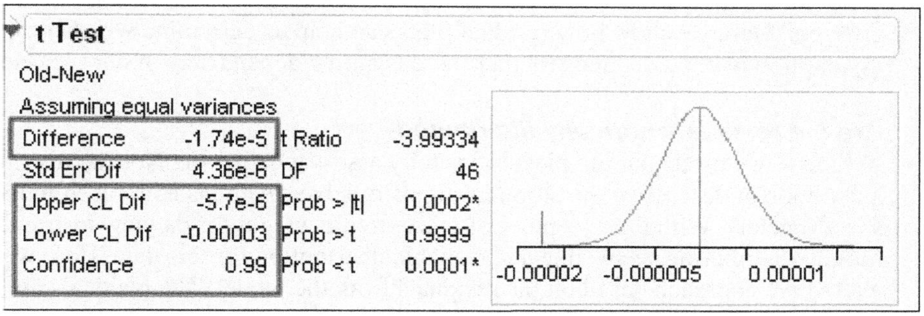

Confidence Intervals for Each Mean

Recall from Step 2 that we are also interested in knowing whether the average performance of each mass spectrometer is close to the reference standard for the atomic weight of silver of 107.86815 amu. Using the confidence intervals for each mean shown in Figure 5.29 we can quickly check this. In order to test the hypotheses stated above, we need to see if these intervals contain the reference value 107.86815. If they contain this value, then we can conclude that the mass spectrometer, on average, is not different from the reference value. On the other hand, if the intervals do not contain this value, then we conclude that, on average, the mass spectrometer is biased and not capable of meeting the reference value.

Figure 5.29 99% Confidence Intervals for Each Mass Spectrometer Mean

Means and Std Deviations

Level	Number	Mean	Std Dev	Std Err Mean	Lower 99%	Upper 99%
New	24	107.8681538	0.000013	2.6665e-6	107.8681463	107.8681613
Old	24	107.8681364	0.000017	3.45e-6	107.8681267	107.8681460

The 99% confidence interval for the old mass spectrometer is (107.8681267, 107.8681460), and since this does not contain the reference value 107.86815, we reject H_0 in favor of H_1. In other words, a bias with respect to the reference value, of about of 0.000136 amu, exists in the old mass spectrometer. For the new mass spectrometer, the 99% confidence interval (107.8681463, 107.868163) contains the reference value 107.86815 providing little evidence to say that a bias exists for this instrument.

Residuals Checks

The validity of our analysis results depends on the three assumptions of independence, normality, and homogeneity of the two mass spectrometers variances. We have already seen that the data does not provide enough evidence to say that the variances are different. Here we show how residual plots can help us determine whether the normality assumption is satisfied, and whether there is a time dependency in our measurements.

Are the residuals normally distributed?

In Step 4, we used quantile plots to visually assess whether the Normal distribution does a good job at describing the atomic weights of silver readings coming from the two mass spectrometers. With the exception of one extreme weight for the new instrument, the normal distribution seems like a reasonable distribution for this data. Here we provide an alternative approach based on the residuals from the model. The residuals are calculated by subtracting from each response value its respective group mean. They represent what is left in our data after we extract the signal, and therefore should behave like white noise.

The residuals can be saved from the **Fit Y by X** platform to the JMP data table by selecting **Save > Save Residuals** from the drop-down menu shown in Figure 5.30. This adds a new column to the end of the JMP table labeled **Ag Weight centered by Mass Spectrometer**.

Figure 5.30 Saving Residuals in Fit Y by X Platform

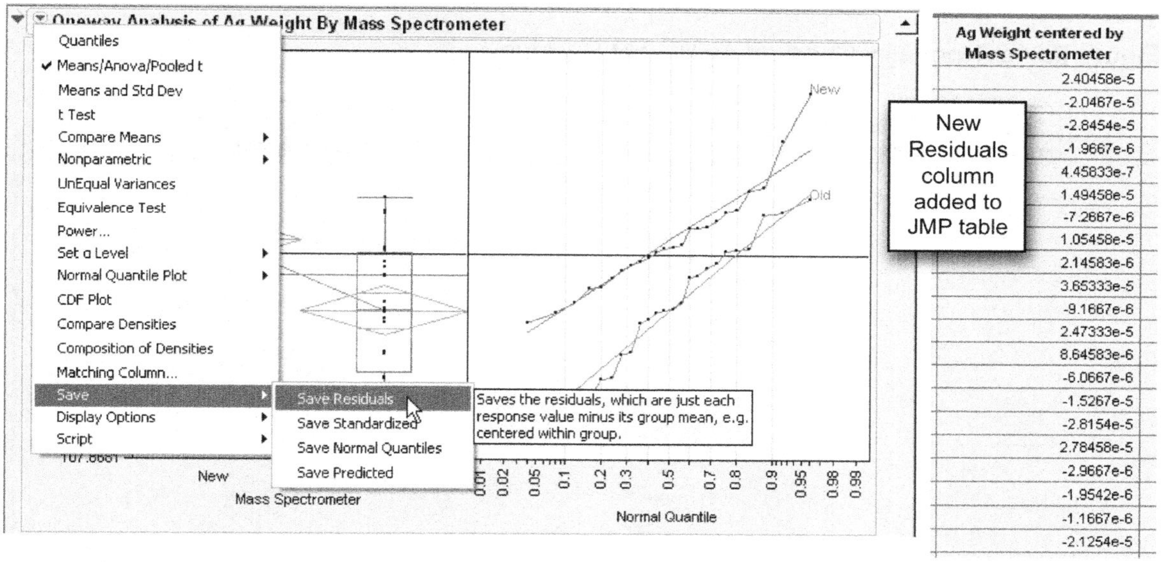

We next select **Analyze > Distribution** from the main menu, place the new column in the **Y, Columns** box, and click **OK**. This produces a histogram of the residuals. If we have not previously changed the default options for the Distribution platform, then only a histogram of the data, along with the quantiles and some summary statistics, is displayed.

To check the normality assumption we use a normal quantile plot of the residuals. If all the points fall within the confidence bands (dotted lines), then we can assume that the normality assumption holds. We do this by clicking on the red triangle next to the response name **Ag Weight centered by Mass Spectrometer**, and selecting **Normal Quantile Plot** from the drop-down menu. All the points are right on top of the centerline of the plot (Figure 5.31a), which indicates that the normal distribution does a very good job at describing the silver weight data. We can also fit a normal distribution to the data by selecting **Fit Distribution > Normal**, again by clicking on the red triangle next to the response name **Ag Weight centered by Mass Spectrometer**. A normality test can be performed if you select **Goodness of Fit** from the drop-down menu (red triangle) next to the **Fitted Normal** name, shown also in Figure 5.31a. The p-value for the Shapiro-

Wilks W test is 0.8764 > 0.05, which indicates that we do not have enough evidence to say that the Normal distribution is not appropriate for our data.

Of these two approaches for verifying the normality assumption, we prefer to use the normal probability plot because it is a straightforward way of visually checking for normality. Even the histogram can sometimes mislead us, as you will see in the case study of Chapter 6.

Figure 5.31a Normality Check of Ag Weight Residuals

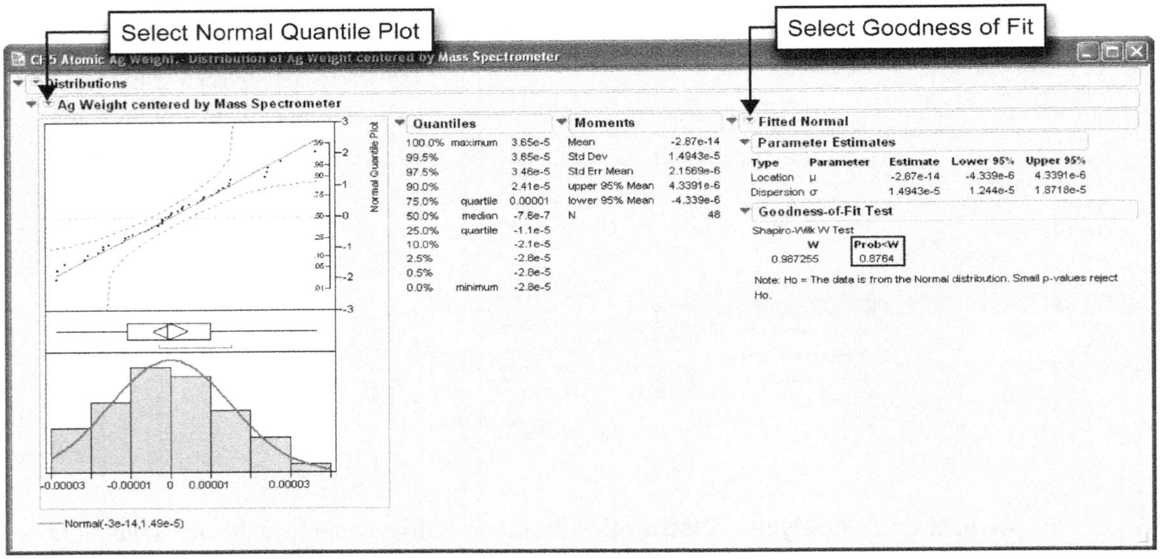

Are the residuals related to time order?

Another useful residuals plot is a plot of the residuals versus time order. This plot helps us determine whether there is some type of time dependency in our observations. In this study the Randomization provides the time order. If we plot the residuals versus the calibration mix, we can detect systematic drifts that might have occurred during the study. Figure 5.31b shows a plot of the Ag Weight residuals versus row order, generated using **Graph > Overlay Plot** by leaving the **X** field blank. The plot has been enhanced by connecting the points, adding a Y-axis reference line at 0, and by using different plotting symbols and colors for the two mass spectrometers. No apparent trend or drift is shown in this plot.

Figure 5.31b Ag Weight Residuals Versus Time Order

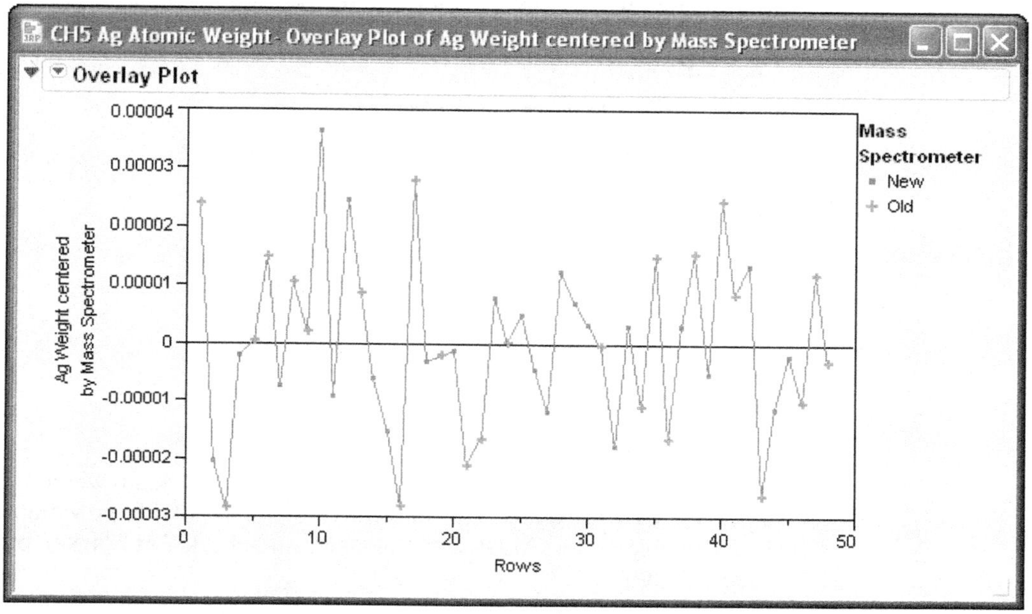

The enhancements to Figure 5.31b can be done as follows:

1. To connect the points click the red triangle next to **Overlay Plot**, and select **Connect Thru Missing** from the drop-down menu.

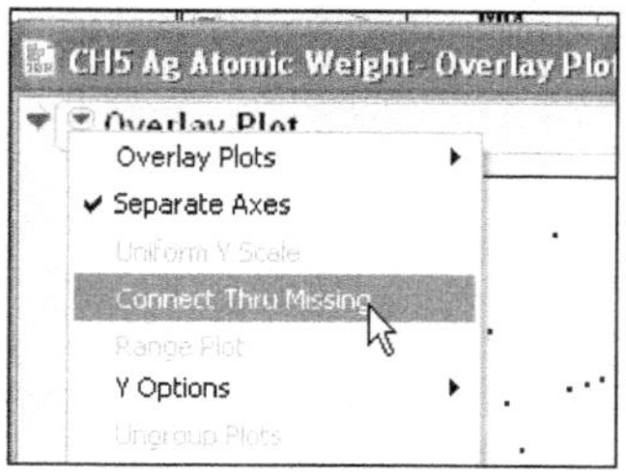

2. To add a Y-axis reference line at 0, double-click the Y-axis to bring up the Y Axis specification window. Now enter `0` in the box next to the left of the **Add Ref Line** button, and click this button to include this value in the box to the left. Click **OK.**

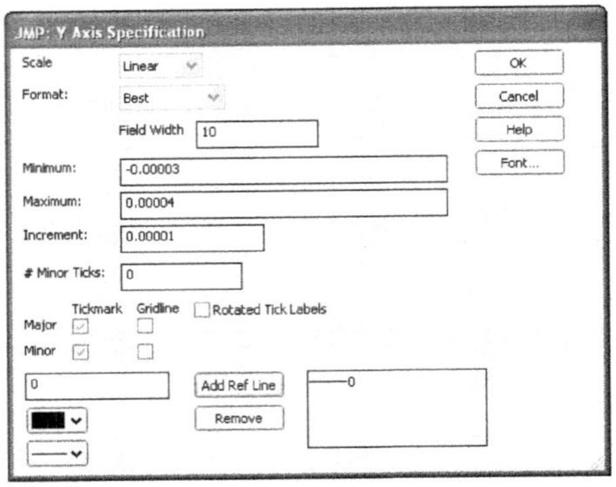

3. To use different plotting symbols and colors, right-click anywhere within the plot and select **Row Legend.** This launches a dialog box with the variable names in the JMP table. Select Mass Spectrometer and check **Set Marker by Value**. Then click **OK**.

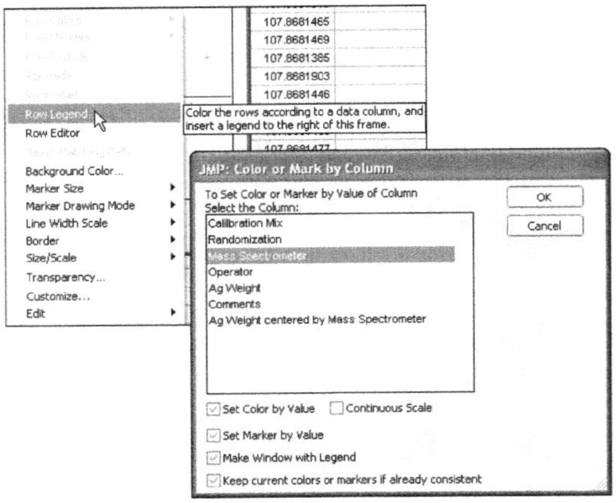

4. To enlarge the marker size, once again, right-click within the plot and select **Marker Size** and **Medium**.

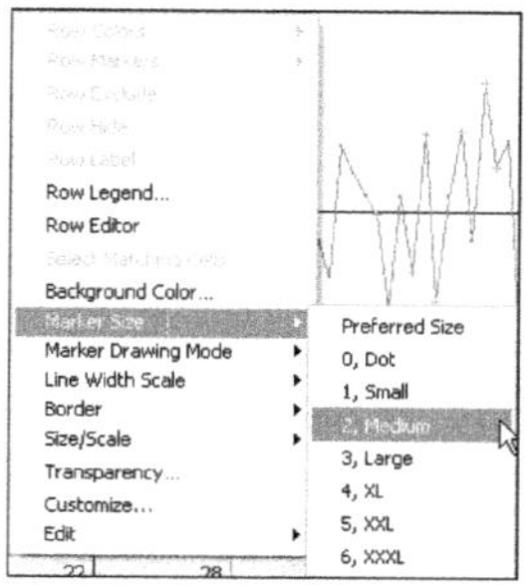

Step 6: Summarize your results with key graphs and summary statistics

In this step we review the results and output of the analyses performed in Step 5. The goal is to select the key graphs and tables that facilitate the presentation of what we learned and the discoveries we made. They should help answer our uncertainties, give us insight into the differences between the two mass spectrometers, enable us to make claims about the future performance of the spectrometers, and quantify the uncertainty in our results. We also introduce two other graphs that help the presentation of results.

Box Plots for Comparing the Performance of the Two Mass Spectrometers

As we discussed in Chapter 2, box plots are a great tool for visually comparing data coming from two or more populations. Figure 5.32a is a condensed version of the output from **Means/Anova/Pooled t** (Figures 5.26, 5.28, 5.29), which contains one key graph and all the important results. The top part of Figure 5.32a shows box plots for each of the two mass spectrometers as well as a reference line at 107.8681, the overall mean of the data. From this graph we can see that the old mass spectrometer gives Ag weight determinations that are lower than those coming from the new mass spectrometer.

Figure 5.32a Ag Weight Summary Results

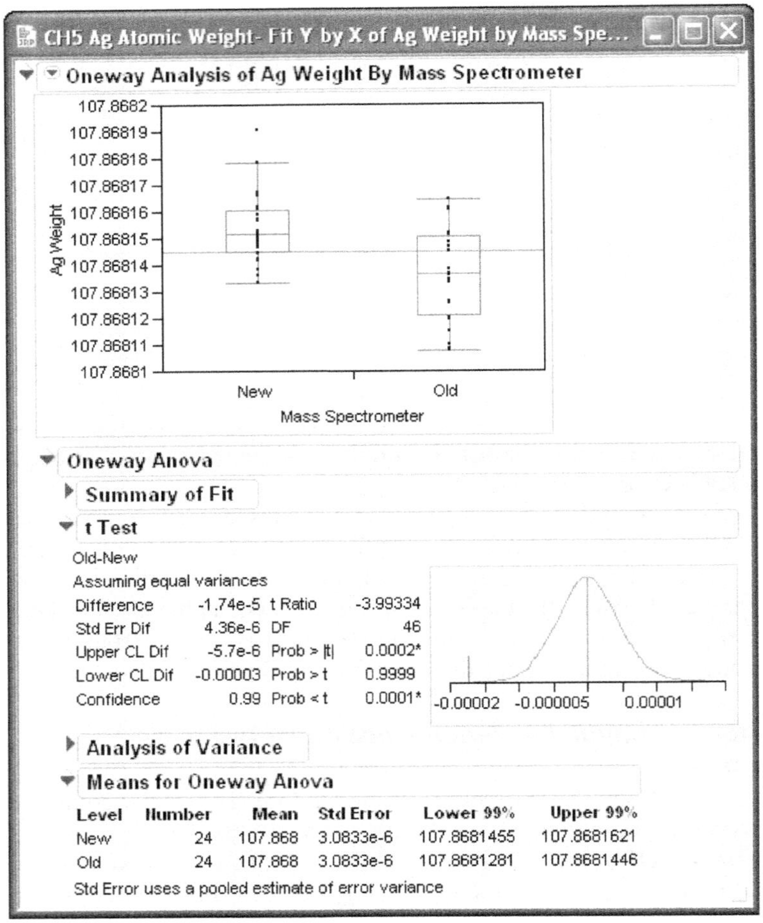

Confidence Interval for the Estimated Difference

The middle section of Figure 5.32a shows the results of the *t*-test in terms of the difference between the **Old–New** mass spectrometers, as well as a 99% confidence interval for the difference. These results show that the following is true:

- The estimated, statistically significant, difference is −1.74E-5 amu.
- At the 99% confidence level the confidence interval is (−3E-5 ; −5.74E−6). This shows that the data supports the claim that the drop in Ag weight when using the old spectrometer can be as large as 3E-5 amu. This information could be important if we need to make claims about the performance of old the

spectrometer in terms of how much worse its performance can be as compared to the new.

- In the previous bullet, we noted that the drop in Ag Weight can be as large as 3E-5 amu. What about the drop with respect to the reference value? As compared to the reference value of 107.86815 amu, the parts per million (ppm) change in the atomic weight of silver (when going from the old to the new spectrometer) is (–3E-5/107.86815 × 1E6 ; –5.7E–6 /107.86815 × 1E6) = (–0.28 ppm ; –0.05 ppm).
- The width of the 99% confidence interval (–5.74E–6 – –3E–5 = 2.43E-5) is about 1.4 times the size of the estimated difference 1.74E-5. The width of the interval gives us a sense for the uncertainty of our estimated difference, with narrow intervals giving more precise estimates.
- The estimated **Std Err Dif** can be used with the **Sample Size and Power** calculator to determine how many samples we would need to improve the precision in the estimate of the difference.

Confidence Intervals for Each Spectrometer Average Ag Weight

The last section of Figure 5.32a shows the 99% confidence intervals for the individual Ag weight averages. We can see that the reference value of 107.86815 is inside the new spectrometer interval, while it is outside the interval for the old spectrometer. This provides a quick and visual way of demonstrating that the old spectrometer has a bias while the new does not.

Prediction Intervals for the Future Performance of the New Spectrometer

The results have shown that the new spectrometer is unbiased, with respect to the reference value, and repeatable. What claim can we make about its future performance? Let's say that we are interested in knowing what the average atomic weight of silver will be if we measure 10 samples in the new spectrometer. Prediction intervals for the mean and standard deviation of the next 10 future samples are what we need.

To generate predictions intervals, select **Analyze > Distribution** to first generate histograms of Ag Weight **by** Mass Spectrometer (see Figure 5.32b.) In the resulting output click the right triangle next to **Ag Weight** for the histogram corresponding to the new spectrometer. Select **Prediction Interval** from the drop-down menu and enter 10 in the **Enter number of future samples** box. A section of the output is shown in Figure 5.32c.

Figure 5.32b Generating 95% Prediction Intervals by Spectrometer

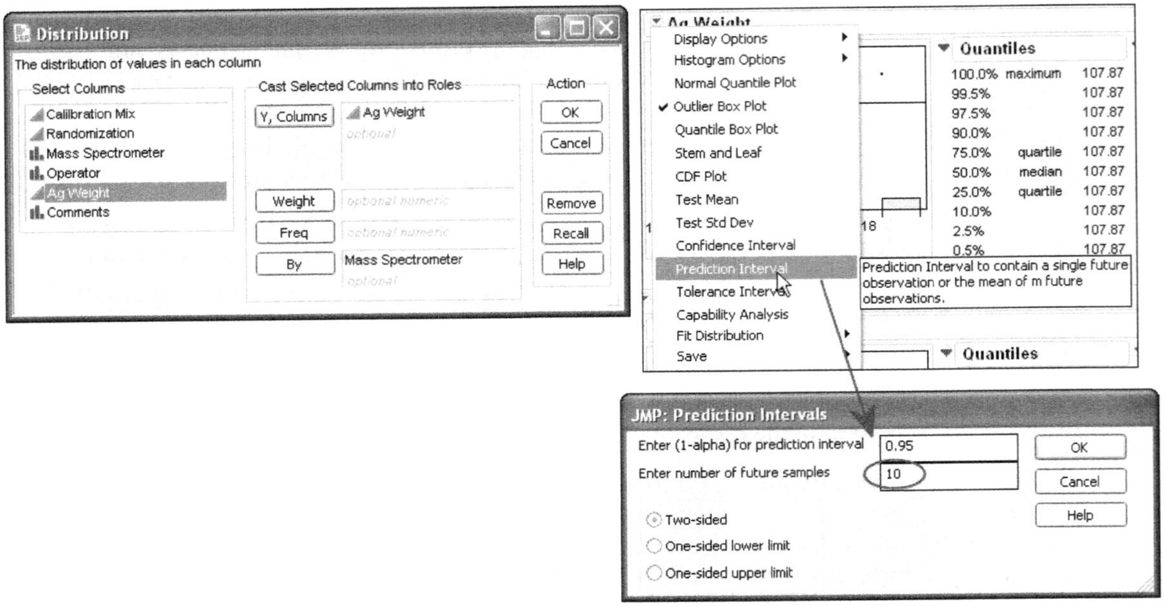

The prediction intervals (as shown in Figure 5.32c) show that we should, with 95% confidence, expect the average Ag atomic weight of 10 future calibration mixes to be between 107.8681436 and 107.8681639, and the standard deviation to be between 0.000006860 and 0.000021589, when weighted with the new spectrometer. This is valuable information since our customers require that the atomic weight of silver be measured with very low uncertainty.

Figure 5.32c 95% Prediction Intervals for the New Spectrometer

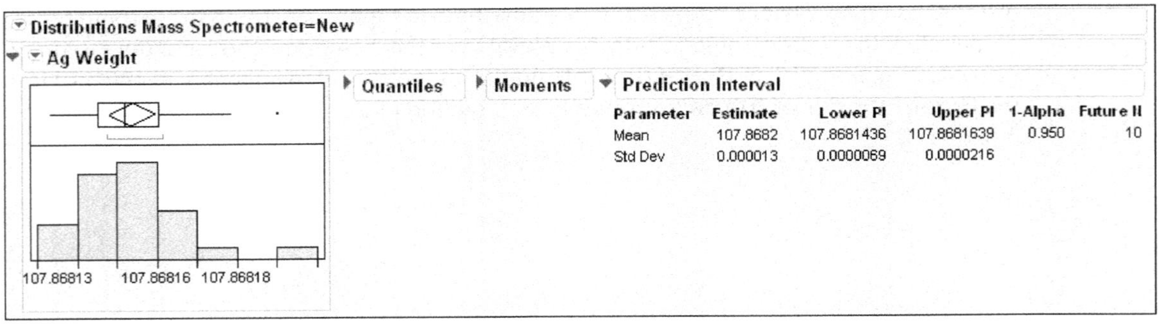

Individual Measurements Control Chart with Phase Variable

A good way to compare the results of the two mass spectrometers side by side is by means of an individuals control chart with a Phase variable. In order to create this chart we first need to sort our data by the Mass Spectrometer and the Calibration Mix. Select **Tables > Sort** from the main menu, select Mass Spectrometer, and then click **By**, then select Calibration Mix and click **By**. Figure 5.33 shows the data sorted by Mass Spectrometer and Calibration Mix. After the data has been sorted, do you notice anything unusual about it? It now appears that Joe ran all of the calibration mixes with the new mass spectrometer while Ann ran all of the calibration mixes with the old mass spectrometer. We shall revisit this observation in Step 7.

Figure 5.33 JMP Table Sorted by Mass Spectrometer and Calibration Mix

Sorted CH5 Ag Weight

Sorted CH5 Ag Weight
Source

Columns (6/0)
Calilbration Mix
Randomization
Mass Spectrometer
Operator
Ag Weight
Comments

Rows
All rows 48
Selected 0
Excluded 0
Hidden 0
Labelled 0

	Calilbration Mix	Randomization	Mass Spectrometer	Operator	Ag Weight	Comments
1	1	0.7383483	New	Joe	107.8681568	
2	2	0.1013341	New	Joe	107.8681465	
3	3	0.64111233	New	Joe	107.8681572	
4	4	0.30007301	New	Joe	107.8681785	
5	5	0.28825689	New	Joe	107.8681446	
6	6	0.26861059	New	Joe	107.8681903	
7	7	0.44118789	New	Joe	107.8681526	
8	8	0.52128954	New	Joe	107.8681494	
9	9	0.47379164	New	Joe	107.8681616	
10	10	0.48424924	New	Joe	107.8681587	
11	11	0.82668622	New	Joe	107.8681519	
12	12	0.748961	New	Joe	107.8681486	
13	13	0.60319225	New	Joe	107.8681419	
14	14	0.67006953	New	Joe	107.8681569	
15	15	0.42085807	New	Joe	107.8681508	
16	16	0.77808219	New	Joe	107.8681672	
17	17	0.36879486	New	Joe	107.8681385	
18	18	0.06289465	New	Joe	107.8681518	
19	19	0.63147828	New	Joe	107.8681662	
20	20	0.79547272	New	Joe	107.8681424	
21	21	0.65322803	New	Joe	107.8681360	
22	22	0.02507942	New	Joe	107.8681333	
23	23	0.63604165	New	Joe	107.8681610	
24	24	0.3592237	New	Joe	107.8681477	
25	25	0.0556518	Old	Ann	107.8681079	
26	26	0.42658282	Old	Ann	107.8681344	
27	27	0.09377345	Old	Ann	107.8681513	
28	28	0.4587308	Old	Ann	107.8681197	
29	29	0.0016095	Old	Ann	107.8681604	
30	30	0.25508308	Old	Ann	107.8681385	
31	31	0.41512423	Old	Ann	107.8681642	
32	32	0.48042018	Old	Ann	107.8681365	
33	33	0.45576197	Old	Ann	107.8681151	

Next select **Graph > Control Chart > IR** to bring up the **IR Control Chart** dialog box. Now select Ag Weight as the **Process** variable, Calibration Mix as the **Sample Label**, and Mass Spectrometer as the **Phase** variable. By entering a variable in the **Phase** field, separate control limits will be calculated for each value of the variable **Mass Spectrometer**. We also want to see how well each mass spectrometer meets the standard reference value of 107.86815. Once the chart is generated, we do this by double-clicking on the Y-axis and entering the reference value of `107.86815` in the box to the left of **Add Ref Line** button, and then clicking on this button to include this value in the box to the right, and then clicking **OK**.

> **JMP Note 5.4**: The data needs to be sorted by the **Phase** variable before generating the phase control chart.

Figure 5.34 shows the resulting output where we can quickly see from the individual measurements (top) chart that Ag weight numbers determined by using the old mass spectrometer are lower than those determined by using the new mass spectrometer. The reference line at the reference value of 107.86815 also shows that the old mass spectrometer is biased toward lower numbers, while the bias of the new mass spectrometer is small. The moving range (bottom) chart shows that the data spread from the two mass spectrometers is similar, but the control limits are wider for the OLD spectrometer.

Figure 5.34 IR Phase Chart for Ag Weight with Reference Line @107.86815

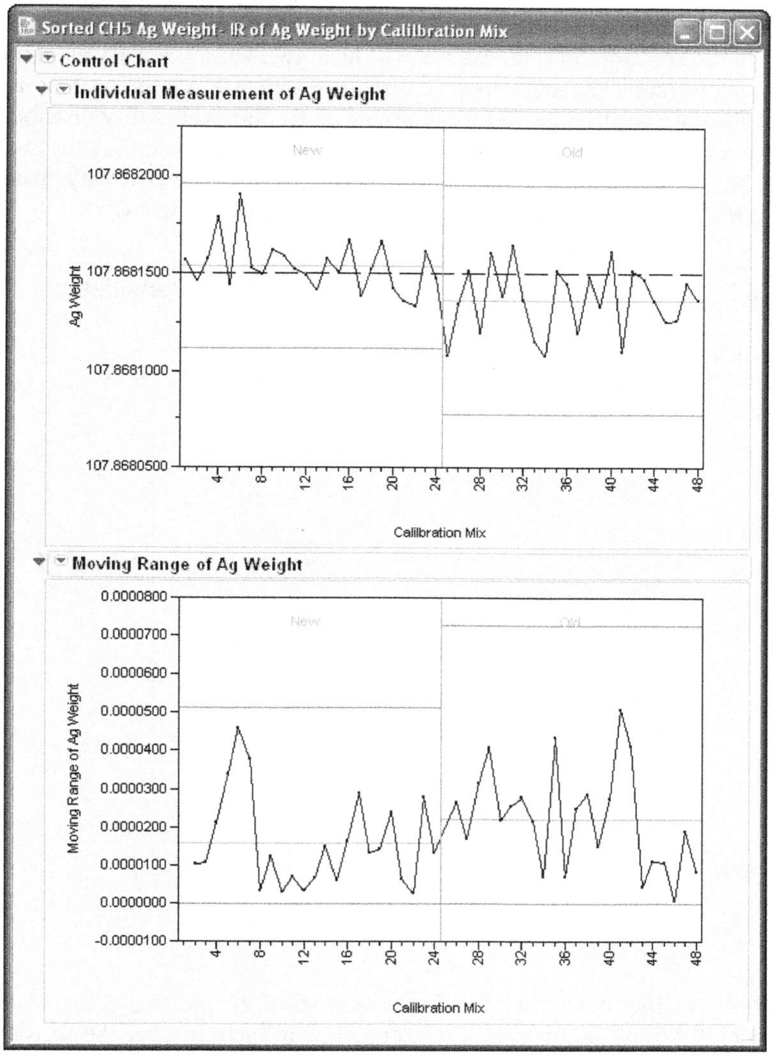

Comparing the Two Mass Spectrometers Densities

One nice tool within the **Fit Y by X** platform is the **Compare Densities** option. This option generates a plot with kernel density estimates for each of the levels of the factor defined by the X variable. In our case we will have two kernel density estimates—one for each mass spectrometer. Kernel density estimators approximate the probability density function of the data without having to assume a particular probability distribution. This gives us a quick visual way to compare the two mass spectrometers. After the **Fit Y by X** platform has been launched, click the red triangle next to the **Oneway Analysis Ag Weight by Mass Spectrometer,** and select **Compare Densities.**

Figure 5.35. Comparing the Two Mass Spectrometers Densities

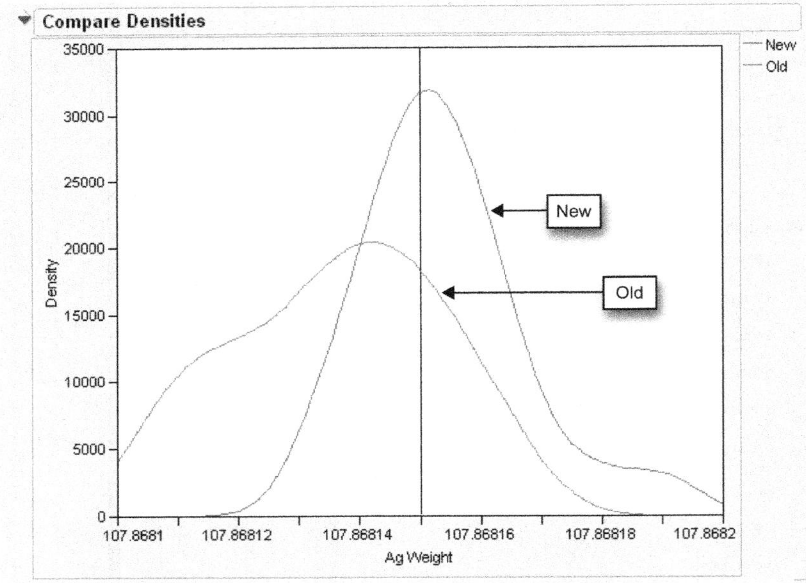

Figure 5.35 shows the two kernel density estimators of the old and new mass spectrometers densities, along with a reference line at the reference Ag weight of 107.86815. We can see, as we saw in Figure 5.34, how close the new mass spectrometer (red line) is to the reference value. However, in this plot it is easier to see how close its density is around this value. The peak of the old mass spectrometer density is to the left of the reference value, reflecting the bias in the instrument.

> **JMP 8 Note 5.1:** In JMP 8 the **Compare Densities** option is now found within the **Densities** menu of the **Fit Y by X** platform: **Fit Y by X** > **Densities**> **Compare Densities.**

Step 7: Interpret the results and make recommendations

After the analysis is done, it is very easy to get lost in the statistical output or caught up in the statistical jargon of significance and *p*-values. That is why in Step 6 we try to select just the key graphs and tables that can contain the information we need to report our findings. After that we need to translate our statistical findings according to the context of the situation at hand. We ran this study to compare a newly purchased mass spectrometer to the current one, and were interested in three main questions:

1. Do the new and old mass spectrometers produce similar results in terms of the atomic weight of silver?
2. Are the mass spectrometers biased when compared with a standard reference material?
3. Do the new and old mass spectrometers have similar precision?

Based on our analysis we found that, statistically, the average performance of the new mass spectrometer is higher than that of the old mass spectrometer by about 1.74E-5 amu. Is this difference also practically significant? The mass spectrometric analytical error is 1.37E-5 amu, which indicates that the difference between the spectrometers is also practically significant. We also discovered that the old mass spectrometer was, on average, producing results that were lower than the standard reference material. For the new mass spectrometer, no statistically significant difference was detected between its average performance and the standard reference value. The standard deviations were found to be similar between the two mass spectrometers, and can be pooled to provide an overall estimate of the calibration mixes variation of about 1.51E-5 amu.

We conclude the following:

1. The new and old mass spectrometers produce different results in terms of the atomic weight of silver by about 1.74E-5 amu.
2. The old mass spectrometer produces lower atomic weight results when compared with the standard reference material.
3. The new and old mass spectrometers have similar precision, at about 1.51E-5 amu.
4. In addition, the average of 10 future calibration mixes measured on the new mass spectrometers is expected, with a 95% confidence, to be between 107.8681436 and 107.8681639.

With respect to statements 1 and 2, we need to determine what is different between the two mass spectrometers in order to help explain the bias between them, and the bias between the old mass spectrometer and the standard reference material. As we saw in Step 6, one interesting observation was that Joe ran all the calibration mixes on the new mass spectrometer, while Ann ran all the calibration mixes on the old mass spectrometer (Figure 5.33). In other words, differences between operators are indistinguishable from differences between the spectrometers, so it is impossible to statistically determine whether the bias between the spectrometers is due to the spectrometers themselves, or it is due to the difference in the two operators.

In conclusion, the new mass spectrometer appears to be working well and should be brought online. The old mass spectrometer needs to be further investigated for its apparent calibration problems. In order to dismiss an operator effect, Joe should run some of the calibration mixes using the old mass spectrometer to see if the bias is due to the spectrometer or the operator.

5.5 Testing Equivalence of Two Materials, Processes, or Products

In this section we will provide a brief overview of how we can conduct a test of equivalence in order to prove that the difference in average performance of the two mass spectrometers is equal within a given bound. The two-sample significance test presented in this chapter assumes that the null hypothesis is true and looks for evidence to suggest otherwise. But what if we want to "prove" that the two means are the same? Although we cannot prove that the means are equal, we can prove that they are *equivalent* up to a certain value. This test is called an equivalence test.

In JMP, an equivalence test between two means is done using the **Equivalence Test** option in the **Fit Y by X** platform as is shown in Figure 5.36. When this option is selected, a dialog box is displayed where we need to enter a value that represents the smallest practical difference between the two means (equivalence bound). That is, if the difference in the two means does not exceed this value, then we would consider them to be equivalent.

Figure 5.36 Test of Equivalence to Compare Two Means

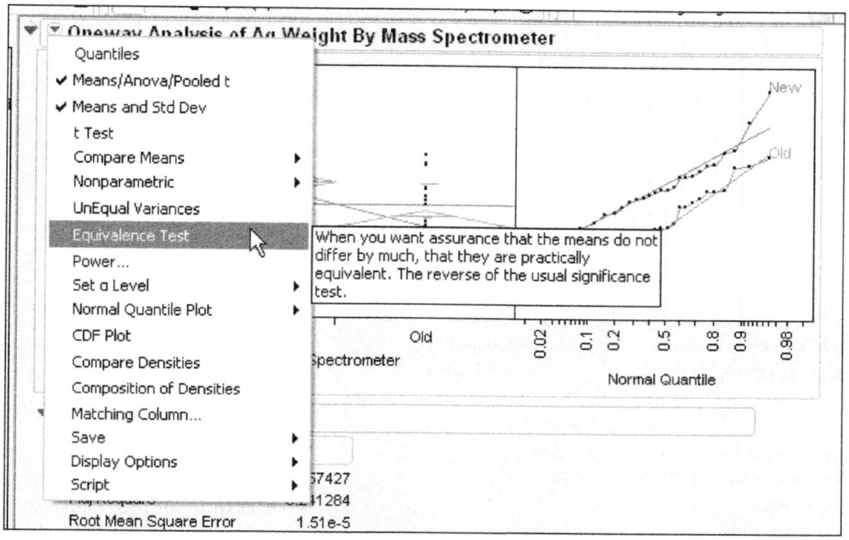

An equivalence test is performed by using two one-sided *t*-tests, and is commonly referred to as the TOST approach. Let's select 0.00001 amu to be the equivalence bound. For this example, we want to know if $|\mu_{New} - \mu_{Old}| < 0.00001$, or equivalently, $-0.00001 < (\mu_{New} - \mu_{Old}) < 0.00001$. The two one-sided hypotheses that we need to carry out are as follows:

a) H_0: $\mu_{New} - \mu_{Old} \leq -0.00001$
H_1: $\mu_{New} - \mu_{Old} > -0.00001$

b) H_0 $\mu_{New} - \mu_{Old} \geq 0.00001$
H_1: $\mu_{New} - \mu_{Old} < 0.00001$

The output for the TOST is shown in Figure 5.37. The *t*-statistics for the hypotheses shown above are both derived using the estimated difference in the two means, the equivalence bound, and the standard error of the difference. The test statistic for the first hypothesis (a) can be found in the JMP output next to the label **Lower Threshold** and is -1.699 = (-0.00001741 + 0.00001) / 0.0000044. Similarly, the test statistic for the hypothesis shown in (b) is labeled the **Upper Threshold** and is = -6.287 = (-0.00001741 – 0.00001) / 0.0000044.

Figure 5.37 Dialog Box and Output for Test of Equivalence with Equivalence Bound of 1E-5

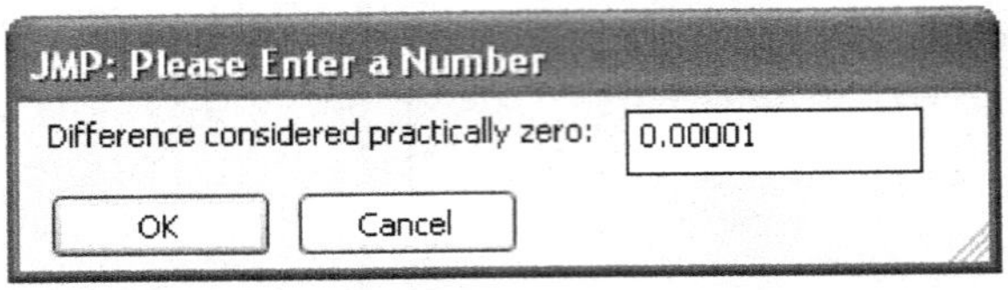

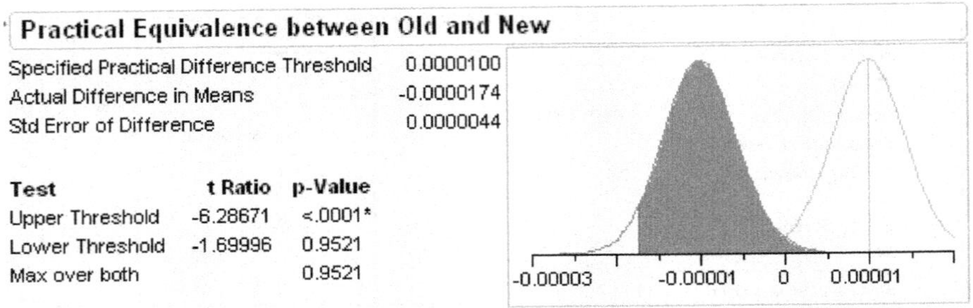

The *p*-values for both test statistics are obtained from a Student's *t*-distribution with 46 degrees of freedom. The *p*-values for (a) and (b) are 0.9521 and < 0.0001, respectively. The JMP output also selects the maximum *p*-value and displays it next to the label **Max over both**. These values suggest that we were not able to reject the null hypothesis in (a), but were able to reject the null hypothesis for (b). In order to prove equivalence within the specified value, we must reject both null hypotheses. Since the estimated difference between the means is -0.0000174 is actually less than -0.00001, we should not be surprised that we could not prove that it was greater than -0.00001.

What would happen if we used a larger value for the equivalence bound? Figure 5.38 shows the JMP output for 0.0001; i.e., we will assume that the two mass spectrometers are equivalent if their difference is to the nearest 4th decimal place. The results show that the mass spectrometers are equivalent to 1E-4 amu. However, since the mass spectrometric analytical error is 1.37E-5, this is not a reasonable equivalence bound and was shown only for illustrative purposes.

Figure 5.38 Test of Equivalence for Equivalence Bound of 1E-4

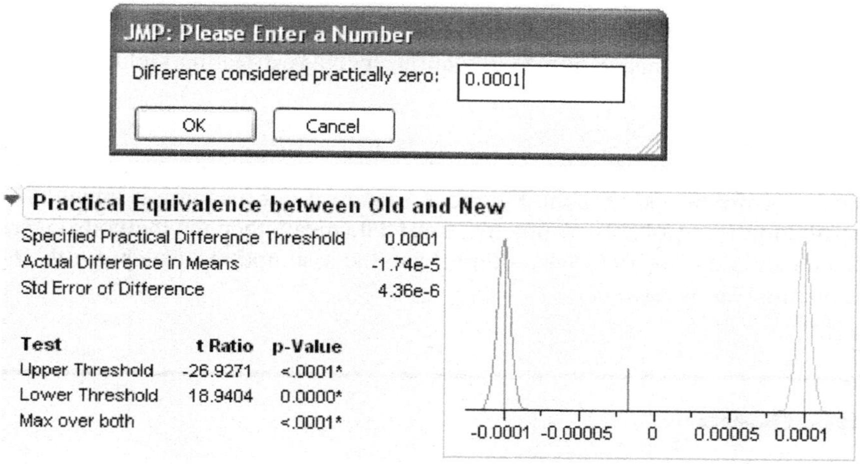

5.6 Summary

In this chapter we learned how to use significance tests to provide insight into how the average performance, and performance variation, of two products, processes, or materials compare with each other. We showed how to set up a comparative study and provided step-by-step instructions for how to use JMP to perform sound statistical analysis and make data-driven decisions. We also illustrated two additional analyses, which included how to analyze matched pairs of experimental units, and how to prove equivalence, up to a given bound, of the average performance of two populations. Some of the key concepts and practices that we should remember from this and previous chapters are shown below.

- Be aware of additional factors that were not accounted for in the study but could impact the results, such as confounding two factors, e.g., Operator and Mass Spectrometer.
- Include tests for comparing both the mean and the variation of the two populations.
- Do not forget to clearly define the population of interest, the experimental unit, and how the study needs to be executed so as to conduct the appropriate analysis, for example, matched pairs versus two-sample *t*-test.
- Use the scientific method to make sure that the statistics are helping us support or refute our hypothesis.

- *Always translate the statistical results back in the context of the original questions.*
- Use probability models in place of sampling statistics to get better estimates of the performance of the population, such as estimating yield loss, and prediction intervals.

We made reference to the assumptions for a one-sample significance test and provided some tools for checking them. For the most part, if our response is quantitative and its distribution is reasonably symmetric around its mean, then the methods presented in this chapter should be robust to departures from the assumptions, except for the *F*-test for comparing variances.

5.7 References

Brown, T.L., H.E. LeMay, and B.E. Bursten. 1994. *Chemistry – The Central Science.* 6th ed. New Jersey: Prentice Hall.

Powell, L.J., T.J. Murphy, and J.W. Gramlich. 1982. "The Absolute Isotopic Abundance and Atomic Weight of a Reference Sample of Silver." *Journal of Research National Bureau of Standards (U.S.)* 87: 9-19.

Chapter 6

Comparing the Measured Performance of Several Materials Processes, or Products

Chapter Goals

- Become familiar with the terminology and concepts needed to compare average performance and performance variation of several materials, products, or processes.
- Use the **Fit Y by X** platform in JMP to compare average performance and performance variation of three or more populations.
- Learn how to determine, statistically, if any of the populations are different from each other using analysis of variance (ANOVA).
- Use multiple comparisons to identify significant and important differences between pairs of averages.
- Translate JMP statistical output in the context of a problem statement.
- Present the findings with the appropriate graphs and summary statistics.

6.1 Problem Description

What do we know about it?

A local residential developer has been building a housing development of 100 semi-custom homes with basements, each built on about one-acre lots. There have been four phases of development and shortly after the fourth and last phase of the development was completed, the developer started to receive a number of complaints for premature cracking in the cement sidewalks leading from the driveways to the front doors. Upon further investigation, it was found that most of the houses completed in the third phase of development exhibited long longitudinal cracks at the top of their sidewalks.

The developer promptly contacted your engineering firm, a company that specializes in residential and commercial architectural designs and land surveying, to help determine the extent of the problem, the likely causes for the cracking, and the potential solutions to repair the damage. As a civil engineer working for this company you were put in charge of this project. After reviewing the design specifications for the sidewalks you concluded that the cement specifications were correctly identified for this type of application. Also, the developer's early records for the first phase indicated that the correct cement was ordered from the subcontractor who made the sidewalks.

What is the question or uncertainty to be answered?

Based on prior experience, you know that there is a potential for large variations in the cement properties, such as compressive strength and tensile strength, even for a specified grade such as the one used in the homes. Since the development was completed in several phases, it is possible that different cement suppliers, or cements with different characteristics, were used to complete the sidewalks. If the properties are significantly different, this could be the cause for the premature cracking in the sidewalks. In fact, further investigation revealed that four different suppliers were used for the four different phases.

How do we get started?

One of the main objectives for this study is to determine whether the cracking of the sidewalks was due to cement that did not meet the specified requirements that were needed for the application. There were four phases in the development, each with a different cement supplier, and the cracking showed up in houses that were built in the third phase of development, but not on the other three phases. You would like to check the properties of the cement that were used in each phase to determine whether they meet the specified requirements, and whether they are similar to each other.

In order to get started, we need to address the following areas:

1. Describe the requirements in terms of a statistical framework that enables you to determine, with a certain level of confidence, that the cement properties from the four phases do not differ significantly from each other or the required values.
2. Be able to translate the statistical findings in a way that clearly conveys the message to your client without getting lost in the statistical jargon.

What is the appropriate statistical technique to use?

A key cement property, compressive strength, will be used to compare the cement used in each of the four phases of the development. We will use a *one-way analysis of variance (ANOVA)* for comparing the average performance and performance variation of the four suppliers. In Section 6.3 we review the key concepts that are relevant to one-way ANOVA tests of significance, in Section 6.4 we introduce a step-by-step approach for performing a one-way ANOVA using JMP, and in Section 6.5 we show how to perform equivalence tests for three or more populations.

6.2 Key Questions, Concepts, and Tools

Several statistical concepts and tools that were introduced in Chapter 2 are relevant when comparing the measured performance of several materials, processes, or products. Table 6.1 outlines these concepts and tools, and the Problem Applicability column helps to relate these to the comparison of the four cement suppliers.

Table 6.1 Key Questions, Concepts, and Tools

Key Questions	Key Concepts	Problem Applicability	Key Tools
Is our sample meaningful?	Inference from a random and representative sample to a population	In our problem description, we see that the cracking sidewalks were all constructed during the same timeframe (phase), and most likely using the same cement suppler. We suspect that the cement used to build the other phases of the development have cement properties that differ significantly from the "cracking" phase. We need to make sure that we can obtain cement samples from each of the suppliers that are representative of the cement used during construction.	Sampling schemes and sample size calculator
Are we using the right techniques?	Probability model using the normal distribution	The properties of cement and their test methods are well known in this industry. The compressive strength is an important material property, which can contribute to cracking. Compressive strength measurements are continuous and should be randomly distributed around the target compressive strength of the given cement formulation. The normal distribution should provide an adequate description of the data, but we need to check this assumption.	Normal quantile plots to check distributional assumptions
What risk can we live with?	Decision theory	Some insight into the problem can be gained by studying the cement properties of the different suppliers. However, there are many factors that contribute to cracking, and it can be difficult to definitively isolate the root cause for the premature cracking of the sidewalks. The risk levels for the design and analysis of this study reflect our likelihood of making the wrong claim a small percent of the time.	Type I and Type II errors, statistical power and confidence

(*continued*)

Table 6.1 *(continued)*

Key Questions	Key Concepts	Problem Applicability	Key Tools
Are we answering the right question?	One-way analysis of variance (ANOVA)	We will be evaluating cements from three or four different suppliers. We will be looking for significant deviations in compressive strength, as measured by average and variation in the cement used in Phase 3, as compared with the other phases. This can be accomplished by completing a one-way ANOVA, followed by multiple comparisons, to compare the six pairs of four averages. We can use a Brown-Forsythe test to compare the variances of the different suppliers.	ANOVA, Brown-Forsythe for the variances, and multiple comparison tests
How do we best communicate the results?	Visualizing the results	We will be comparing four suppliers to each other and looking for differences in their average performance and performance variation. This can be visualized very effectively with a box plots and summary statistics.	Box plots and confidence intervals

6.3 Overview of One-way ANOVA

6.3.1 Description and Some Applications

One-way ANOVA, or one-way analysis of variance, is a statistical technique that can be used for comparing the average performance of several materials, processes, or products. The term *one-way* indicates that we are dealing with one factor with multiple levels, or "treatments," that we want to compare. In other words, we would like to know whether the observed differences in our data can be explained by the different treatments. In our example the factor is the development phase, with four levels corresponding to the four phases in the housing development that we want to compare in terms of the compressive strength of the cement.

Analysis of variance? Why do we analyze variances in order to compare means? The term analysis of variance (ANOVA) describes the method we use to compare the averages corresponding to the levels of the one-way factor. The total variance is partitioned into two components: one variance component due to the signal, which is a function of the differences *between* the factor level averages, and one variance component due to the noise, which reflects the combined variation *within* the factor levels. The global test of significance to determine whether the observed differences in response averages are due to the factor levels is the *signal-to-noise ratio* of these two components. Note that since the noise reflects the combined variation of the levels, we need to make sure that the levels variation is similar before we can pool them into a single estimate.

Some examples from science and engineering where we could use a one-way ANOVA approach include the following:

- Determining if there is a difference in the average metal thickness of wafers in four different zones of a deposition tube.
- Determining if there is an operator bias among three different technicians who use the same set of hand calipers to measure the thickness of a soft plastic tube at a quality checkpoint.
- Finding alternate source suppliers for a key raw material that exhibits properties similar to the raw material supplied by the current source supplier.
- Studying the operational equipment effectiveness of five different braiders for electrical conductors in a high-volume manufacturing area.
- Selecting from three proposals the most effective way to set up a die cutting process, in terms of reaching a steady state in the least amount of time and with a minimum number of wasted parts.

For the above studies, we can passively collect data under the three or more conditions (levels of our factor), or we might intentionally alter the levels of the causal factor, hoping to see a signal in our response. In either case, we collect a number of samples representing each level of our factor under study. Measurements taken on these samples are then be used to compare the factor levels average performances and performance variations.

What are the criteria by which we declare three or more items (populations) different from each other? Similar to the one- and two-sample tests of significance test, we look for differences in both the means and variances among the different populations by using appropriate test statistics. However, one distinct difference between a one-way ANOVA and methods discussed in the previous chapters is the claims that we can make following a statistically significant result. When using a two-sample *t*-test, we can conclude that the two population means in our study are different from one another. For a one-way ANOVA, a significant test statistic merely indicates that factor levels are different from each other. If we want to determine which, and how, factor levels are different from each other (for example, which cement suppliers differ from the one used in Phase 3), then we must do further analyses that are referred to as multiple comparisons.

In the following sections we discuss what is involved in an analysis of variance:

- Section 6.3.2 describes how a one-way ANOVA can be used to compare the average performance of three or more populations, outline the assumptions needed for this test statistic, and what to do if they are violated.
- Section 6.3.3 illustrates how to determine whether the ANOVA signal-to-noise ratio test is statistically significant, and which means are different from each other using multiple comparisons.
- Section 6.3.4 explains how to determine whether the population variances for the different factor levels are homogeneous.
- Section 6.3.5 discusses sample size considerations.
- Finally, in Section 6.4, we show how to carry out this type of analysis for comparing the concrete suppliers using JMP.

6.3.2 Comparing Average Performance of Several Materials, Processes, or Products

In the previous chapter, we showed you how to compare the average performance of two materials, processes, or products using a two-sample *t*-test, and the average variation using an *F*-test. For the two-sample scenario we are making a direct comparison between the two population parameters and, in the event of a significant test statistic, we can conclude that the two population parameters are different from each other. In this section we extend these ideas to three or more populations and describe some of the statistical underpinnings of the analysis of variance methodology. To keep things simple we start with an example in which we want to know if there is a difference, in terms of Overall Equipment Efficiency (OEE), between the three molding machines used in the manufacturing floor.

Statistics Note 6.1: The Overall Equipment Efficiency (OEE) is a metric used to assess the performance of production equipment. The OEE is defined as the ratio of the Actual Output/Theoretical Maximum Output, or the product of the Availability Ratio x Performance Ratio x Quality Ratio, and as such is a number between 0 and 100. As we discussed in Chapter 2, even though the OEE is a percentage it possesses a meaningful zero; i.e., an equipment with OEE=0 produces no output. Therefore, OEE can be treated as belonging to a ratio scale and an analysis of variance is appropriate.

In Figure 6.1 the OEE for each machine is represented by a normal curve (normality is an ANOVA assumption that we discuss later). The population representing the overall equipment efficiency from Machine 1 (M1) can be described by a normal distribution with mean μ_1, the population representing the overall equipment efficiency from Machine 2 (M2) can be described by a normal distribution mean μ_2, and the population representing the operational effectiveness from Machine 3 (M3) is described by a normal distribution with mean μ_3. Random and representative samples are taken from machines M1, M2, and M3, the OEE calculated for each machine, and a one-way ANOVA is performed to determine whether there is a statistically significant difference between μ_1, μ_2, and μ_3.

Figure 6.1 Three Populations Defined by Machines M1, M2, and M3

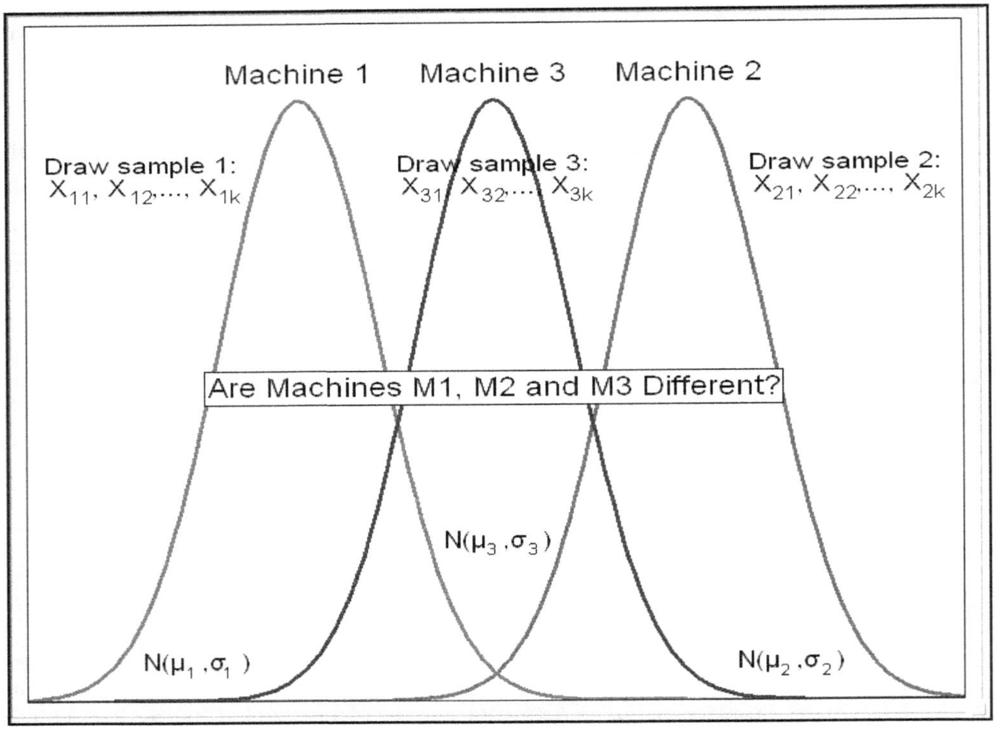

For the one-sample (Chapter 4) and two-sample (Chapter 5) scenarios three sets of null and alternative hypothesis statements are possible. For comparing two population means, recall that we can write a one- or two-sided alternative hypothesis: a) H_1: $\mu_1 \neq \mu_2$, b) H_1: $\mu_1 < \mu_2$ or, c) H_1: $\mu_1 > \mu_2$. For the case of three or more population means, we can no longer write three different sets of alternative hypothesis. Instead, for *k* populations, defined by the k factor levels or treatments, means, $\mu_1, \mu_2, \mu_3, \ldots \mu_k$ we write the hypothesis in the following way:

H_0: $\mu_1 = \mu_2 = \mu_3 \ldots = \mu_k$ (Assume)
H_1: not all μ_i are equal, i=1, 2, ..., k (Prove)

The null hypothesis (H_0) *assumes* that the average performances of all k population means are the same. While the alternative hypothesis (H_1) states that the average performances of the k populations are not equal to each other. If our test of significance favors the alternative hypothesis, then we have *proven*, with statistical rigor, that the average performances are different without making any statements regarding which means are different from each other and the direction of their departure.

Statistics Note 6.2: As was the case with the one- and two-sample test of significance discussed in previous chapters, we assume the means to be similar and set out to prove that they are different. If we do not reject the null hypothesis (H_0), we have not *proven* that the means are the same. However, if we reject the null hypothesis we have enough evidence (proof) to demonstrate, or establish, that not all the means are equal.

The one-way ANOVA F-statistic

The ANOVA F-statistic is a ratio of variances (a signal-to-noise ratio) that is calculated by first partitioning the total variation in our response into two components: one due to the *signal* and one due to the *noise*. The total *sum of squares* (SS) is partitioned into the factor sum of squares (signal), and the error sum of squares (noise). This fundamental equation is expressed as follows:

$$SS_{Total} = SS_{Factor} + SS_{Error}$$

Total Variation = Signal Variation + Noise Variation

The factor sum of squares represents the variation *between* the different levels, or treatments, of the factor, while the error sum of squares represents the variation *within* the different levels. For a factor with k levels, each level having n_i observations, the formulas for the total, factor, and error sums of squares are shown in equations 6.1, 6.2, and 6.3, respectively.

$$SS_{Total} = \sum_{i=1}^{k}\sum_{j=1}^{n_i}(y_{ij} - \overline{\overline{y}})^2, \tag{6.1}$$

where k is the number of factor levels, n_i is the sample size for level i, y_{ij} is the j-th observation within factor level i, and $\overline{\overline{y}}$ is the overall mean.

$$SS_{Factor} = \sum_{i=1}^{k} n_i \left(\overline{y}_i^2 - \overline{\overline{y}}\right)^2, \tag{6.2}$$

where k is the number of factor levels, $\overline{y}_i$ and n_i are the average and sample size for level i.

$$SS_{Error} = \sum_{i=1}^{k}\sum_{j=1}^{n_i}\left(y_{ij} - \overline{y}_i\right)^2 \tag{6.3}$$

where y_{ij} is the j-th observation within factor level i.

Let's illustrate the partitioning of the total variation with an example. Suppose we want to know if the operational effectiveness, as measured by the Overall Equipment Efficiency (OEE), of the three molding machines is similar. During the course of one year, we calculated the OEE for three machines on a monthly basis and ended up with 12 independent readings for each machine for a total of 36 OEE readings.

The data are plotted in Figure 6.2, along with the estimates for their means and standard deviations (Table 6.2).

Table 6.2 OEE Summary Data

Machine	Number of Readings	Mean(OEE)	Std Dev(OEE)
M1	12	74.88	1.4603
M2	12	80.63	0.9423
M3	12	80.19	0.9671

The variation between the OEE averages for the three machines determines the signal for the test statistic. In other words, how much of the total variation in OEE values is due to the three machines? We use equation 6.2 to calculate the machine sum of squares as the sum of the squared deviations from the overall mean (78.5694), of the three machine averages 74.8833, 80.6333, and 80.1917, as follows:

$$SS_{Machine} = 12\times(74.8833-78.5694)^2+12\times(80.6333-78.5694)^2+12\times(80.1917-78.5694)^2 = 245.7439$$

The total sum of squares is calculated from all the 36 OEE readings as follows:

$SS_{Total} = (n - 1)\ s^2 = (36 - 1) \times 2.8747^2 = 289.2564$

Finally, the error sum of squares is easily calculated as the difference between the total sums of squares, SS_{Total}, and the machine sums of squares, $SS_{Machine}$:

$SS_{Error} = SS_{Total} - SS_{Machine} = 289.2564 - 245.7439 = 43.5125$

Figure 6.2 Comparing the Average OEE of Machines 1, 2, and 3

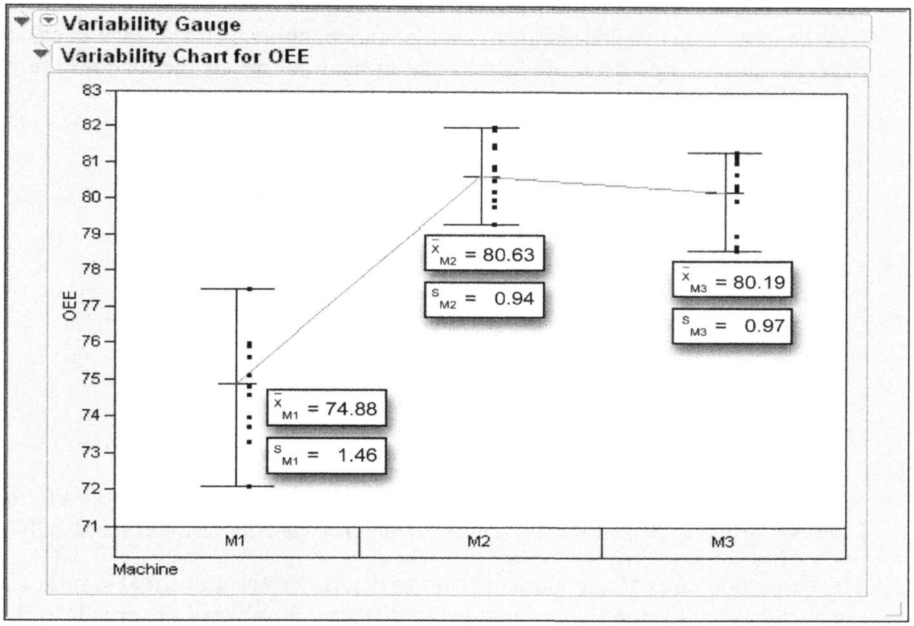

Statistics Note 6.3: For a given machine, the sum of squared deviations from the machine average, of the 12 OEE readings, is a measure of the Machine variation (Table 6.3a).

Table 6.3a Sum of Squares Deviations from Average for SS_{Error}

Source	Sum of Squared Deviations
M1	23.4567
M2	9.7667
M3	10.2892
Error (=M1+M2+M3)	43.5125

Table 6.3a clearly shows that the error sum of squares is a pooled measure of noise since it is the sum of the variations from each of the machines. However, it is easier to calculate the error sum of squares as the difference of the total minus the machine sums of squares.

The total variation 289.2564 is decomposed into a contribution from the machine and the error as shown in Table 6.3b.

Table 6.3b Sum of Squares Decomposition

Source	Sum of Squares
Machine	245.7439
Error	43.5125
Total	289.2564

We can clearly see that machine is the largest contributor to the total variation in the data contributing about 85%. However, is this enough to declare our factor as statistically significant?

The sum of squares alone is not enough to determine the statistical significance of our factor since they are computed using different number of observations. In order to determine the statistical significance of the factor, we must first standardize the sums of squares by their corresponding degrees of freedom, and then calculate a signal-to-noise ratio (a test statistic) and its probability distribution in order to obtain the p-value. The F-statistic is the test statistic for a one-way ANOVA, and is calculated as

$$F - statistic = \frac{Signal}{Noise} = \frac{\frac{SS_{Factor}}{df_{Factor}}}{\frac{SS_{Error}}{df_{Error}}} = \frac{\frac{SS_{Factor}}{(k-1)}}{\frac{SS_{Error}}{(n-k)}} = \frac{MS_{Factor}}{MS_{Error}} \quad (6.4)$$

A *mean square* (MS) is a sum of squares divided by the appropriate degrees of freedom, and represents the variation contribution per degree-of-freedom. The mean squares for the factor and the error are obtained by dividing their sums of squares by the corresponding degrees of freedom df_{Factor} and df_{Error}. The F-statistic is the signal-to-noise ratio of the MS_{FACTOR} to the MS_{ERROR}. For a one-way ANOVA, df_{Factor} is $(k-1)$ one less than the number of levels for the factor, and df_{Error} is $(n-k)$, the number of observations minus the number of factor levels. This gives another decomposition: $df_{Factor} + df_{Error} = n-1$, which is the number of observations minus 1.

Degrees-of-freedom decomposition: dfFactor + dfError = n-1, the number of observations minus 1.

Statistics Note 6.4: For normally distributed data, the MS_{FACTOR} and the MS_{ERROR} are independent, each distributed as a Chi-square. The ratio of two independent Chi-squares is distributed as an *F*-statistic.

For the OEE example, $df_{Machine} = 3 - 1 = 2$; $df_{Error} = 36 - 3 = 33$ and the F-statistic is calculated as $(245.74 / 2) / (43.5125 / 33) = 93.19$. This tells us that the signal in our data, coming from the differences between machines, is 93.19 times the noise in the data, and will most likely result in a statistically significant effect. As a confirmation we need to calculate the p-value associated with this test statistic. The ANOVA signal-to-noise ratio follows an F-distribution, with numerator and denominator degrees of freedom, $df_{Factor} = k - 1$ and $df_{Error} = n - k$, respectively. As it was the case with the t-statistic in Section 4.3.2, the p-value = 2.6645E-14 is the area under the F-distribution, which corresponds to our calculated F-statistic, with 2 and 33 degrees of freedom, and is shown graphically in Figure 6.3 (the x-axis is in the log scale to show detail). For a significance level of 5%, since this p-value is less 0.05, we reject the null hypothesis, H_0: $\mu_1 = \mu_2 = \mu_3$ in favor of the alternative, H_1: $\mu_1 \neq \mu_2 \neq \mu_3$. We therefore conclude that the average machines OEE performances are different. For a description of what a p-value is see Section 4.3.2.

Statistics Note 6.5: The *F*-test of significance is an omnibus type test in the sense that it indicates only that the average performances are different; it does not tell us where the differences are. To find where the differences are coming from we need to do multiple comparisons (as shown in Section 6.3.3) to compare each mean against the others. For our cement example comparing four suppliers, the one-way ANOVA tests whether the compressive strength of the cement is impacted by the different suppliers, but it will not tell us which suppliers have better or worse strength properties.

Figure 6.3 F-distribution (log scale) for OEE Test Statistic

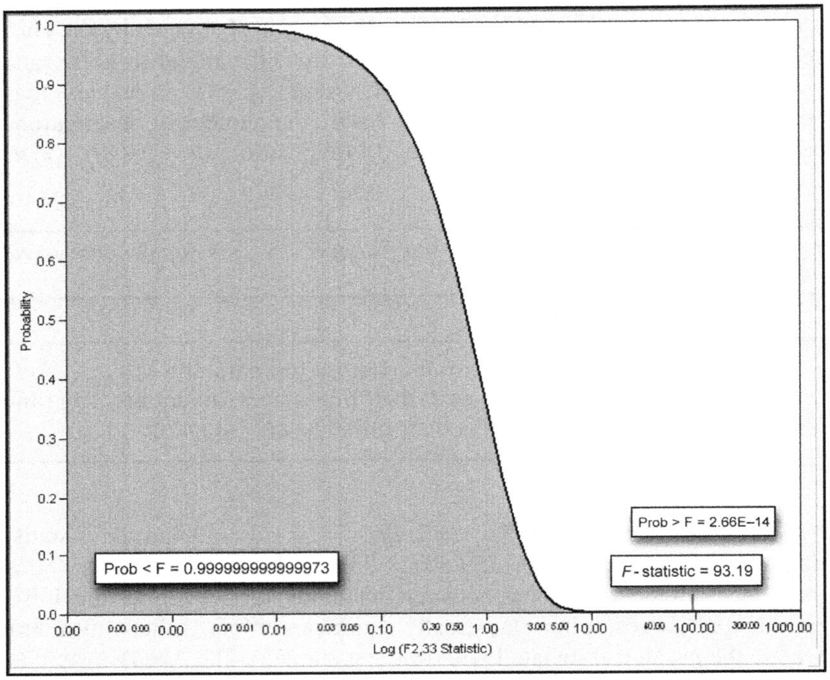

The sum and mean squares calculations are summarized in an ANOVA table, as is shown in Table 6.4. This table makes it easy to see the sum of squares contributions of the factor and the error as well as the calculated *F*-statistic and its corresponding *p*-value.

In the planning stages of our study, it is a good practice to fill out the first two columns (Source and DF) of the ANOVA table. Think of this as an accounting exercise for keeping track of how we are "spending" our *n* observations.

Table 6.4 ANOVA Table Layout

Source	DF	Sum of Squares	Mean Square	F Ratio	Prob > F (P-value)
Factor	k – 1	SS_{Factor}	$MS_{Factor} = SS_{Factor} / (k-1)$	MS_{Factor} / MSE	Prob > F Ratio
Error	n – k	SS_{Error}	$MSE = SS_{Error} / (n-k)$		
Total	n – 1	SS_{Total}			

The ANOVA table for the OEE example is shown in Table 6.5.

Table 6.5 ANOVA Table for OEE

Source	DF	Sum of Squares	Mean Square	F Ratio	Prob > F (P-value)
Machine	2	245.74	122.87	93.1864	< 0.001
Error	33	43.51	1.32		
Total	35	289.26			

JMP Note 6.1: The one-way ANOVA for testing multiple population means can be found under the **Analyze > Fit Y by X** platform. Since JMP selects the analysis technique appropriate for the data, it is important that the factor modeling type is set to ordinal or nominal, and not continuous. After selecting **Fit Y by X**, a dialog box appears and we need to identify the **X, Factor** and **Y, Response** in our data set, as is shown in Figure 6.4. Note the modeling type for machine is set to nominal, as is indicated by the red unordered bars icon next to the label.

Figure 6.4 One-way ANOVA JMP Dialog Box for the OEE Data

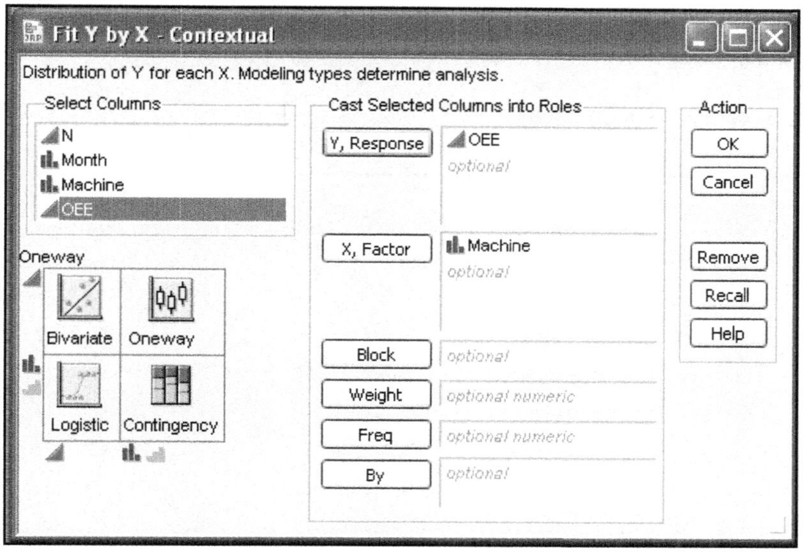

If the default preferences have not been changed, when you click **OK** a simple plot of the data appears. This plot displays the data values for each of the levels of the factor (three machines in this case) and a horizontal line representing the overall average of the data. A full listing of the options can be obtained by clicking on the red triangle next to the **One-way ANOVA** banner at the top of the graph, as shown in Figure 6.5. While there are many options available in this platform, the ones that are discussed in this chapter are provided in Table 6.6.

Figure 6.5 Options for One-way ANOVA Platform

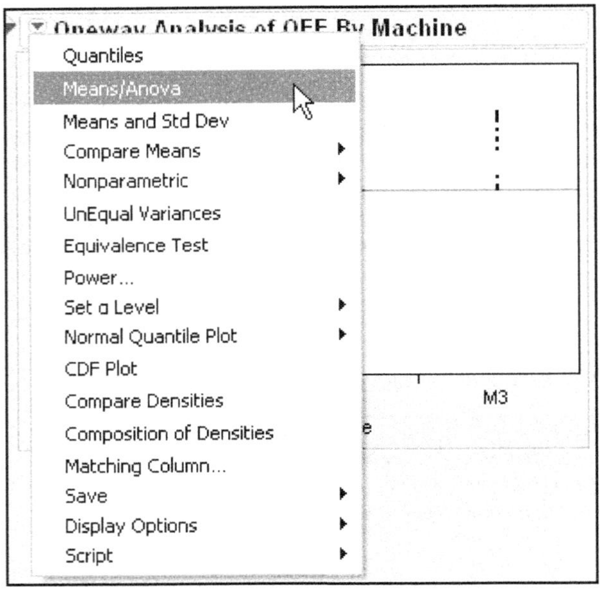

Table 6.6 One-way ANOVA Topics Illustrated in Chapter 6

Topic	Description	Figure Reference
Means and Std Dev	Table for the means and their confidence intervals and standard deviations for each factor level. Note that these confidence intervals are slightly different from the ones produced in Means / Anova because they do not use a pooled estimate of the variance.	Figure 6.23 Figure 6.25
UnEqual Variances	Tests that the variances are equal among the factor levels using several methods, for example, Brown-Forsythe and Levene. Also produces a graph and a table of standard deviation estimates for each factor level. Welch's ANOVA test statistic is also included at the bottom of the output, which should be used in place of the Means / Anova F-test if the variances are found to be different.	Figure 6.26 Figure 6.27
Means/Anova	Conducts a one-way ANOVA analysis and outputs the summary of the fit, the analysis of variance table, and means and confidence intervals for each factor level. This choice is appropriate when the homogeneous variance assumption has been met. **Means diamonds** are also overlaid on the initial graph of the data.	Figure 6.29 Figure 6.30
Set α Level	Used to change the significance level, α. This affects some of the output in the one-way ANOVA platform, such as the confidence intervals for the means.	Figure 6.31
Compare Means	Includes several multiple comparisons methods to perform pairwise comparisons of factor level means, such as Tukey and Dunnett's tests. A **comparisons circles plot** is added to the initial graph of the data that can be used for visually comparing all factor level means to each other to find out which ones are different. Note, the tests are not adjusted to control the overall error rate.	Figure 6.32 Figure 6.33
Normal Quantile Plot	Produces a normal quantile plot next to the initial graph of the data. This plot can be used to quickly check the normality for each factor level, as well as compare their means (vertical positions) and variances (slopes).	Figure 6.35a
Save	Saves the residuals to the JMP table in order to check the model assumptions with other platforms.	Figure 6.34
Equivalence Test	Conducts equivalence tests for all two-pair combinations of factor level means in order to determine whether they are equivalent within a given equivalence bound. Note that the tests are not adjusted to control the overall error rate.	Figure 6.41

In order to generate the ANOVA table shown in Table 6.5, to compare the OEE for three different machines, select the **Means/Anova** option.

This produces the output shown in Figure 6.6. The **Means Diamond** plot is at the top of the JMP output and gives us a visual display of our data and some insight into how different the means are for the different factor levels. A quick significance test can be done using the overlap marks within the means diamonds. If the overlap marks of two diamonds do not overlap, it indicates that there might be a statistically significant difference between the levels of the factor. Note that this is true only if the sample sizes are equal for all the levels of the factor under study. For this reason, and because sometimes it is difficult to judge visually if the overlap marks overlap, we prefer to use the *F*-statistic and its *p*-value to make a determination about the hypothesis outcome. The one-way ANOVA tabular output begins with a **Summary of Fit**.

Figure 6.6 JMP One-way ANOVA Output for OEE Example

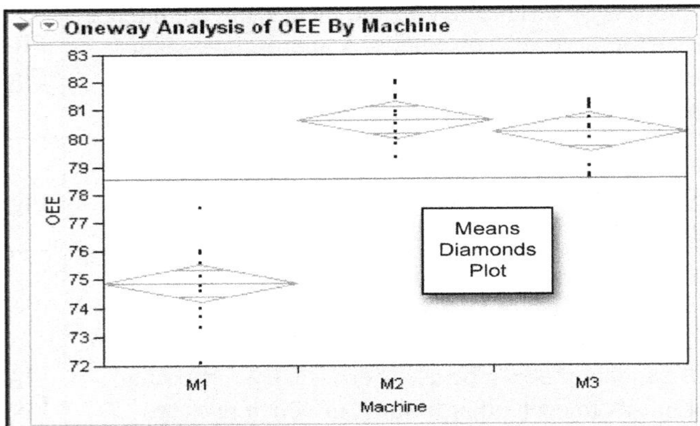

Oneway Anova

Summary of Fit

Rsquare	0.849571
Adj Rsquare	0.840454
Root Mean Square Error	1.148286
Mean of Response	78.56944
Observations (or Sum Wgts)	36

Useful information

Analysis of Variance

Source	DF	Sum of Squares	Mean Square	F Ratio	Prob > F
Machine	2	245.74389	122.872	93.1864	<.0001*
Error	33	43.51250	1.319		
C. Total	35	289.25639			

F-statistic and *p*-value

Means for Oneway Anova

Level	Number	Mean	Std Error	Lower 95%	Upper 95%
M1	12	74.8833	0.33148	74.209	75.558
M2	12	80.6333	0.33148	79.959	81.308
M3	12	80.1917	0.33148	79.517	80.866

Std Error uses a pooled estimate of error variance

The *Rsquare* value is a statistic between 0 and 1 that describes the amount of variation in the response that is due to the factor. The Rsquare is calculated by taking the ratio of $SS_{Machine} / SS_{C. Total}$. We can multiply this number by 100 to interpret it in a percentage scale. For our OEE example, 84.96% of the variation in OEE is due to the different machines and 15.04 is unexplained or due to the error. The *Root Mean Square Error* (RMSE), which is 1.148, is a pooled estimate of the experimental error (under the assumption of equal variances), and it is calculated by taking the square root of the *Mean Square Error* (MSE), which is 1.319, in the ANOVA table. Finally, the *Mean of Response*, which is 78.569, is the mean of all observations, and the *Observations (or Sum Wgts)* is the total number of observations in the study (36 OEE readings).

The **Analysis of Variance** table is presented next with the five columns of Table 6.4. The **Source** column includes entries for our factor machine, the error term, and the total variation, which is labeled C.Total. The degrees of freedom for each entry are shown in the next column and they match the calculations in Table 6.5. The sum of squares and mean square for each term are provided in columns 3 and 4 of the JMP output and, once again, match the hand calculations in Table 6.5. Finally, the *F*-statistic and its corresponding *p*-value are given in columns 5 and 6. If the *p*-value is less than our pre-specified significance level of 0.05 it will have an asterisk (*) next to it, indicating a statistically significant result.

The last section of Figure 6.6 shows the sample means for each of the factor levels and their confidence intervals. There is a note at the bottom of the output indicating that the confidence intervals are calculated using a pooled estimate (the RMSE) of the standard deviations from all three machines.

ANOVA Assumptions

The partitioning of the total variation into two components is an algebraic property of the data that does not depend on any distributional assumptions. In other words, the ANOVA calculations can be performed with any set of continuous data. However, in order to make valid statistical inferences when comparing the average performance of multiple materials, processes, or product, three assumptions must be met:

1. Independent observations between and within the populations defined by the factor levels
2. Normally distributed observations
3. Homogeneous variances of the populations defined by the factor levels

1. Independence

Observations are independent if the probability of the outcome of any observation in the sample is not affected by the outcome of any other observation in the sample. In the case of the one-way ANOVA, this assumption should hold within the levels of the factor and between the levels of the factor. The best way to verify this assumption is through knowledge of the way the data was collected. A violation of the independence assumption can result in a biased error term, and therefore impact the F-statistic and p-value obtained from the F-distribution. If the observations are positively correlated the estimate of error (noise) tends to be inflated, i.e., the estimate of error is larger than it should be, and, in turn, making the signal-to-noise ratio smaller than it should be. The best defense is to set up the study appropriately and use randomization.

2. Normality

In order for the F-statistic to follow an F-distribution both the numerator and denominator mean squares should follow a Chi-square distribution. This, in turn, requires the data to be normally distributed. The normality assumption for a one-way ANOVA is easy to check using a **Normal Quantile Plot** option in the one-way ANOVA output window. Another approach to check the normality of the data is to save the residuals to the JMP data table, and then use the **Analyze > Distribution** to check the normality with a normal quantile plot and a goodness of fit test. The good news is that the F-statistic is fairly robust against departures from normality. The F-test, however, can be sensitive to large departures from normality, in particular if the sample sizes for the different levels are seriously unbalanced.

As we described in Chapter 2, certain responses tend to follow a particular distribution. For example, physical measurements like length, pressure, and temperature can be modeled using a normal distribution; time-to-failure using a Weibull distribution; a Poisson distribution can approximate defect counts, while pass or fail data is suited for a Binomial distribution. If there is prior knowledge that the response is not normally distributed then the analysis should be adjusted accordingly. In this chapter we deal with responses that can be modeled with a normal distribution.

3. Homogeneous variances

The denominator of the F-statistic, MSE, is a pooled estimate of the variances of the populations defined by the factor levels. If these variances are not homogeneous, then the MSE is not a good estimate of the noise, particularly if the sample sizes for the different levels are unbalanced. This is why it is important to check this assumption *before* we perform the ANOVA. If the variances are not similar a special form of ANOVA, Welch's ANOVA, needs to be used.

The assumption of homogeneous variances can be tested quite easily in JMP by using the **UnEqual Variances** option in the one-way ANOVA output window. Figure 6.7 shows the output for the **UnEqual Variances** option.

Figure 6.7 Unequal Variances Output for OEE

Tests that the Variances are Equal

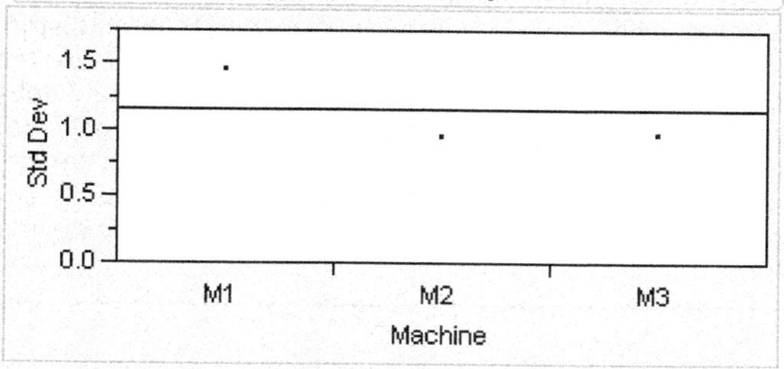

Level	Count	Std Dev	MeanAbsDif to Mean	MeanAbsDif to Median
M1	12	1.460282	1.133333	1.133333
M2	12	0.942273	0.783333	0.783333
M3	12	0.967150	0.776389	0.758333

Test	F Ratio	DFNum	DFDen	Prob > F
O'Brien[.5]	1.6930	2	33	0.1996
Brown-Forsythe	1.2265	2	33	0.3063
Levene	1.2218	2	33	0.3077
Bartlett	1.3643	2	.	0.2555

Welch Anova testing Means Equal, allowing Std Devs Not Equal

F Ratio	DFNum	DFDen	Prob > F
69.5034	2	21.4	<.0001*

The Std. Dev. by Machine plot shows the standard deviation of machine M1 (1.46) to be somewhat higher than the standard deviations of the other two machines (≈0.95). JMP provides four different tests to test the hypothesis that the machine variances are equal. For the OEE data the *p*-values for all four tests are greater than the significance level of 0.05, indicating that we do not have enough evidence to say that the variances are not homogeneous. If the unequal variance test suggests that the population variances are not homogeneous, we must use a different form of the test statistic in order to compare the average performance. The details for the homogeneity tests, as well as Welch's ANOVA, are included in Section 6.3.4 and are not discussed here.

Statistics Note 6.6: If the sample sizes for the factor levels are equal or nearly equal, and not too small (> 5), the ANOVA *F*-statistic is quite robust to departures from normality, and somewhat robust to mild departures from the homogeneous variances assumption, provided that the distribution is not skewed. For the case of heterogeneous variances the *F*-test tend to be liberal, i.e., likely to reject the null hypothesis when it should not (when the large sample sizes correspond to the levels with the small variances), and conservative, i.e., unlikely to reject the null hypothesis when it should (when the small sample sizes correspond to the levels with the small variances). Also, the *F*-statistic is not robust to violations of the independence assumption.

6.3.3 Multiple Comparisons to Detect Differences between Pairs of Averages

For the one-way ANOVA the *F*-test answers the question: Are there any differences between the average performance of different materials, products, or processes? If the answer is yes, however, the *F*-test does not tell us where the differences are. The differences can be due to pairs of averages being different from each other, or functions of the averages being different from each other. For example, one can compare the average compressive strength from Phase 3 with the average compressive strength from Phases 1, 2, and 4; i.e., average (Phase 3) vs. average (Phase 1, Phase 2, and Phase 4).

Let's look at the simple case of comparing each pair of means. For the OEE example we have three machines (M1, M2, and M3) given three possible pairwise comparisons, namely, M1 vs. M2, M1 vs. M3, and M2 vs. M3. As the number of levels increases so do the number of pairwise comparisons. For a factor with five levels we have $5\times(5-1)/2 = 10$ pairwise comparisons. In general for a factor with k levels we have $k\times(k-1)/2$ pairwise comparisons, corresponding to the number of combinations of two elements taken from k elements or $k!/[(k-2)!\times 2!]$. Since the number of multiple comparisons increases as the number of factor levels increases, the likelihood of finding a significant difference by chance alone also increases. The significance level of the pairwise comparisons test needs to be adjusted to take this into account.

Statistics Note 6.7: Recall that when comparing two means, a Type I error is declaring a statistically significant difference when in fact there is no difference between the means. As the number of *simultaneous* pairwise comparisons increases so does the cumulative probability, or *Overall Type I error rate*, of finding a statistically significant difference between a pair of means when there is none. It can be shown that for m simultaneous tests the probability of making at least one Type I error is as follows:

Probability(making at least one Type I error in m simultaneous tests) =

$$1 - \text{Probability(making no Type I)}^m =$$
$$1 - (1 - \text{Probability(making a Type I)}^m)$$
$$\textit{Overall Type I error rate} = 1 - (1 - \alpha)^m$$

To illustrate, for a factor with five levels we have 10 pairwise *simultaneous* comparisons. If for each pairwise comparison we use an individual Type I error rate $\alpha = 0.05$, then the overall Type I error rate is not 5% but $1 - (1 - 0.05)^{10} = 0.4013$. In other words, 40% of the time we will be declaring at least one pairwise comparison statistically significant just by chance alone!

Some multiple comparisons methods enable us to perform simultaneous comparisons, while keeping the overall Type I error rate at a pre-specified level. JMP offers several multiple comparisons tests that are accessible in the one-way ANOVA platform using the **Compare Means** option. The choices are shown in Figure 6.8 and are briefly described below:

- **Each Pair, Student's t** – As the name suggests, this method computes individual pairwise comparisons using Student's t-test on all the means. Since there is no protection for controlling the overall Type I error, α, we recommend using another multiple comparisons test that controls for the overall α.
- **All Pairs, Tukey HSD** – HSD stands for Honestly Significant Difference test, also known as the Tukey-Kramer HSD test. This test controls for the overall Type I error rate, α, and when all factor levels sample sizes are equal, it is an exact α level test. If the factor levels sample sizes are not equal, then α will be on the conservative side. When all pairwise comparisons are desired this is the recommended approach because it controls the overall Type I error rate.
- **With Best, Hsu MCB** – This test determines if each mean is either the maximum mean or minimum mean from some unknown maximum or minimum. This test can be useful, for example, when we know the desired performance of a material, product, or process and want to determine whether any of the populations in the study exceeds this desired performance.

- **With Control, Dunnett's** – This test is used to determine whether the other means are statistically different from the mean that is identified as the *control*. This test can be useful when we want to compare several new materials, products, or processes to a standard material, process, or product.

Figure 6.8 Multiple Comparison Tests for Means in JMP

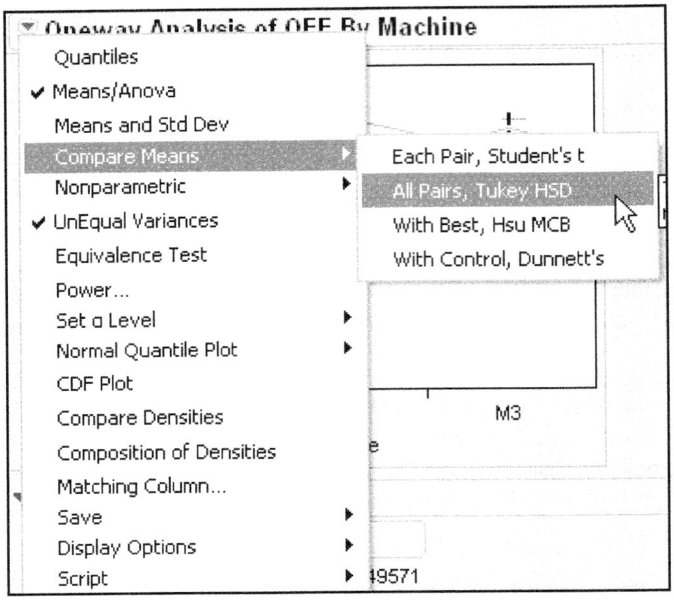

All Pairs, Tukey HSD

The global *F*-test for machine in Table 6.5 was statistically significant, which indicates that there are differences in the average OEE performance for the three machines. For the three machines we have three pairwise comparisons ({M1, M2}, {M1, M3}, and {M2, M3}) that can be performed. The **All Pairs, Tukey HSD** output for this example is shown in Figure 6.9. At the top of the output we have the overall Type I error rate, **Alpha**, and the quantile of the studentized range statistic, **q***. The quantile q* is used to scale the Least Significant Differences (LSD), and depends on the number of levels, the number of samples per level, and Alpha. For the OEE example, we see that q* = 2.45379 and alpha= 0.05.

The first matrix in the output is called the *Difference matrix,* which is not part of the default output, but can be easily obtained by selecting the **Difference Matrix** option from the red triangle in the title at the top of this window. The population means in this matrix are ordered from largest in value to smallest in value, 80.63 (M2), 80.19 (M3), and 74.88 (M1). Each cell in the matrix is the difference between the corresponding row and column. The diagonal entries are therefore 0, and off-diagonal entries are symmetric, but with different signs. For example, the matrix entry of –5.3083 in row M1 and column M3 is calculated as mean M1 – mean M3, or 74.88 – 80.19; while the entry in row M3 and column M1 is 80.19 – 74.88 = 5.3083. Looking at this matrix we see that the largest differences in means occur between machines M1 and M2 (|5.75|) and machines A and C (|5.3083|), while the smallest difference in means (|0.4417|) occurs between machines M2 and M3. We will use a test of significance procedure to determine whether these differences are statistically significant.

Figure 6.9 JMP Output for Tukey's HSD Multiple Comparison Test for OEE

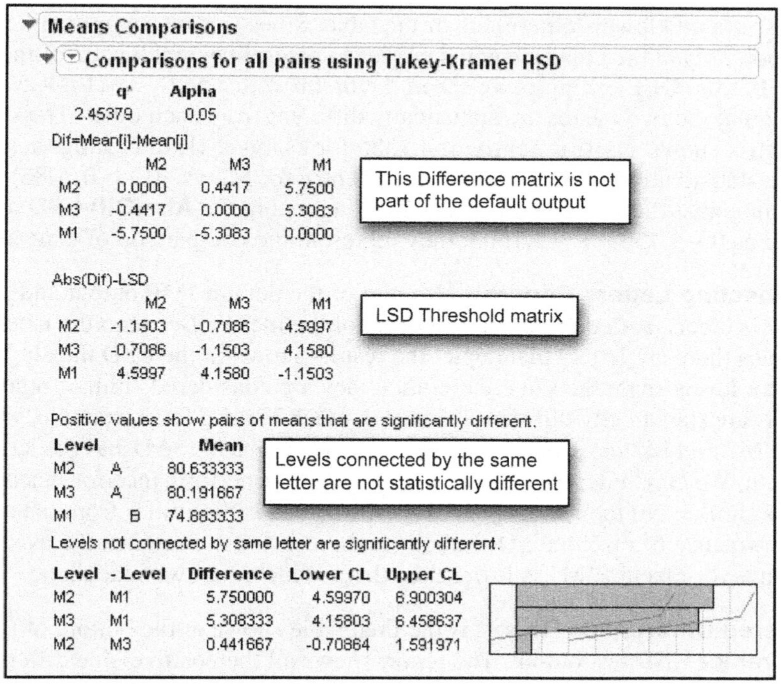

Means Comparisons

Comparisons for all pairs using Tukey-Kramer HSD

q*	Alpha
2.45379	0.05

Dif=Mean[i]-Mean[j]

	M2	M3	M1
M2	0.0000	0.4417	5.7500
M3	-0.4417	0.0000	5.3083
M1	-5.7500	-5.3083	0.0000

This Difference matrix is not part of the default output

Abs(Dif)-LSD

	M2	M3	M1
M2	-1.1503	-0.7086	4.5997
M3	-0.7086	-1.1503	4.1580
M1	4.5997	4.1580	-1.1503

LSD Threshold matrix

Positive values show pairs of means that are significantly different.

Level			Mean
M2	A		80.633333
M3	A		80.191667
M1		B	74.883333

Levels connected by the same letter are not statistically different

Levels not connected by same letter are significantly different.

Level	- Level	Difference	Lower CL	Upper CL
M2	M1	5.750000	4.59970	6.900304
M3	M1	5.308333	4.15803	6.458637
M2	M3	0.441667	-0.70864	1.591971

The **LSD Threshold Matrix** appears next in the JMP output and is labeled **Abs(Dif)-LSD**. This matrix shows the absolute value of the actual difference in the means minus the Least Significant Difference (LSD) needed to indicate a statistically significant difference. The LSD is a studentized range statistic multiplied by the standard error of the difference of the means being compared. When the sample size for each of the k levels is the same (for example, n), the LSD becomes the following:

$$LSD = q_{\alpha,k,k(n-1)}\sqrt{\frac{2MSE}{n}}$$

For the OEE example (shown in Figure 6.6) k=3, n=12, MSE = 1.139, and q=q^*=2.453, given:

$$LSD = q_{0.05,3,33}\sqrt{\frac{2\times 1.319}{12}} = 2.45379\times 0.46886 = 1.1503$$

If the **Abs(Dif)-LSD** entry is positive, then the two means are significantly different, and if the entry is negative then they are not. The greatest differences are typically in the upper-right and lower-left corners of the table. Since this matrix is symmetric, we need to look only at the upper diagonal entries to determine which pairs of means are different. For the OEE example, we see in the first row that M2 vs. M1 is 4.5997 > 0, indicating that the two means are statistically different from each other. The second row of this matrix shows a positive entry of 4.1580 for M3 vs. M1, indicating that these two means are statistically different. However, the entry for M2 vs. M3 (–0.7086) is negative so we assume that they are not different. The diagonal in the **Abs(Dif)-LSD** matrix are calculated as 0 – 1.1503 = – 1.1503; they represent the comparison of a mean with itself.

The **Connecting Letters Report** is also part of the default JMP output and shows the factor levels labels, sorted in decreasing order of the means. Between the labels and the mean values there are letters that depict the results shown in the LSD threshold matrix. If the factor levels share the same letter, then they are considered similar; otherwise, the two means are statistically different from each other. For the OEE example, we see that machines M2 and M3 have an "A" in common, while machine M1 has the letter "B" in this column. We conclude, once again, that the average performance for machines M2 and M3 is similar, but their average OEE performance is dissimilar from the average OEE performance of machine M1. In both of these cases, we see that the average OEE performance for machine M1 is lower than that for the other two machines.

The **Ordered Differences Report** is the final table shown at the bottom of the JMP **Tukey-Kramer HSD** test output. The report shows all the positive-side differences with their confidence interval in sorted order. If any confidence interval contains 0, then we conclude that the difference between the two means is not statistically significant, while if it does not contain 0, we conclude that the difference is statistically significant. The first row of the **Ordered Differences Report** in Figure 6.9 shows the confidence interval for the difference (5.75) between machine M2 and machine M1 as (4.5997, 6.9003), which does not contain 0. The graph to the right of the report shows bars corresponding to the pairwise difference and the overlaid confidence. Confidence intervals that do not fully contain their corresponding bar indicate a pair of means that are significantly different from each other.

When the **All Pairs, Tukey HSD** test option is selected the initial plot displayed at the top of the one-way ANOVA output is augmented with circles that provide a graphical way of performing the Tukey's HSD test. Several versions of this plot are shown in Figure 6.10 in order to illustrate the comparisons for each of the three factor levels. In general, circles that do not intersect (or that intersect slightly so that the outside angle of intersection is less than 90 degrees) indicate which means are different from each other. For borderline cases, rather than relying on the visual comparison (sometimes it is hard to tell if the outside angle is less than 90 degrees), it is better to click a circle to see which circles are statistically significant from it. These are the ones with a thick gray pattern.

The upper left plot shown in Figure 6.10 shows the results of clicking on the lowest circle, corresponding to machine M1. The color of the selected circle, and all other circles representing similar population means, turn red; while the circles representing population means which are statistically different from the selected one turn gray. The corresponding factor level name (and all the factor levels with similar means) are bold and red on the x-axis; while factor levels with statistically different means are colored gray.

The plot at the bottom left-hand side of Figure 6.10 shows the results of clicking on the circle corresponding to machine M2. Note how the x-axis label M3 is also colored red, indicating that the average performance of machine M2 and M3 is similar. The circle and x-axis label for machine M1 is gray, indicating that its OEE average performance is different from the other two machines. The plot to the right shows what happens when machine M3 is selected. The green diamonds in all these plots represent individual 95% confidence intervals around the mean.

Figure 6.10 Visual Means Comparison for Tukey's HSD Test

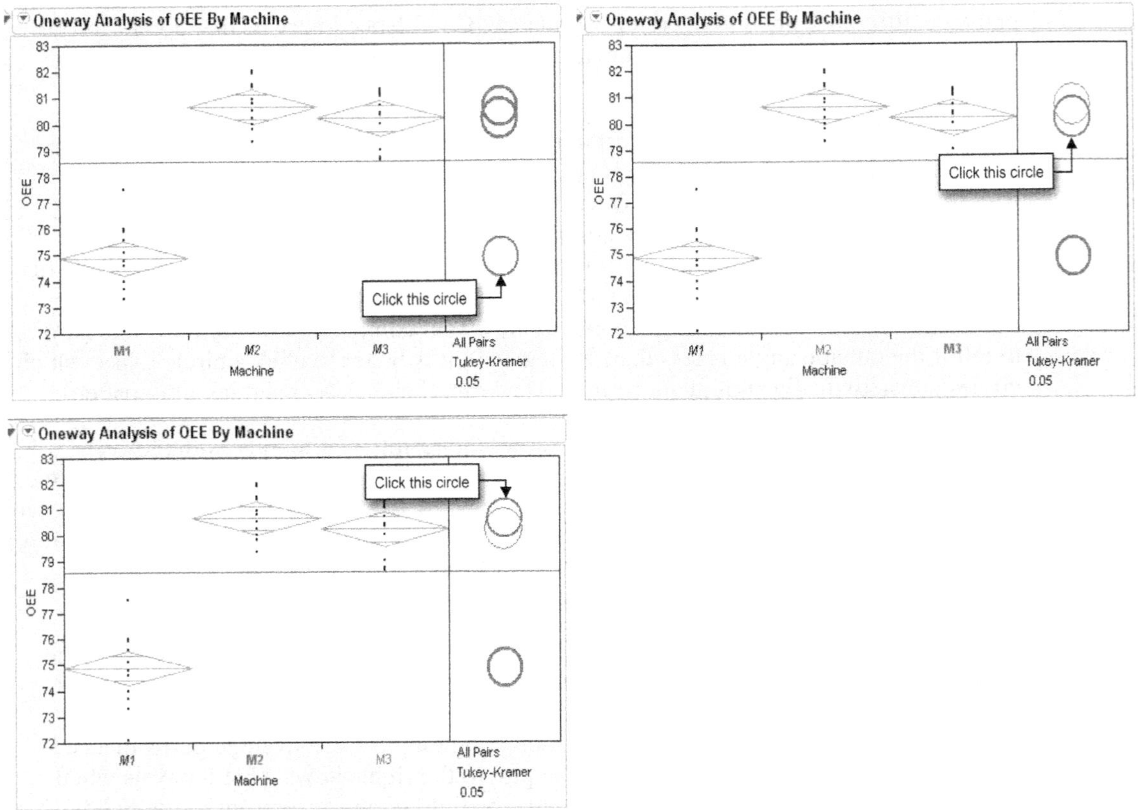

For the OEE example the comparison circles are a quick and graphical way of simultaneously testing, at the 0.05 significance level, the following three hypotheses:

1. H_0: Average Performance Machine M1 = Average Performance Machine M2
   H_1: Average Performance Machine M1 ≠ Average Performance Machine M2
2. H_0: Average Performance Machine M1 = Average Performance Machine M3
   H_1: Average Performance Machine M1 ≠ Average Performance Machine M3
3. H_0: Average Performance Machine M2 = Average Performance Machine M3
   H_1: Average Performance Machine M2 ≠ Average Performance Machine M3

The graphs in Figure 6.10 indicate that we have enough evidence to reject the null hypothesis, H_0, for comparisons (1) M1 vs. M2, and (2) M1 vs. M3, but not for the comparison (3), M2 vs. M3.

Statistics Note 6.8: In order to save time, you might be tempted to skip the ANOVA analysis and jump right into the multiple comparisons tests. Remember however, that the *F*-test is an omnibus test that tells us only whether the means are significantly different from each other. If the *F*-test is not significant, and we skip it, we might find differences that are not real. In other words, we must first detect that there is a difference among the levels of our factor (significant *F*-test), and only then can we carry out multiple comparisons to find out where the differences are coming from.

With Control, Dunnett's

Another class of comparisons involves comparisons with a control. Suppose that machine M2 is the "standard" machine and we would like to compare the OEE performance of machines M1 and M3 to M2. In this case we have only two comparisons: M1 vs. M2 and M3 vs. M2. For a factor with *k* levels, this results in $(k-1)$ comparisons.

When you select **Compare Means > With Control, Dunnett's** from the drop-down list at the top of the output window, you will be prompted to select the factor level representing the control for the comparisons unless you have a row selected in the data table, in which case JMP uses that value as the control. For the OEE example, machine M2 is selected as the control, as shown in Figure 6.11. Figure 6.11 also shows the output which includes the Difference Matrix and the LSD Threshold matrix, just like Tukey's HSD. But unlike Tukey's HSD test, Dunnett's LSD threshold matrix includes *p*-values for each comparison. We look for positive values in this matrix, or *p*-values that are less than our chosen value for α. It is no surprise that machine M1 is different from our control, while machine M3 is not. The machine averages are included in Figure 6.11 for reference.

Figure 6.11 Dunnett's Test for Multiple Comparisons for OEE Example

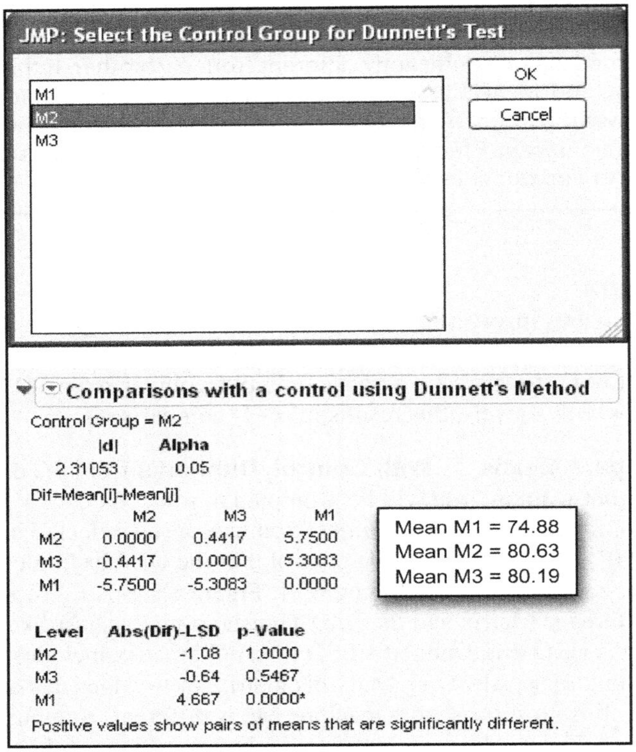

6.3.4 Comparing the Performance Variation of Several Materials, Processes, or Products

As it was the case with the two-sampl *t*-test, *homogeneity of variances* is also a key assumption underlying the validity for a one-way ANOVA test to compare several means. In fact, this is the first test we must conduct *before* we perform the ANOVA. In section 6.3.2, we described how the population variances get pooled into the error sum of squares (SSE) that is used as the noise term in the denominator of the *F*-statistic. In order to be able to pool these variances into one error term, we must make sure that they are similar to each other. We check this assumption by means of a significance test. The null and alternative hypotheses are written as follows:

H_0: $\sigma_1 = \sigma_2 = \ldots = \sigma_k$ (Assume)
H_1: not all σ_i are equal, $i=1, 2, \ldots, k$ (Prove)

> *In order to be able to pool these variances into one error term, we must make sure that they are similar to each other.*

This is very similar to the hypothesis for comparing several means and the outcome interpretation is also similar. If we reject the null hypothesis in favor of the alternative hypothesis, then we will not know which population variances are different from each other, only that they are not homogeneous. This additional information about differences in the performance variation for the different levels of our factor provides us with insight into which population can help us minimize the overall variation.

Test for Comparing Performance Variation

Several tests have been proposed for comparing the performance variation of three or more processes, products, or materials. JMP offers four such tests: O'Brien, Brown-Forsythe, Levene, and Bartlett. Several of these test statistics are based on transforming the response in order to get a measure of dispersion for each of the populations, and then conducting a one-way ANOVA on this transformed response. For example, to construct the Brown-Forsythe test we first determine the median value for each factor level group, and then subtract this median value from each response value in that group. A one-way ANOVA is then conducted using the absolute value of this transformed response. The corresponding F-test for the ANOVA serves as the significance test for the homogeneity of variances hypothesis. In other words, if the calculated p-value for the factor is less than our chosen value for α, then we conclude that the population variances are not homogeneous. As opposed to the comparisons of means, there are no known methods for comparing all of the pairwise variances combinations.

The selection of an appropriate test for testing the hypothesis of homogeneous variances depends on the underlying distribution of the data. Bartlett's test has accurate Type I rates and optimal power when the underlying distribution is normal, but it becomes very inaccurate for the slightest departures from normality. For this reason, we do not recommend this test. The Levene test is similar to the Brown-Forsythe test mentioned above but it uses the absolute deviations from mean instead of the median. The problem with the Levene test is that absolute deviations from the mean tend to be highly skewed, which violates the normality assumption of the ANOVA that is used to perform the test on the absolute mean deviations. Therefore, the Levene test is not as robust as the Brown-Forsythe test. Of the tests offered by JMP we recommend the Brown-Forsythe test because simulation studies (Conover, Johnson, and Johnson 1981, for example) have suggested that this test is robust in regards to departures from normality while retaining good statistical power.

The Brown-Forsythe Test

Although we can easily generate the Brown-Forsythe test in JMP, here we are going through the steps manually, using the OEE example, to show what calculations are done behind the scenes when we select the **UnEqual Variances** option in the one-way ANOVA output. You will see that the Brown-Forsythe test is an ANOVA run on a transformation of the data.

The hypothesis statement is shown below. A rejection of the null hypothesis implies that we have violated the homogeneous assumption for the one-way ANOVA to compare the machine's means, as well as, at least one machine has a more (or less) desirable performance regarding its variability.

H_0: $\sigma_{M1} = \sigma_{M2} = \sigma_{M3}$ (Assume)
H_1: not all σ_{Mi} are equal, i=1, 2, ..., k (Prove)

For each level of the factor machine we need to calculate the median. The summary statistics for M1, M2, and M3, calculated using **Table > Summary**, are shown in Figure 6.12a.

Figure 6.12a Descriptive Statistics for OEE Example

	Machine	N Rows	Mean(OEE)	Median(OEE)	Std Dev(OEE)	N(OEE)
1	M1	12	74.8833333	74.95	1.46028225	12
2	M2	12	80.6333333	80.65	0.9422732	12
3	M3	12	80.1916667	80.35	0.96714983	12

The Brown-Forsythe transformation requires subtracting the group median from each observation in the group, and then taking the absolute value. For example, we will subtract 74.95 from each OEE value for machine M1 and then take the absolute value. A partial listing of the transformed data is shown in Figure 6.12b.

Figure 6.12b Partial Listing of Transformed Data

Month	Machine	OEE	Median	Transformed OEE
JAN	M1	76	74.95	1.05
JAN	M2	79.8	80.65	0.85
JAN	M3	80.7	80.35	0.35
FEB	M1	73.3	74.95	1.65
FEB	M2	79.3	80.65	1.35
FEB	M3	81.1	80.35	0.75
MAR	M1	75.6	74.95	0.65
MAR	M2	81.5	80.65	0.85
MAR	M3	81	80.35	0.65
APR	M1	74.6	74.95	0.35
APR	M2	79.3	80.65	1.35
APR	M3	81.3	80.35	0.95
MAY	M1	74.8	74.95	0.15
MAY	M2	82	80.65	1.35
MAY	M3	80	80.35	0.35
JUN	M1	75.1	74.95	0.15
JUN	M2	81.4	80.65	0.75
JUN	M3	80.3	80.35	0.05
JUL	M1	75.9	74.95	0.95
JUL	M2	80.5	80.65	0.15
JUL	M3	80	80.35	0.35

In order to generate the Brown-Forsythe test statistic and associated p-value, we carry out a one-way ANOVA using the transformed response. We do this by selecting the **Analyze > Fit Y by X** platform with Transformed OEE as the response, and Machine as the factor. The resulting ANOVA table has an F-statistic for machine equal to 1.2265 and a p-value of 0.3063 (Figure 6.12c). Since the p-value = 0.3063 is greater than α=**0.05**, we do not reject the null hypothesis that all variances are similar. In other words, we *assume* that the average deviation from each group's median value is similar, and therefore we assume that their variances are similar or homogeneous.

Figure 6.12c ANOVA for Transformed OEE Data

Analysis of Variance

Source	DF	Sum of Squares	Mean Square	F Ratio	Prob > F
Machine	2	1.055000	0.527500	1.2265	0.3063
Error	33	14.192500	0.430076		
C. Total	35	15.247500			

Conducting a Brown-Forsythe test in JMP is easy. Just select the **UnEqual Variances** option from the Oneway Analysis menu (the red triangle at the top of the output window as shown in Figure 6.13).

Figure 6.13 Invoking the Unequal Variances Test in JMP

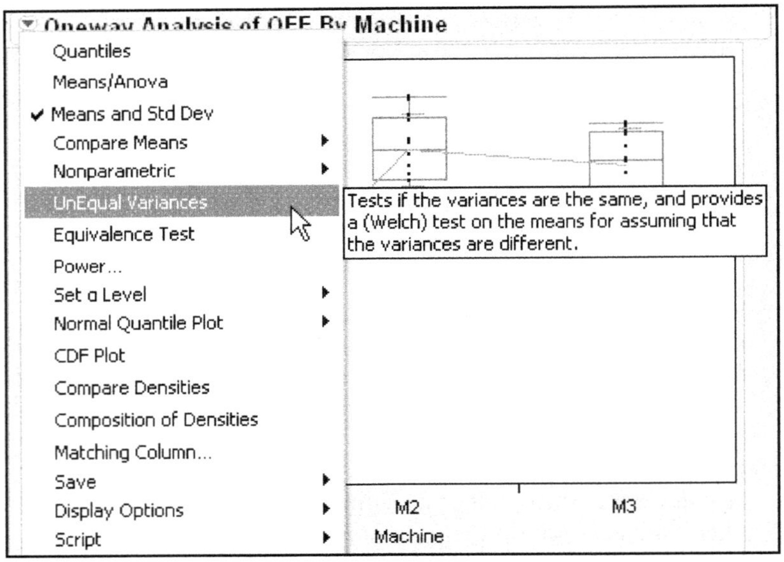

The output for the OEE example is shown in Figure 6.14 with a Brown-Forsythe test statistic (second entry) of 1.2265 and a *p*-value of 0.3063. The output also shows the other three tests mentioned before. Some authors, for example, Milliken and Johnson (1984), recommend using a 1% significance level when conducting homogeneity of variance tests.

Figure 6.14 JMP Output for UnEqual Variances for OEE Data

Tests that the Variances are Equal

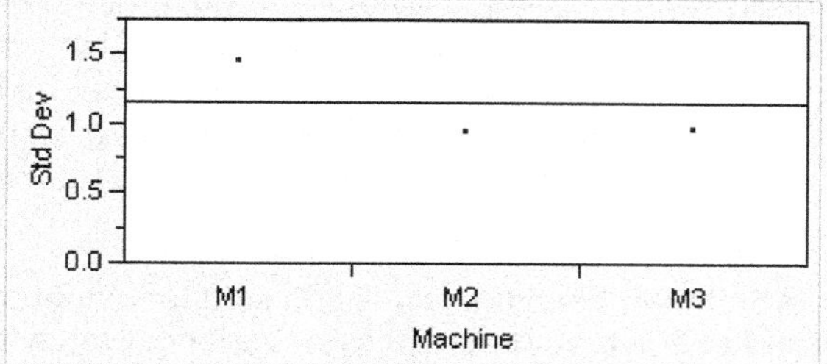

Level	Count	Std Dev	MeanAbsDif to Mean	MeanAbsDif to Median
M1	12	1.460282	1.133333	1.133333
M2	12	0.942273	0.783333	0.783333
M3	12	0.967150	0.776389	0.758333

Test	F Ratio	DFNum	DFDen	Prob > F
O'Brien[.5]	1.6930	2	33	0.1996
Brown-Forsythe	1.2265	2	33	0.3063
Levene	1.2218	2	33	0.3077
Bartlett	1.3643	2	.	0.2555

Welch Anova testing Means Equal, allowing Std Devs Not Equal

F Ratio	DFNum	DFDen	Prob > F
69.5034	2	21.4	<.0001*

What if the variances are heterogeneous?

Although we did not fail the homogeneous variance check for the OEE example, there might be situations where we get a significant test statistic and conclude that the variances among the factor levels are not homogeneous. If this happens then at least two of the factor levels have variances that are not similar. In JMP there is currently no test to determine which variances resulted in a significant test statistic. However, we can compare the standard deviations, using the plot at the top of the **Unequal Variances** report (Figure 6.14) to get a feel for which ones are different. For example, if we are trying to choose a factor level that optimized the average performance in some way, it will be useful to know if it has a lower variance as well.

Welch's ANOVA is based on the usual ANOVA *F*-test, but the group means are weighted by the group variances to take into account their variances. The formula can be found in the Statistics and Graphics Guide within the **Help** menu. The Welch ANOVA is calculated anytime we select the **UnEqual Variances** option. The numerator degrees of freedom for the *F*-test are $k - 1$, the same as the regular ANOVA. The denominator degrees of freedom, however, are approximated using the variances for each group (Statistics and Graphics Guide), and are usually non-integer.

The Welch's ANOVA output for the OEE example is shown in Figure 6.7, and at the bottom of Figure 6.14. The F Ratio is 69.50, as compared to 93.18 of the regular ANOVA, and the non-integer denominator degrees of freedom is 21.4 vs. 33 in the regular ANOVA (Figure 6.6.). Milliken and Johnson (1984) recommend using the Welch's ANOVA whenever the homogeneity of variance test is rejected at the 1% level.

> **JMP Note 6.2:** The JMP Brown-Forsythe test is adjusted for singleton zero residuals. For the Brown-Forsythe homogeneity of variance test, with odd-number sample sizes within levels of the independent variable, one residual will be equal to zero. This artificially dampens the variance estimate for that group. JMP replaces the zero by the minimum non-zero value if there is only one zero residual. Otherwise, the zeros are kept. This adjustment is recommended in Theory of Rank Tests by Jaroslav Hajek and Zbynek Sidak. (Cited in JMP FAQ # 2078 at www.jmp.com/support/faq/.)

6.3.5 Sample Size Calculations

In order to conduct a one-way ANOVA to compare three or more populations, we must define our *sampling plan*. This plan consists of the *sample size*, i.e., how many samples, as well as the *sampling scheme*, i.e., how we select the experimental units from the larger populations. In this section we show how to perform sample size calculations to determine how many samples we need in our study. The sampling scheme depends on the situation at hand but a *random sampling* scheme, in which each sample has an equal chance of being selected, is typical for these types of studies.

> *Sample plans consist of the sample size and the sampling scheme.*

The sample size calculator in JMP can be used to help determine appropriate sample sizes for testing multiple population means and it can be found under **DOE > Sample Size and Power** (Figure 6.15). For testing three or more means using a one-way ANOVA, we select **k Sample Means** from the **Sample Size and Power** window. The user needs to provide several inputs (as shown in Figure 6.16), which are similar to the inputs required to calculate sample sizes for conducting one and two-sample test of hypothesis. These steps are described in detail next.

Figure 6.15 Accessing Sample Size and Power Calculations in JMP

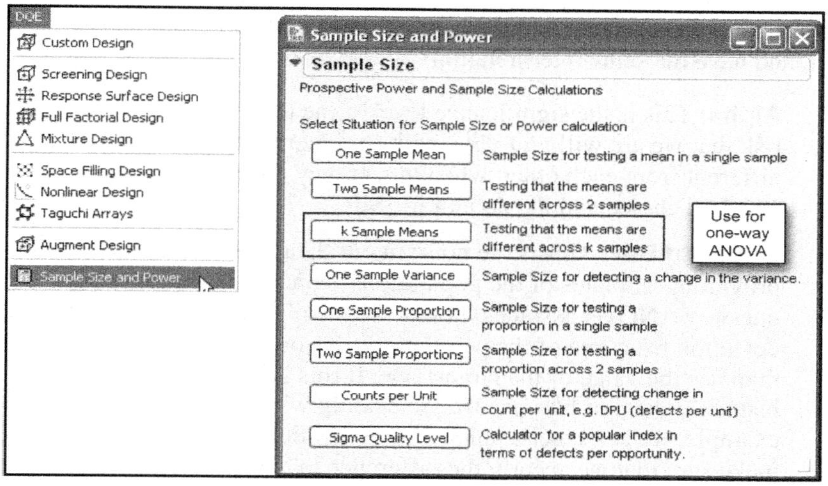

Figure 6.16 Inputs for Sample Size Calculations for *k* Sample Means

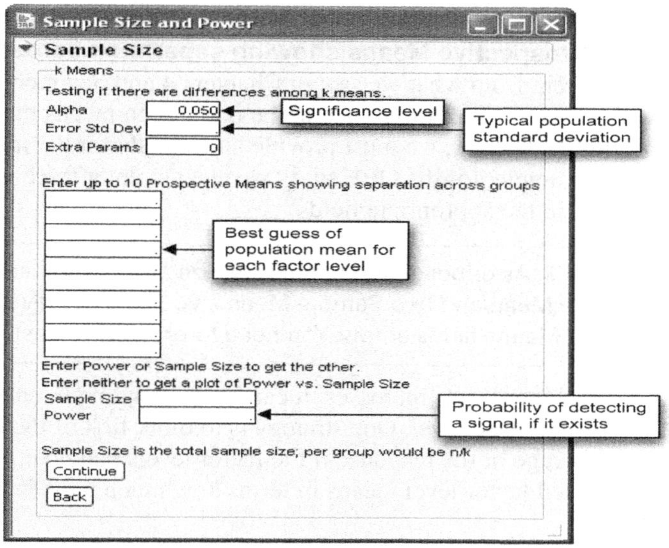

k Sample Means (Average Performance of Three or More Populations)

The following inputs are required for sample size calculations in JMP. Several of these inputs are identical to the ones discussed in Chapters 4 and 5 (for example, Alpha and Power, and have the same interpretation).

1. **Alpha**: This is the significance level of the test, which defines the amount of risk that we are willing to live with for incorrectly stating that the *k* means are different from each other, when in fact they are not. An alpha of 0.05 is the standard choice and the default in JMP.
2. **Error Std Dev**: This is the noise in our data. Historical data is helpful for providing estimates of the populations' standard deviations. Once again, a one-way ANOVA assumes these to be equal in which case we use the standard deviation from one of the populations. If you suspect that they are not equal, than use the value of the largest one. If this is not possible because we have no historical data, or because we are dealing with a new process or material, for example, we can enter a value of 1 in this field. If we do this then we need to make sure that we specify the difference to detect in terms of multiples of this unknown standard deviation.
3. **Extra Params**: This field is used for studies involving more than one factor (multiple-factor experiments), which is a topic not covered in this book. We will leave this zero for the *k* sample test for comparing the means.
4. **Enter up to 10 Prospective Means showing separation across groups**: This entry is different from what we saw in Chapters 4 and 5, which required us to provide just the smallest difference in the population means that we want to detect. For this dialog box, we must provide an estimate for the mean of each factor level. For example, for the OEE study we need to enter three means (one for each machine) in the appropriate fields.

> **JMP Note 6.3:** As opposed to the Sample Size and Power calculator for One Sample Mean and Two Sample Means, you cannot leave the Prospective Means fields empty. You need to provide at least 2.

As the number of factor levels increases, it can be challenging to enter the expected averages for each level. One strategy is to think first of the grand mean, i.e., the overall average of the response if the factor levels had no effect, and think of the expected factor level means in terms how much they deviate from the grand mean.

5. **Sample Size**: This is the minimum sample size recommended for the study. We need to collect measurements on this many experimental units (EUs) in order to be able to detect the practical differences defined in input 4 with a specified power. For a one-way ANOVA, this number represents the total sample size for the study. We divide this number by the number of factor levels, *k*, in order to get the number of required samples (experimental units) for each population.
6. **Power**: The probability of detecting a difference between the average performance of the *k* populations, when in fact the average performance are different from each other. Because it is a probability, the power takes on values between 0 and 1. We normally try to have a high power; i.e., a high probability of detecting a difference. We recommend starting with a power of at least 0.8 and adjusting it up or down as needed. As we mentioned in Chapters 4 and 5, don't be too greedy. The more power you want, the more samples you will need!

As it was the case with the **One Sample Mean** and **Two Sample Means** calculators, we can leave the **Sample Size** and **Power** fields empty and JMP will output a Sample Size vs. Power curve.

For the OEE example, the average OEE is expected to be around 80%, with an error of 2%. We believe that machine 1 will have a lower than average OEE by about 4%, that machine 2 will have an average OEE, and that machine 3 will have a somewhat higher at around 82%. For a power of 90% the sample size calculator (shown in Figure 6.17) indicates that we need a total of 11.77 experimental units (EUs), or 4 EUs per machine.

As we have discussed before, the concept of an experimental and observational unit is crucial to carrying out a meaningful significance test. The *experimental unit (EU)* is the smallest unit that is affected by the "treatment" or condition we want to investigate, while the *observational unit (OU)*, is the smallest unit on which a measurement is taken. Remember, observational units are not replications, if more than one observational unit is taken per experimental unit, the total number of measurements will increase, but the total sample size for our experiment will not. For example, we need 4 samples (EUs) per machine in our study. If we decide to measure each experimental unit four times, then we will end up with 16 measurements (OUs) per machine, but only 4 EUs per machine.

Figure 6.17 Sample Size for One-way ANOVA

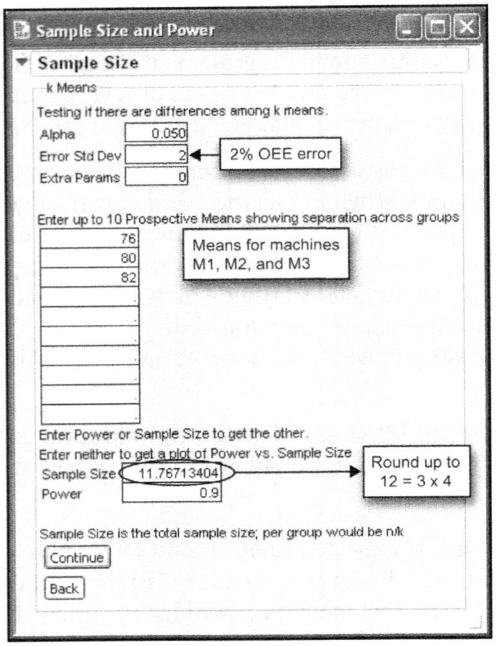

If we want to explore the relationship between power and sample size then we can leave the **Sample Size** and **Power** fields blank and click **Continue** to have JMP output a Sample Size vs. Power curve as shown in Figure 6.18. The resulting graph shows the Error Std. Dev., 2, Alpha, 0.05, and the Difference in Means, which is calculated as the square root of the square deviations from the overall mean. For the OEE data the mean of the three prospective means is 79.33, and the Difference in Means is equal to the following: $\sqrt{(76-79.33)^2 + (80-79.33)^2 + (82-79.33)^2} = 4.3205$.

Using the crosshair tool in this graph shows that for a sample size of 12 the power is about 91%, while for 15 samples, 5 per machine, the power is about 97.3%. Beyond 18 samples no gains are obtained. Sample Size vs. Power graphs are very useful because they enable us to evaluate the risk of a given sample size.

Figure 6.18 Power vs. Sample Size Output for 3 OEE Sample Means

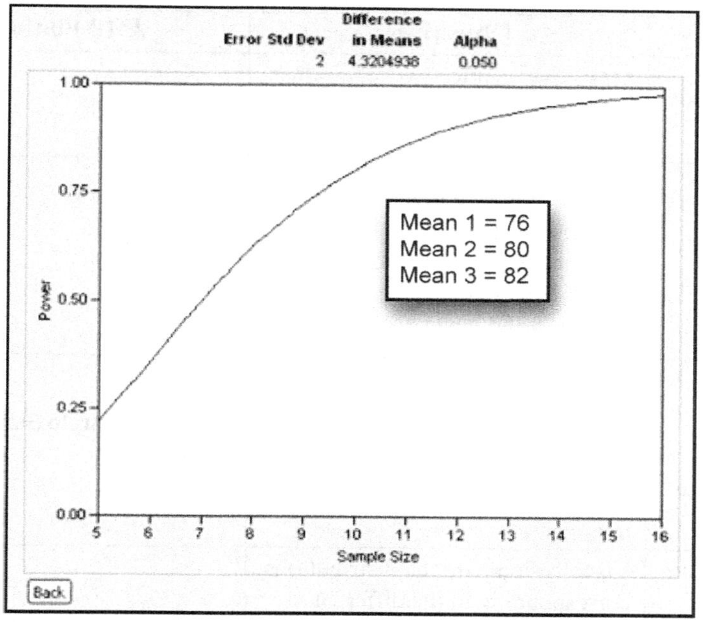

Caution: The values that we input into the various fields of the JMP sample size calculator help us to get an idea of the number of experimental units that we need in order to detect the specified differences between the means with some degree of confidence. However, they do not guarantee that we are going to be able to exactly detect the specified difference between the *k* materials, products, or processes and should be used as a guideline. In other words, the information that JMP provides is only as good as the input information that you give. If error std. dev. is really greater (or lower) than 2, these numbers will be off. If the estimated means are incorrect, these numbers will be off.

6.4 Step-by-Step JMP Analysis Instructions

For the cracking sidewalks situation described in Section 6.1 we will illustrate how to conduct a one-way ANOVA to compare the compressive strength of the different cement suppliers used in the four phases. As we have done in other chapters, we follow the general 7-step procedure shown in Table 6.7. The first two steps should be stated before any data is collected or analyzed, and they do not necessarily require the use of JMP. However, Steps 3 through 6 will be conducted using JMP and the output from these steps will help us complete Step 7.

Table 6.7 Step-by-Step JMP Analysis for Cement Supplier Comparison

Step	Objectives	JMP Platform
1. Clearly state the question or uncertainty.	Make sure that we are attempting to answer the right question with the right data.	Not applicable.
2. Specify the hypotheses of interest.	Is there evidence to indicate that the cements (suppliers) used in each phase are different from each other? Is there evidence to say that they meet the lower specification?	Not applicable.
3. Determine the appropriate sampling plan and collect the data.	Identify how many samples, experimental units, and observational units are needed. State the significance level, α, i.e., the level of risk we are willing to take.	• **DOE > Sample Size and Power**
4. Prepare the data for analysis and conduct exploratory data analysis.	Visually compare the distributions corresponding to the different cements. Look for outliers and do a preliminary check of assumptions (normality, independent observations, and similar variances).	• **Analyze > Distribution** • **Analyze > Fit Y by X**
5. Perform the analysis to verify your hypotheses, and answer the questions of interest.	How far are the compressive strengths averages from the overall average, and from each other? What about the variances? Check the analysis assumptions to make sure your results are valid.	• **Analyze > Fit Y by X > Unequal Variances, Means / ANOVA** • **Compare Means > All Pairs, Tukey HSD** • **Graph > Overlay Plot**
6. Summarize the results with key graphs and summary statistics.	Find the best graphical representations of the results that give us insight into our uncertainty. Select key output for reports and presentations.	• **Analyze > Fit Y by X** • **Graph > Control Chart > IR Chart**
7. Interpret the results and make recommendations.	Translate the statistical jargon into the problem context. Assess the practical significance of your findings.	Not applicable.

Step 1: Clearly state the question or uncertainty

Shortly after the fourth, and last, phase of a housing development of semi-custom homes was completed, and after an unusually cold winter, the developer started to receive a number of complaints for premature cracking in the cement sidewalks leading from the driveways to the front doors. Upon further investigation, it was found that most of the houses completed in the third phase of development exhibited long longitudinal cracks at the top of their sidewalks. Since there is a potential for large variations in the cement properties, such as compressive strength and tensile strength, even for a specified grade such as the one used in the homes, we would like to reexamine the cement batches used in the various phases of construction.

One of the main objectives for this study is to determine whether the cracking of the sidewalks is due to using cement that does not meet the specified properties needed for the application. The compressive strength of the cements provided by the four cement suppliers will be used as performance criteria, and compared with the stated requirements for the application. The required compressive strength for a residential sidewalk should be a minimum of 2900 pounds per square inch (psi).

Step 2: Specify the hypotheses of interest

From the problem description, we would like to compare the compressive strengths of the cement used in the four different phases of the development, thus yielding a study with one factor (Phase) and four levels (1, 2, 3, and 4). We are using the factor Phase as analogous to Supplier since each phase in the development is associated with a particular supplier. We are looking for evidence to be able to claim that there is, in fact, a difference in the compressive strength performance, both in average and variation, of the cements used in the four phases of the development. As was the case in Chapters 4 and 5, this comparison involves evaluating both the means and the variances defined by the four phases.

For a one-way ANOVA, the hypothesis statement involving the means can be written as follows:

H_0: $\mu_{\text{Phase 1}} = \mu_{\text{Phase 2}} = \mu_{\text{Phase 3}} = \mu_{\text{Phase 4}}$ (Assume)
H_1: not all the μ_i are equal, i=1, 2, 3, 4 (Prove)

We assume that all of the average compressive strengths are equal unless proven otherwise. A significant test result will not indicate how they differ, just that they do.

In addition to knowing whether the compressive strengths differ, we would also like to make sure that they meet the minimum requirement of 2900 pounds per square inch (psi). If we want to show this using a significance test, then we should use a one-sided alternative hypothesis, which demonstrates that the mean is greater than the minimum value. This can be accomplished by testing the following hypothesis (shown below for Phase 3), for each of the four means:

H_0: $\mu_{\text{Phase 3}} \leq 2900$ psi (Assume)
H_1: $\mu_{\text{Phase 3}} > 2900$ psi (Prove)

Statistics Note 6.9: We prove the alternative hypothesis in the sense of *establishing by evidence a fact or the truth of a statement* (The Compact Oxford English Dictionary).

Finally, we must also determine whether variances of the compressive strengths used in the four phases are homogeneous or similar. We need to check this assumption first (see Section 6.3.4) to insure the validity of the ANOVA *F*-statistic. This homogeneity of variances hypothesis is shown below:

H_0: $\sigma_{\text{Phase 1}} = \sigma_{\text{Phase2}} = \sigma_{\text{Phase 3}} = \sigma_{\text{Phase 4}}$ (Assume)
H_1: not all the $\sigma_{\text{Phase i}}$ are equal, i=1, 2, 3, 4 (Prove)

As was the case for the two-sample *t*-test for comparing two means, if we reject the null hypothesis of homogeneity of variances, then we need to use another form of the test statistic, such as Welch's ANOVA, to determine whether there is a difference among the compressive strengths of the four phases. In addition, knowing that the variances among the populations are different also provides insight into how well the cement used in each phase of construction can meet the minimum specification limit.

Step 3: Determine the appropriate sampling plan and collect the data

In this section we determine how many experimental units are needed to carry out our study, so that we can adequately compare the compressive strength of the cement used in the four phases of the development. What is the experimental unit (EU) for our study? Compressive strength is tested using cement briquettes that are dried for up to 14 days; therefore, one cement briquette is our experimental unit for this study. Since we will be measuring the compressive strength of the briquette, the briquette is also the observational unit (OU). A cement briquette is representative of one batch of cement.

Decide on the level of risk

In order to determine the number of samples required for our study, we need first to decide the level of risk, α, that we are willing to take when using a given sample size. While it is very tempting to use the popular, and default, value of $\alpha = 5\%$, we should consider and interpret our choice in the context of our study. For the hypothesis involving the comparison of the four phase means, if we select $\alpha = 0.05$, then 5% of the time, by chance alone, we will conclude, in error, that at least one of the cements has a different average compressive strength than the rest. A similar interpretation can be made for our hypothesis involving the four population variances. What are the implications of committing this type of error in our study? Should we be more conservative, or more liberal, in our choice of α? For the mean hypothesis, we will stick with the standard choice $\alpha = 0.05$, and select α=0.01 for the variance hypothesis. Note that selecting a significance level of $\alpha = 0.05$ (5%) is the same as setting our confidence level to 95%.

We also have four sets of hypothesis to determine whether the average performance meets minimum compressive strength of 2900 psi. Since we are conducting four *simultaneous* one-sided tests, we need to make an adjustment to our individual Type I error rate in order to maintain an overall Type I error rate of 0.05. If we select $\alpha = 0.01$ for conducting each of the one-sided *t*-tests, then the approximate overall Type I error will be equal to $1 - (1-0.01)^4 = 1 - 0.99^4$ or 3.9%.

Apart from the level of risk we are willing to live with, there are other inputs that are needed in order to calculate the required sample size. In this case we also need to specify an estimate of the noise in the compressive strengths measurements, the prospective means of the four cements, and the desired power for the significance test. These specifications are the inputs for the **Sample Size and Power** calculator in JMP.

Sample size calculation

Select **DOE > Sample Size and Power** in order to get access to the sample size calculator for testing **Three or more Sample Means**. Enter `0.05` for **Alpha**, our choice for the significance level, and `0.8` for Power (we want an 80% chance of detecting a difference if, in fact, the cements have different compressive strengths). Next, we must enter a value for the **Error Std Dev**, which represents the average standard deviation for the compressive strength for the four phases. We can use a historical estimate for this value or enter a "best guess." We can also rely on an estimate of the *coefficient of variation* (CV), which is the ratio of the standard deviation to the mean, to get an estimate for the standard deviation. The coefficient of variation is a normalized measure of noise that is appropriate when dealing with ratio scale data, like the compressive strength. Small values of the CV indicate a population with small amount of variation. Suppose that from previous experience we know that the CV for compressive strength is in the order of 5%. Since we know the lower bound for the compressive strength for this application, we can use the CV formula to solve for the standard deviation as follows:

CV = 100 x (Error std dev / sample mean)
5% = 100 x (Error std dev / 2900) ⇨ Error std dev = 145 psi.

We are using the minimum specification value of 2900 psi instead of a sample mean.

We can use different values of the CV to get a feel for how much the standard deviation changes. For example, for a CV=1% the compressive strength standard deviation is 29 psi, while for a CV=10% it is 290 psi. We can use these values of the standard deviation to calculate different sample sizes and evaluate their applicability.

Before we can calculate the total sample size, we must enter estimates for compressive strength for the four phases into the fields for the **Prospective Means**. At first this might seem like a vicious cycle because we might not know what the compressive strengths should be. However, we can use our subject matter knowledge and hypotheses to guide us. Since the material specifications require a cement with a compressive strength of at least 2900 psi for this application, we will assume that three of the phases have average compressive strength close to this lower specification limit (LSL), and the phase where the cracks occurred has a compressive strength below the lower specification limit. We can enter `3050`, `3000`, `2800`, and `2950` as our **Prospective Means** and then click **Continue** in order to obtain the total number of experimental units recommended for our study. The output is shown in Figure 6.19 and yields a total sample size of 30.40, or, rounding up 30.4/4 = 7.6, eight briquettes per phase. Since we were not quite sure about what to enter for our standard deviation, we can repeat these calculations using an Error Std Dev of 29 (CV = 1%) and 290 (CV = 10%), giving us total sample sizes of 6.4 and 108.78, or 2 and 27 briquettes per phase, respectively.

Figure 6.19 Sample Size and Power for Comparing the Phase Cements

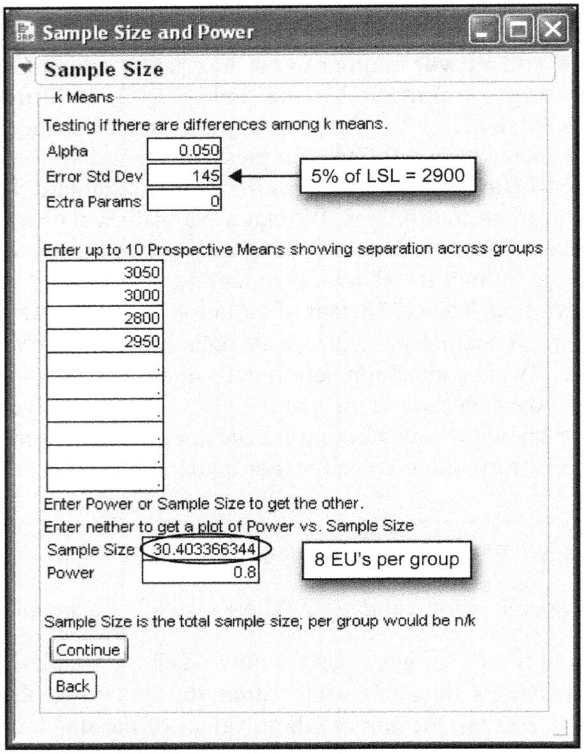

From previous studies that have been conducted, it is typical to use five briquettes in order to get a good estimate of the average compressive strength, and the observed CV has been between 1% and 5%. What should we do? Based on timing and resources, we decide to use five briquettes per phase, resulting in a total of 20 briquettes for our study.

We can use the **Sample Size and Power** calculator to evaluate our choice which, for an **Error Std Dev** of 90 (CV ~ 3%), gives us a 95% power.

It is always a good idea to write out a partial ANOVA table (source and degrees of freedom) in order to understand the partitioning of the degrees of freedom associated with the signal and noise. This shows us how the data will be allocated to the estimation of the signal and the noise, before the data is collected. The partial ANOVA table for our study is shown in Table 6.8 where the factor Phase, with 4 levels, has 3 (= 4–1) degrees of freedom, and the error has 16 = 4 × (5−1) degrees of freedom, which should provide us with a good estimate of noise. As a rule-of-thumb estimates of variation with 12 or more degrees of freedom are reliable.

Table 6.8: Partitioning Degrees of Freedom for Cement Study

Source	DF
Phase	3
Error	16
Total	19

> *Factor DF = Number of Levels − 1*
> *Factor DF = Number of Levels × (Number of Replications − 1)*

Now that we know that we need five cement briquettes from the cement used in each phase of the housing development, how do we obtain them? Standard tests procedures indicate that briquettes should be made from each batch of cement. Luckily, the developer saved several bags of the cement, corresponding to different batches used in each phase of the development, which can be used to make briquettes for this study. A *random sample* of cement will be taken from each of five bags of cement used in Phase 1, and five briquettes, one per bag, will be made, dried, and prepared for testing according to standard procedures. This will be repeated for the three other phases. The 20 briquettes will be tested in a random order in order to help prevent test bias from confusing our findings.

Now that we have determined the sampling plan for comparing the compressive strength of cement used in the four phases, we should carefully construct a data collection sheet that can be used to gather the data from the study, including all the relevant information. For example, the data sheet should always include a column for comments in order to capture any excursions or problems that we might have, or any relevant information captured while conducting the study. This information will provide valuable insights if we observe any outliers or unexpected patterns or trends in the data that can invalidate the assumption of independence. As we mentioned before, checking for independence is not easy. We can, however, ensure that we design and run our studies in a way that minimizes any potential bias. A sample of the compressive strength data is provided in Table 6.9. The entire data set can be found in Figure 6.20.

Table 6.9 Sample Data for Comparing the Four Phase Cements

Phase	Briquette	Test Order	Sample Preparation	Compressive Strength	Comments
Phase 1	1	1	Joe	3008.16	
Phase 1	2	19	Joe	3075.61	
Phase 1	3	7	Joe	3073.36	
Phase 1	4	14	Joe	2991.34	
Phase 1	5	11	Joe	3037.51	
Phase 2	1	12	Joe	3244.39	
Phase 2	2	13	Joe	3242.79	
Phase 2	3	5	Joe	3185.29	

Step 4: Prepare the data for analysis and conduct exploratory data analysis

The first thing that we need to do is read our data into JMP and make sure that it is in the right format for the **Fit Y by X** analysis platform. The data is read into JMP from its original source, an Excel file, by selecting the **Open Data Table** icon in the JMP Starter window and selecting the **Excel Files (*.XLS)** for file types. Once the correct filename is selected, we click the **Open** icon and the data is converted to a JMP data table. Note that if we want to save the data in this JMP format, we need to select **File > Save** and enter a new name and location for the JMP table.

The JMP table for our study is shown in Figure 6.20. This is a similar layout to what is required for a two-sample *t*-test. Each row contains all information pertaining to one briquette, and one column for each of the variables in the study. On the left-hand side of the table we can identify how the **Modeling Type** is set for each column variable. The Phase should be specified as nominal and have a red unordered bars icon, while our response (the compressive strength) should be specified as continuous and have a blue right triangle as a symbol. The compressive strength is recorded in psi (pounds per square inch). The compressive strength is also shown in MPa (Mega Pascals), which is the International System of Units (SI units).

Figure 6.20 JMP Table for Comparing Compressive Strength

CH6 Cement Compressive Strength

CH6 Cement Compressive Stre
Notes C:\Documents and Settings
Control Chart

Columns (6/0)
Phase
Briquette
Test Order
Compressive Strength (psi)
Compressive Strength (MPa)
Comments

	Phase	Briquette	Test Order	Compressive Strength (psi)	Compressive Strength (MPa)	Comments
1	Phase 1	1	1	3008.15896	20.7405313	Joe prepared and tested all samples
2	Phase 1	2	19	3075.61122	21.2055984	
3	Phase 1	3	7	3073.3583	21.1900651	
4	Phase 1	4	14	2991.34394	20.6245958	
5	Phase 1	5	11	3037.50682	20.9428778	
6	Phase 2	1	12	3244.39111	22.3692951	
7	Phase 2	2	13	3242.79246	22.3582728	
8	Phase 2	3	5	3185.28935	21.9618027	
9	Phase 2	4	4	3244.72604	22.3716044	
10	Phase 2	5	20	3222.61417	22.2191483	
11	Phase 3	1	2	2777.12836	19.147631	
12	Phase 3	2	16	2803.15209	19.3270584	
13	Phase 3	3	3	2785.15629	19.2029816	
14	Phase 3	4	10	2775.10642	19.1336902	
15	Phase 3	5	15	2774.83922	19.1318479	
16	Phase 4	1	18	2995.32189	20.6520228	
17	Phase 4	2	6	3037.6613	20.9439428	
18	Phase 4	3	9	3067.17029	21.1474002	
19	Phase 4	4	17	3065.28973	21.1344342	
20	Phase 4	5	8	3060.16864	21.0991255	

Now that our data is in JMP it is a good idea to take a quick look at the data for outliers, typos, or any other unusual results. We also need to check the assumptions for the *F*-test used in a one-way ANOVA. Namely, we need to check that the underlying populations can be described by a distribution that closely follows the normal distribution; that the samples are independent; and that the variances for each of the phases are similar.

In order to check our data for outliers, check the normality of the samples, and determine whether the variances are homogeneous, we will launch the **Fit Y by X** platform. (See Figure 6.21.)

Steps for Launching the Fit Y by X Platform

1. From the primary toolbar, select **Analyze > Fit Y by X**. A dialog box appears that enables us to make the appropriate choices for our data.
2. Select Compressive Strength (psi) from the **Select Columns** window and then click **Y, Response**. This populates the text box to the right of this button with the name of our response.
3. Select Phase from the **Select Columns** window and then click **X, Factor**. This will populate the text box to the right of this button with the name of the factor under investigation.
4. Click **OK**.

Figure 6.21 Fit Y by X Dialog Box for Cement Comparison

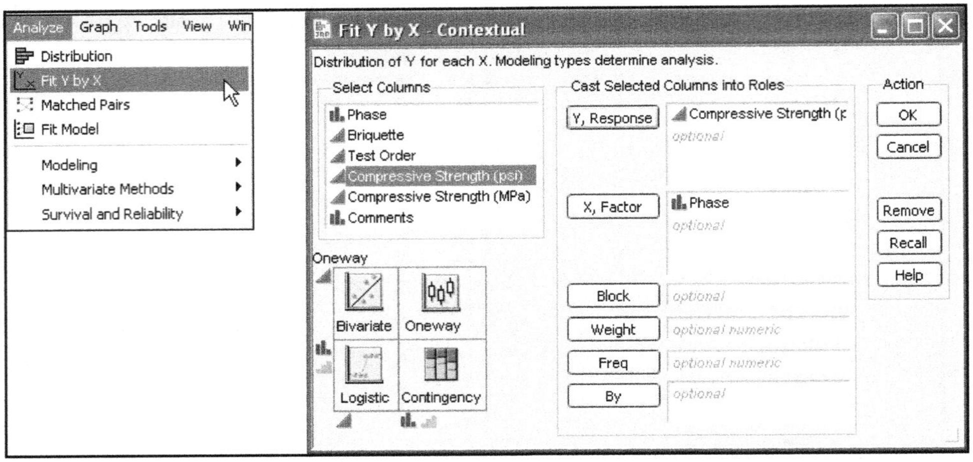

The default JMP output is shown in Figure 6.22, with Compressive Strength (psi) by Phase plotted on the y-axis and the values for the four phases plotted on the x-axis. We can enhance the appearance of this plot by turning on some additional features, which can be accessed by clicking on the red triangle at the top of the plot next to the label Oneway Analysis of Compressive Strength (psi). By selecting **Display Options** we can turn on the desired features one by one. A more efficient way to turn on multiple display options is to hold down the ALT key and click the same red triangle at the top of the plot. This will launch the menu shown in Figure 6.23, which enables us to select multiple features by clicking in the boxes of interest. Finally, if we want these options to apply to all analyses we conduct in this platform, then we can permanently set them from the **JMP Starter > Preferences** icon shown in Figure 6.24.

Figure 6.22 Default Graph from Fit Y by X Platform Comparing Compressive Strength

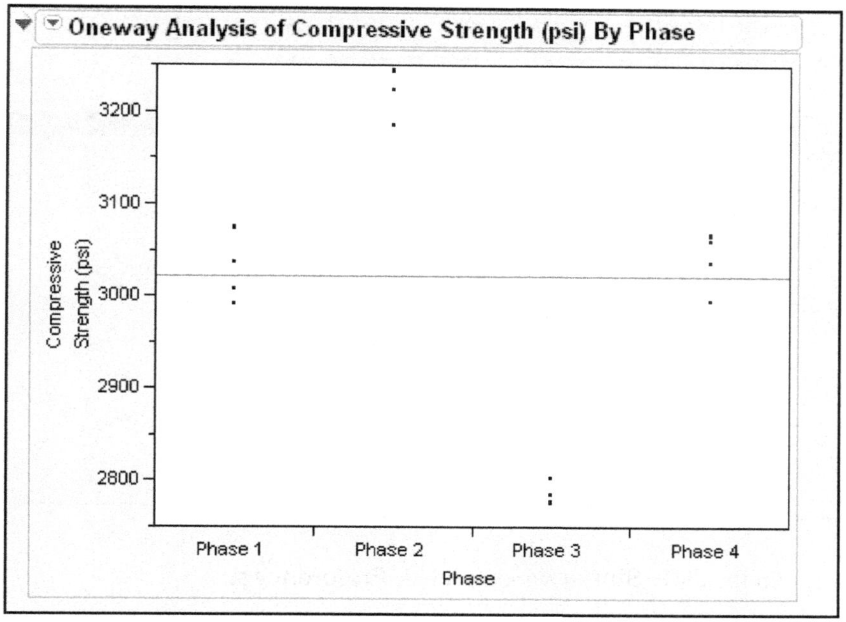

Figure 6.23 Options for Changing Plot Features in Fit Y by X

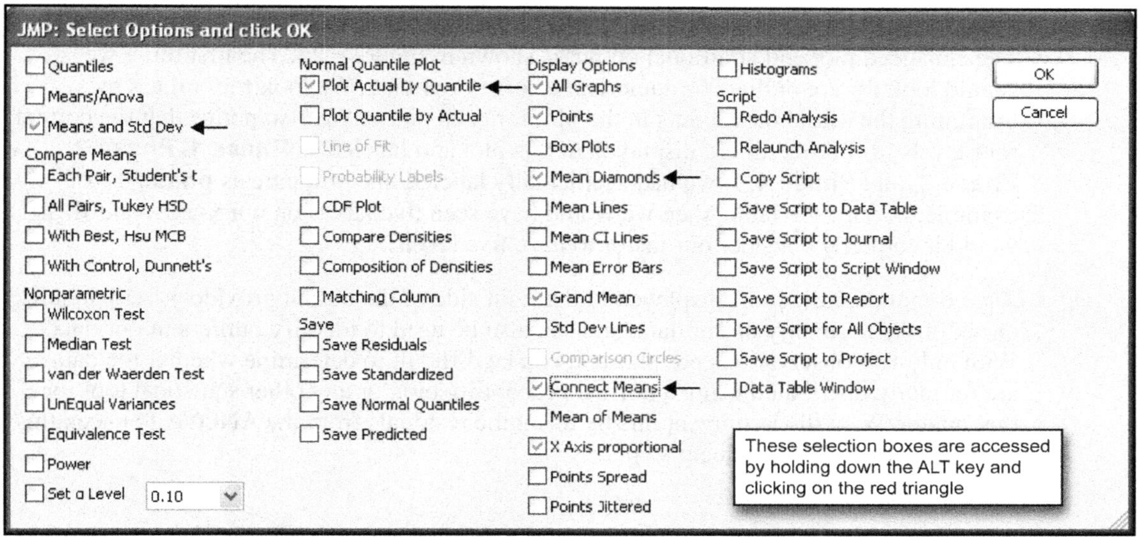

Figure 6.24 JMP Menu for Setting System Preferences for the Oneway Platform

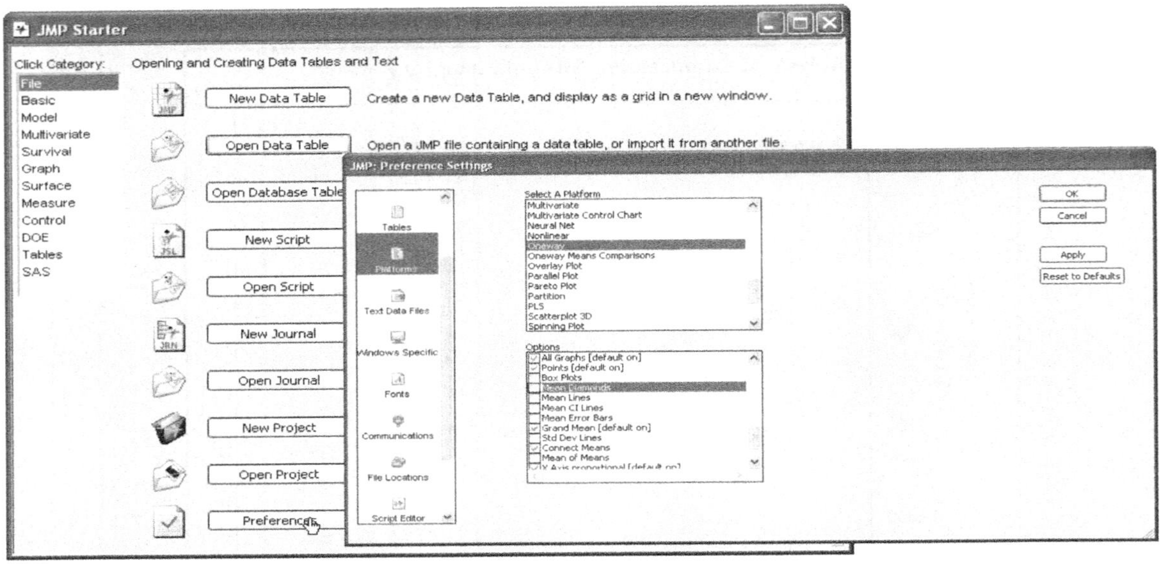

1. In the JMP Starter window, click **Preferences**.
2. In the JMP: Preference Settings dialog box, select **Platforms**.
3. From the **Select a Platform** list, select **Oneway**.
4. From the **Options** list, select the desired options.

The enhanced plot and additional output is shown in Figure 6.25. The first thing we should look for are outliers or numerical errors in our data. We look for outliers by examining the actual data points in the plot for each phase. We also notice that the correct four levels for our factor are displayed in this plot and labeled as **Phase 1**, **Phase 2**, **Phase 3**, and **Phase 4**. If we had accidentally labeled one briquette as **phase 1**, for example, in our JMP table, then we would have seen five labels on our x-axis, and JMP would incorrectly consider our factor to have five levels.

The normal quantile plots displayed to the right side of the output provide us with a quick check of the normality of our data, and can also be used to identify outliers in our data. With only five observations per phase, it can be difficult to determine whether the data are normally distributed using a normal probability plot, or any other statistical tool for that matter. We will circumvent this by using the residuals from the ANOVA to check this more formally in a subsequent step.

Figure 6.25 Enhanced Plot for Comparing Compressive Strength

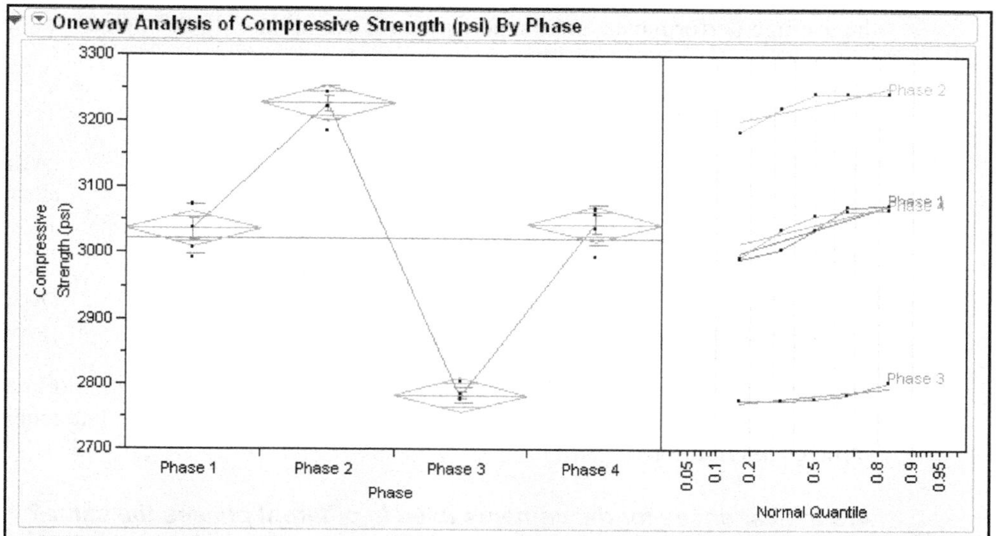

Means and Std Deviations

Level	Number	Mean	Std Dev	Std Err Mean	Lower 95%	Upper 95%
Phase 1	5	3037.20	37.8454	16.925	2990.2	3084.2
Phase 2	5	3227.96	25.5951	11.446	3196.2	3259.7
Phase 3	5	2783.08	11.9803	5.358	2768.2	2798.0
Phase 4	5	3045.12	30.2282	13.518	3007.6	3082.7

The plot clearly shows the separation of the compressive strength among the cement used in the four phases, with Phases 1 and 4 overlapping quite a bit, and close to the overall average of 3023 psi. Note that the overlap marks cannot be used for multiple comparisons since they have not been adjusted for that purpose.

The sample means and standard deviations of the compressive strength measurements for each phase are shown below the plots. We can see that the average compressive strength for Phases 1, 2, and 4 are higher than the lower specification of 2900 psi. However, the average compressive strength of the briquettes representing the cement used in Phase 3, 2783.08 psi, is lower than the lower specification. The sample standard deviations of the four phases range from 11.98 psi up to 37.85 psi. However, it is hard to determine whether this will result in a violation of our homogeneous variance assumption (we will check this assumption in Step 5).

Based on our exploratory analysis of the data we would speculate that the means will be different, and this is most likely caused by the lower mean from Phase 3. It is not evident if the means from Phases 1, 2, and 4 will be found to be different from each other and if the standard deviations will be statistically different. In Step 5 we compute significance tests for comparing the means and variances.

Step 5: Perform the analysis to verify your hypotheses

Recall from Step 2 that there are two sets of hypotheses that we want to test, to compare the average performance and the performance variation of the cements used in the four phases of the development:

Average performance

H_0: $\mu_{Phase\ 1} = \mu_{Phase2} = \mu_{Phase3} = \mu_{Phase\ 4}$ (Assume)
H_1: not all the μ_i are equal, i=1, 2, 3, 4. (Prove)

Performance variation

H_0: $\sigma_{Phase\ 1} = \sigma_{Phase\ 2} = \sigma_{Phase\ 3} = \sigma_{Phase\ 4}$ (Assume)
H_1: not all the $\sigma_{Phase\ i}$ are equal, i=1, 2, 3, 4 (Prove)

These hypotheses can be tested using the **Fit Y by X** platform. We should first test the performance variation hypothesis since the ANOVA hinges on the homogeneity of variance assumption.

Are the variances for the cements used in different phases the same?

This question is easily answered by clicking on the small red triangle above the plots and selecting **Unequal Variances** from the drop-down list, as is shown in Figure 6.26. Notice the tooltip that is automatically displayed and describes the purpose of the selection.

Figure 6.26 Unequal Variance Selection from Fit Y by X

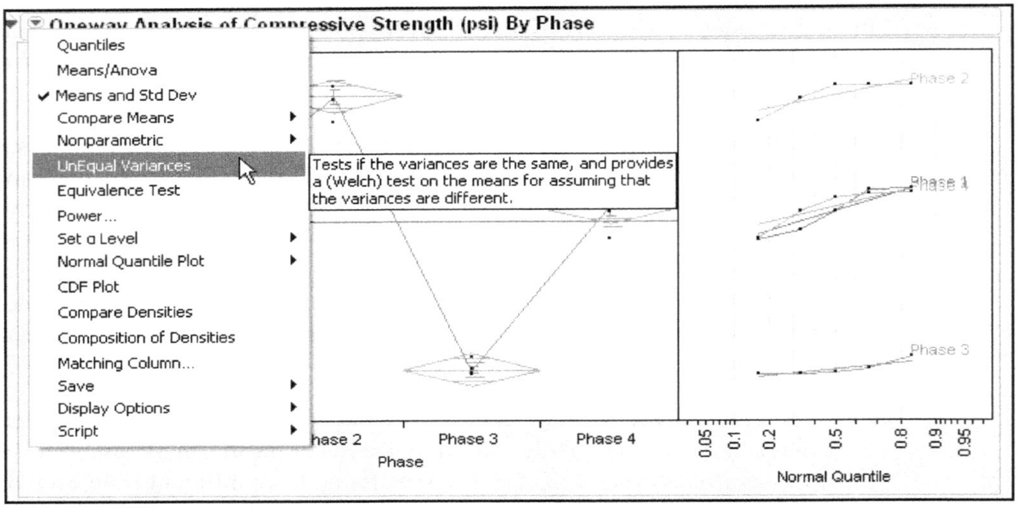

The test results are outputted in a table below the table for the means and standard deviations, as shown in Figure 6.27. As was discussed in Section 6.3.4, we recommend using the Brown-Forsythe test statistic for comparing the variances (Conover et al. 1981). The calculated Brown-Forsythe F-statistic is 1.918 with a p-value (Prob > F) equal to 0.1674, obtained from an F-distribution with 3 (DFNum) and 16 (DFDen) degrees of freedom. For this F-statistic the numerator *degrees of freedom* correspond to our factor levels and the denominator *degrees of freedom* to our experimental error (shown in Table 6.8), and the p-value represents, assuming that the variances are equal, how likely it is to observe an F-statistic of 1.918 or larger. The Brown-Forsythe results indicate that there is a 16.74% probability of obtaining an F-statistic as large or larger than 1.918, by chance alone. Given that the p-value is greater than 0.01 we do not have enough evidence to say that the variances are not equal. As usual, we are not proving that they are equal but only that we cannot say that they are different.

Figure 6.27 Output for Unequal Variances Test

Tests that the Variances are Equal

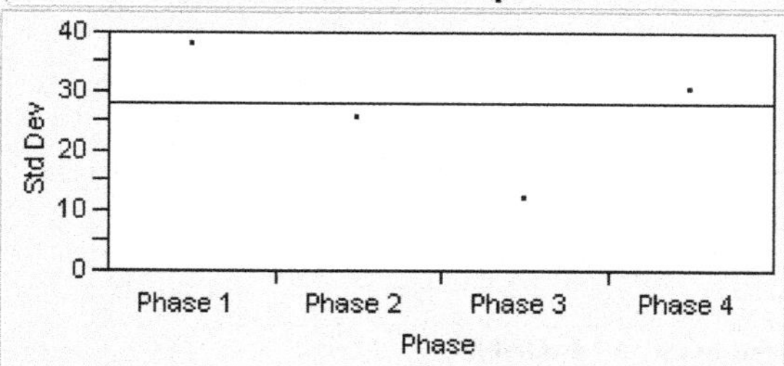

Level	Count	Std Dev	MeanAbsDif to Mean	MeanAbsDif to Median
Phase 1	5	37.84540	29.95552	35.76290
Phase 2	5	25.59515	19.20870	16.56246
Phase 3	5	11.98030	8.86217	8.07694
Phase 4	5	30.22821	22.90462	20.91959

Test	F Ratio	DFNum	DFDen	Prob > F
O'Brien[.5]	1.2538	3	16	0.3235
Brown-Forsythe	1.9180	3	16	0.1674
Levene	1.9147	3	16	0.1679
Bartlett	1.3679	3	.	0.2505

Warning: Small sample sizes. Use Caution.

Welch Anova testing Means Equal, allowing Std Devs Not Equal

F Ratio	DFNum	DFDen	Prob > F
421.9117	3	8.0105	<.0001*

Is the average performance of the cements similar for all phases?

As was noted earlier, the test statistic and *p*-value for testing the following hypotheses depend upon the outcome for the test of the four variances:

$$H_0: \mu_{Phase\ 1} = \mu_{Phase\ 2} = \mu_{Phase\ 3} = \mu_{Phase\ 4}$$
$$H_1: \text{not all the } \mu_i \text{ are equal, } i = 1, 2, 3, 4$$

If the four variances are shown to be different, then we use Welch's ANOVA to test the four means. This output is found below the JMP output for testing the four variances and is shown in Figure 6.28. As we saw above, the *p*-value for testing our variances using the Brown-Forsythe statistic is 0.1674, which is > 0.05, indicating that the data does not provide enough evidence to say that the four standard deviations are different, and therefore we do not need to use Welch's ANOVA.

Figure 6.28 JMP Output for Welch's ANOVA

Tests that the Variances are Equal

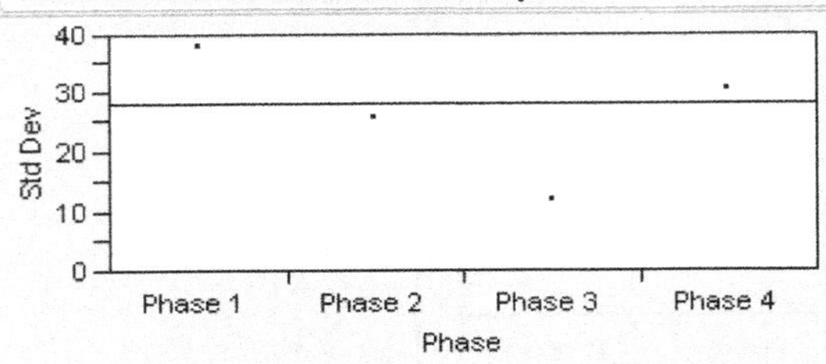

Level	Count	Std Dev	MeanAbsDif to Mean	MeanAbsDif to Median
Phase 1	5	37.84540	29.95552	35.76290
Phase 2	5	25.59515	19.20870	16.56246
Phase 3	5	11.98030	8.86217	8.07694
Phase 4	5	30.22821	22.90462	20.91959

Test	F Ratio	DFNum	DFDen	Prob > F
O'Brien[.5]	1.2538	3	16	0.3235
Brown-Forsythe	1.9180	3	16	0.1674
Levene	1.9147	3	16	0.1679
Bartlett	1.3679	3	.	0.2505

Warning: Small sample sizes. Use Caution.

Welch Anova testing Means Equal, allowing Std Devs Not Equal

F Ratio	DFNum	DFDen	Prob > F
421.9117	3	8.0105	<.0001*

Used to compare the four Phase means when variances are not equal

To obtain the one-way ANOVA *F*-test, click the small red triangle at the top of the window next to the **Oneway Analysis ...** label, and select **Means/Anova**, as shown in Figure 6.29. This is the appropriate choice for when the population variances are assumed equal and can be pooled into one error term.

Figure 6.29 Testing Four Means in JMP Using Means/ANOVA

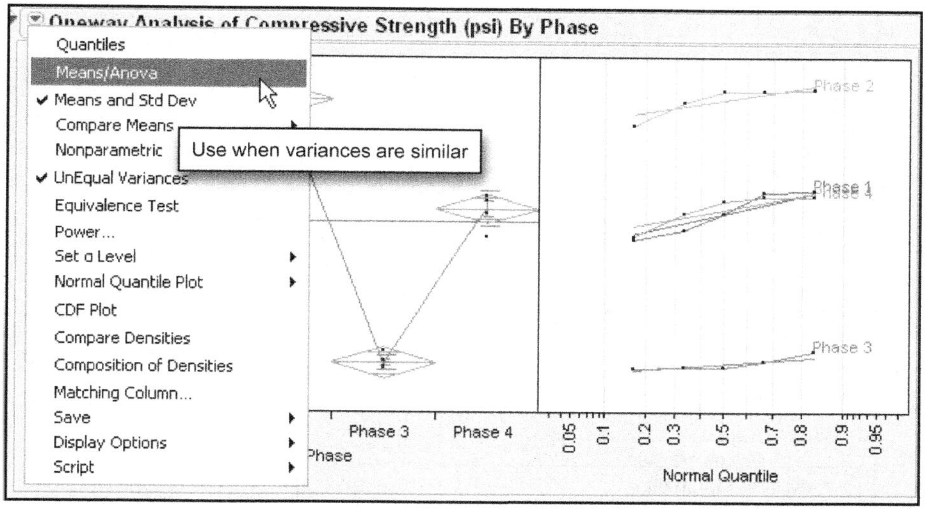

The output for our selection of **Means/Anova** is shown in Figure 6.30. The first several lines of output provide general information about our data, starting with the Rsquare = 0.9755. This number is telling us that 97.55% of the total variation in our compressive strength data is explained by differences in the cement used in the different phases of the development. The Root Mean Square Error of 28.039 is the pooled estimate of our experimental error, and the grand mean for compressive strength is 3023.339. Finally, there are 20 observations used in this analysis.

Figure 6.30 One-way ANOVA Output for Testing Phase Means

Oneway Analysis of Compressive Strength (psi) By Phase

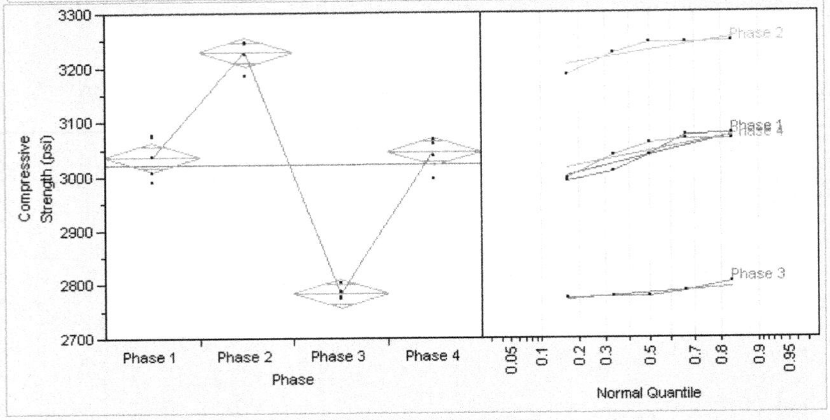

Oneway Anova

Summary of Fit

Rsquare	0.975523
Adj Rsquare	0.970934
Root Mean Square Error	28.03863
Mean of Response	3023.339
Observations (or Sum Wgts)	20

Analysis of Variance

Source	DF	Sum of Squares	Mean Square	F Ratio	Prob > F
Phase	3	501317.19	167106	212.5582	<.0001*
Error	16	12578.63	786		
C. Total	19	513895.82			

Not all the μ_i are equal

Means for Oneway Anova

Level	Number	Mean	Std Error	Lower 95%	Upper 95%
Phase 1	5	3037.20	12.539	3010.6	3063.8
Phase 2	5	3227.96	12.539	3201.4	3254.5
Phase 3	5	2783.08	12.539	2756.5	2809.7
Phase 4	5	3045.12	12.539	3018.5	3071.7

Std Error uses a pooled estimate of error variance

The ANOVA table is below the **Summary of Fit** results. It is good practice to verify the degrees of freedom for the factor, the error, and the total in order to make sure that it matches our study (shown in Table 6.8). The degrees of freedom for Phase should be equal to 3, the number of phases minus one, while the degrees of freedom for error are obtained as the number of phases, 4, times the number of briquettes per Phase minus one, 5 – 1; i.e., 4×(5 – 1) = 16. The total degrees of freedom (19) is one less than the total number of observations in the study or the sum of the Phase and error degrees of freedom. The F Ratio is the test statistic that we use to determine how large the signal, due to differences in the cement used in the different phases, is relative to the noise in our study.

The F ratio = MSFactor / MSError = 167106 / 786 = 212.5582, so the phase signal generated is 212 times larger than the noise. Because this value is so large, we are sure it indicates a statistically significant result. The p-value confirms this; Prob > F is <.0001, which is less than our significance level of 0.05.

The last section of the output shows the factor level means, their standard errors, and 95% confidence intervals. The note at the bottom of the output indicates that the **Std Error uses a pooled estimate of error variance** in accordance with the homogeneity of variance assumption.

> **JMP Note 6.4**: The confidence intervals in the **Means/ANOVA** output are not identical to the 95% confidence intervals shown in the **Means and Std Dev** output, which are calculated using the standard deviations for each factor level rather than the pooled estimate.

The data provides enough evidence to "prove" that not all of the average compressive strengths are the same. However, the ANOVA *F*-test does not enable us to say which one, or which ones, are responsible for this difference. This can be done by performing a multiple comparisons test like Tukey's HSD, which tests all six 4 × (4–1)/2 pairwise comparisons between the four phase averages.

Is the average cement performance for each phase meeting the lower specification?

Recall from Step 2 that we are also interested in knowing whether the average performance of the cement used in each phase is greater than the minimum specification of 2900 psi. This is done by testing four one-sided hypotheses.

H_0: $\mu_{\text{Phase i}} \leq 2900$ psi (Assume)

H_1: $\mu_{\text{Phase i}} > 2900$ psi, for i = 1, 2, 3, 4. (Prove)

Since these are four simultaneous hypotheses with four *p*-values we need to adjust the individual *p*-values for each test in order to maintain an overall Type I error rate less than or equal to 5%. We achieve this by setting each individual significance level, α_{indiv}, to 1%. Using the formula shown in Step 3 the overall Type I error rate is equal to $1 - (1 - 0.01)^4 = 1 - 0.99^4$ or 3.9%, where 4 is the number of simultaneous hypothesis.

> **JMP Note 6.5**: Since JMP calculates two-sided confidence intervals, we need to set the alpha level to 0.02 (Figure 6.31), not 0.01, because this produces a (100 – 0.02/2)% = 99% lower confidence bound, which is what we want. In other words, the lower confidence bound from a two-sided 98% confidence interval is the same as a one-sided 99% confidence lower bound.

Figure 6.31 Confidence Intervals for Each Phase Mean

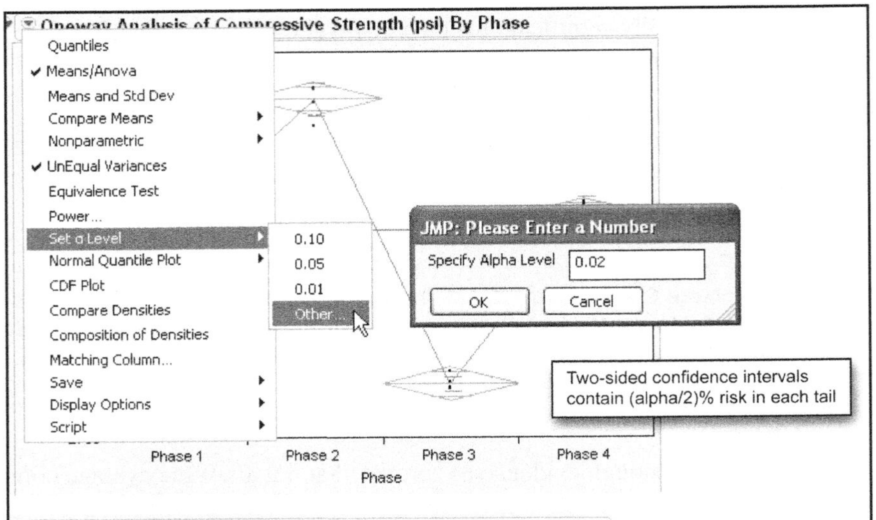

Level	Number	Mean	Std Dev	Std Err Mean	Lower 98%	Upper 98%
Phase 1	5	3037.20	37.8454	16.925	2973.8	3100.6
Phase 2	5	3227.96	25.5951	11.446	3185.1	3270.9
Phase 3	5	2783.08	11.9803	5.358	2763.0	2803.2
Phase 4	5	3045.12	30.2282	13.518	2994.5	3095.8

For each of the four averages, we compare the 98% lower confidence bound, Figure 6.31, to the lower specification of 2900 psi. If the lower bound is greater than 2900, then we have evidence to claim that a given cement is exceeding the lower specification. The 98% lower confidence bounds for Phases 1, 2, and 4 are all greater than 2900 psi, while the lower bound for Phase 3 is not. This confirms our suspicion that the cement used in the Phase 3 of the housing development was substandard.

Pairwise Comparisons of Phase Cement Averages

Multiple comparisons allow us to better understand the differences in average compressive strengths between the four cements. We click the red triangle at the top of the window and select **Compare Means > All Pairs, Tukey HSD**, as shown in Figure 6.32.

Figure 6.32 Multiple Comparison Tests for Phase Means

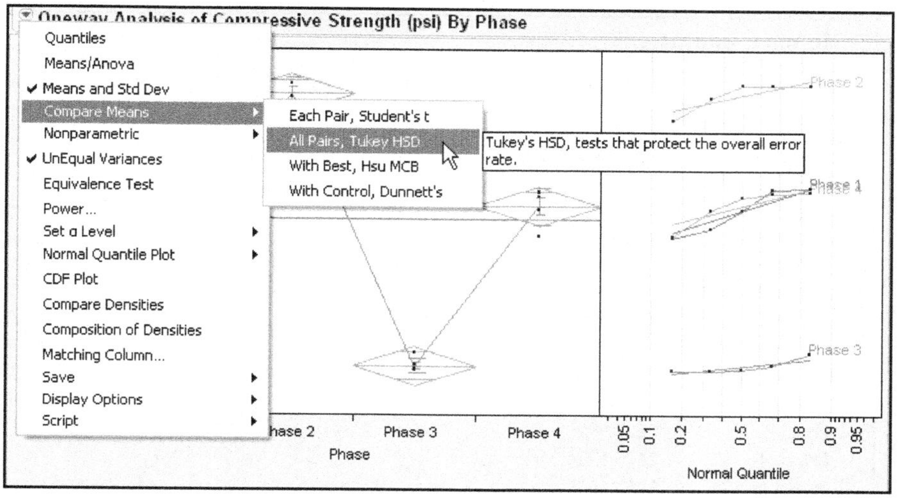

At the top of this output, Figure 6.33, we see that the Alpha is set to 2%. This is because we set it to that value, Figure 6.31, when we compared the individual averages to the minimum requirement limit. In this case, JMP is keeping the overall Type I error to 2% rather than the default value of 5%. In Step 3 we selected 5% as the significance level for the overall ANOVA *F*-test. Here we are using a 2% significance level for the six simultaneous pairwise comparisons.

In the middle section of the output we have a display of each of the phases (**Level**) and their means (**Mean**). Each phase level is accompanied by a letter, with levels sharing the same letter being statistically similar. We can quickly see the following:

- The average cement performance for the cement used in Phase 2 (letter A) is higher than the average performance for the other cements.
- The cements used in Phases 1 and 4 (letter B) have similar average performances.
- The cement used in Phase 3 (letter C) has an average performance lower than the rest.

Figure 6.33 Multiple Comparisons

Means Comparisons

Comparisons for all pairs using Tukey-Kramer HSD

q'	Alpha
3.32749	0.02

Abs(Dif)-LSD

	Phase 2	Phase 4	Phase 1	Phase 3
Phase 2	-59.01	123.83	131.76	385.88
Phase 4	123.83	-59.01	-51.08	203.04
Phase 1	131.76	-51.08	-59.01	195.11
Phase 3	385.88	203.04	195.11	-59.01

Positive values show pairs of means that are significantly different.

Level				Mean
Phase 2	A			3227.9626
Phase 4		B		3045.1224
Phase 1		B		3037.1959
Phase 3			C	2783.0765

Levels not connected by same letter are significantly different.

Level	- Level	Difference	Lower CL	Upper CL
Phase 2	Phase 3	444.8862	385.879	503.8931
Phase 4	Phase 3	262.0459	203.039	321.0529
Phase 1	Phase 3	254.1194	195.112	313.1263
Phase 2	Phase 1	190.7668	131.760	249.7737
Phase 2	Phase 4	182.8403	123.833	241.8472
Phase 4	Phase 1	7.9265	-51.080	66.9335

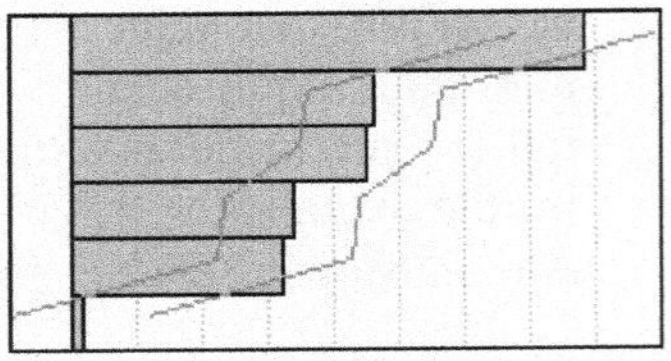

The last section of the output shows the estimated difference between the phase's average performance, as well as a confidence interval for the estimated difference. As we mentioned before, confidence intervals are better at showing the actual differences than a significance test. The first three lines show that the average performance of the cement used in Phase 3 is lower than the one for Phase 2 by about 445 psi; lower than the one in Phase 4 by 262 psi; and lower than the one for Phase 1 by 254 psi. The cement used in Phase 3 did not meet the required specification of 2900 psi, and it was of inferior quality to the cement used in the other three phases.

Residuals Checks

The ANOVA for comparing the compressive strength of the different phases also provides a statistical model that enables us to predict the value of the compressive strength for the cement used in a given phase. The predicted value for a given phase is the average compressive strength of the five briquettes in that phase. Figure 6.31 shows the average compressive strength for Phase 2 to be 3227.96 psi. This is the predicted value for the cement used in that phase. Similarly, the other three averages represent what the statistical model predicts the compressive strength to be for that phase.

Residuals are obtained as the differences between an observation and the value predicted by the model. The first observation in Figure 6.20, corresponding to briquette 1 in Phase 1, is 3008.15896. The predicted average value for Phase 1 is 3037.1959. The residual is then 3008.15896 – 3037.1959 = –29.0369 (Figure 6.34). Residuals represent experimental error, or the variation above and beyond the statistical model. In our example, they represent the variation not accounted for by the different phases. A careful examination of the residuals, using the appropriate plots, is key for checking the ANOVA model assumptions that the observations are independent and normally distributed, with similar variances for each factor level (Montgomery and Runger 2003).

Although the ANOVA is quite robust in regards to departures from normality, and somewhat robust in regards to modest variance heterogeneity, we should check the assumptions of independence, normality, and equal variances to make sure that serious violations have not occurred, so not to invalidate the results of the analysis. We also need to check for outliers since discrepant observations can sometimes cast some doubt on the results of the analysis. Sometimes analysts consider outliers to be observations that need to be discarded. The blind application of throwing away "bad values" should be avoided because, as Box, Hunter, and Hunter (2005) point out, "If their cause can be identified, they can provide important and unanticipated information."

Are the residuals normally distributed?

In Step 4, we used normal quantile plots to visually assess if the normal distribution does a good job at describing the cement briquettes compressive strengths. However, since there were only five briquettes per phase, this was difficult to do. A better method is to save the residuals from the ANOVA to the JMP table, and then use the **Analyze > Distribution** platform to create a normal probability plot, based on 20 briquettes. The ANOVA residuals are calculated by subtracting from each response value its respective group mean. They represent what is left in our data after we extract the signal, and therefore, if the model does a good job at explaining the data, should behave like white noise.

The residuals can be saved from the **Fit Y by X** platform to the JMP data table by selecting **Save > Save Residuals** from the drop-down list shown in Figure 6.34. This adds a new column to the end of the JMP table labeled **Compressive Strength (psi) centered by Phase**. We also save the predicted values, group means, by selecting **Save > Save Predicted.** We will use the predicted values in conjunction with the residuals to create another residual plot.

JMP Note 6.6: If the variances were proven to be heterogeneous then we would have selected **Save Standardized** instead, which are the residuals normalized by the standard deviation for each factor level.

Figure 6.34 Saving Residuals in Fit Y by X Platform

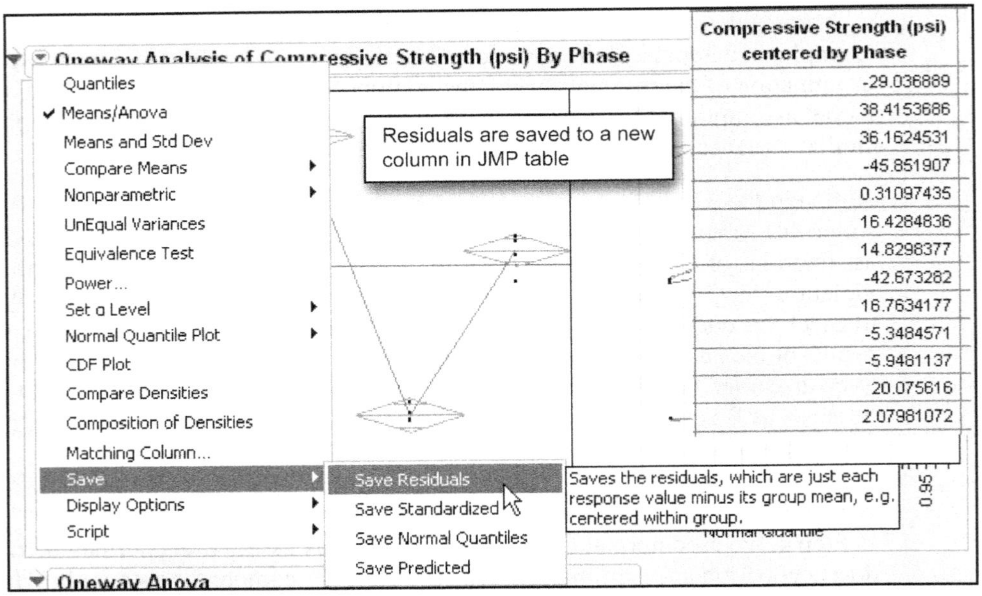

We next select **Analyze > Distribution** from the main menu, and place the new column in the **Y, Columns** box and click **OK**. This produces a histogram of the residuals. If we have not previously changed the default options for the Distribution platform, then only a histogram of the data, along with the quantiles and some summary statistics, will be displayed.

In order to check the normality assumption we select **Normal Quantile Plot** by clicking on the red triangle next to the response name **Compressive Strength (psi) centered by Phase**. This displays a normal quantile plot of the data as shown in Figure 6.35a.

Figure 6.35a Normality Check of Compressive Strength Residuals

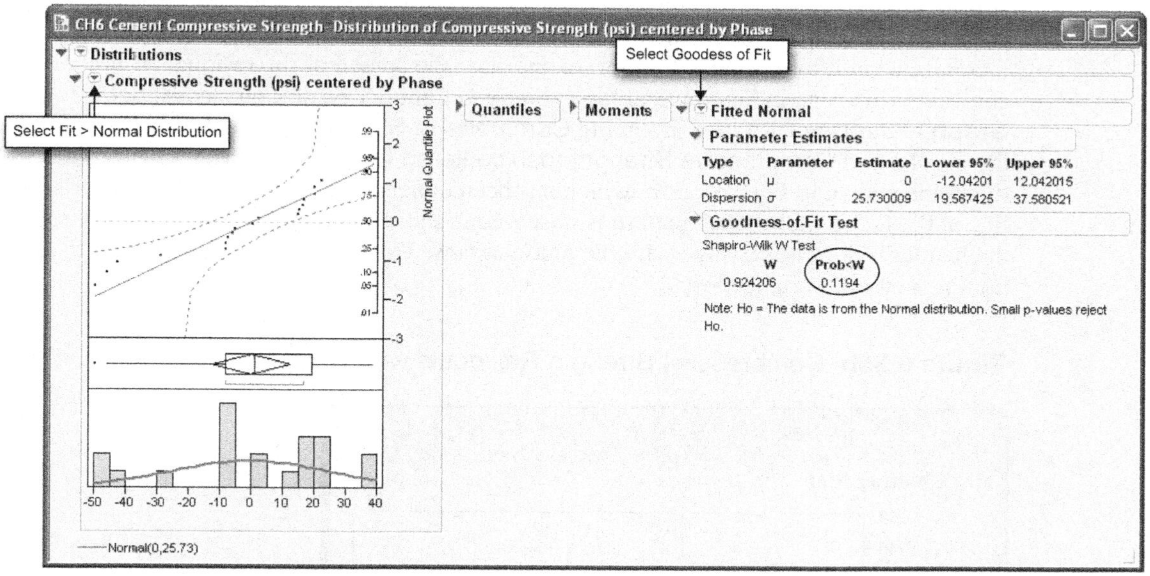

For small samples the deviations from the straight line can be large making the interpretation of the plot difficult (see Sall and Jones 2008). The confidence bands help in these situations, as well as the use of other residuals plots as we show in Chapter 7. Since all of the 20 points fall within the red 95% confidence bands (dotted red lines) of the normal plot we can say, with 95% confidence, that the data can be approximated by a normal distribution.

We can also fit a normal distribution to the data by selecting **Fit Distribution > Normal**, again by clicking on the red triangle next to the response name **Compressive Strength (psi) centered by Phase**. A normality test can be performed if you select **Goodness of Fit** from the drop-down menu (red triangle) next to the **Fitted Normal** name. The output, also shown in Figure 6.35a, shows a normal curve superimposed on the histogram, and the Shapiro-Wilk W test for normality. The W test has a p-value = 0.1194 > 0.05 indicating that we do not have enough evidence to say that the normal distribution is not appropriate for our data.

As we mentioned in Chapter 5, we prefer to use the normal probability plot because it is a straightforward way of visually checking for normality. Even the histogram can sometimes mislead us. If you look at the histogram in Figure 6.35a you will get the impression that the normal distribution does not do a good job at describing this particular data set, when the normal quantile plot shows otherwise. In some case this could be due to the JMP default bin size and axis scaling. You can change the bin size by double-clicking on the X-axis as shown in Figure 2.16.

Does the spread of the residuals increase as the response increases?

A plot of the residual versus the predicted values should show a random distribution of points around zero. If the residuals tend to increase, or decrease, as the predicted values increase, what it looks like a "funnel shape," then the spread of the residuals is not constant. This is an indication that the variances might not be homogeneous. Selecting **Graph > Overlay Plot**, and assigning Compressive Strength (psi) mean by Phase to the X role, and Compressive Strength (psi) centered by Phase to the Y role, produces the graph shown in Figure 6.35b, which has been enhanced by adding a Y-axis reference line at 0. No "funnel shape" pattern is observed in this plot, suggesting that the spread of the residuals does not change with the phase means. This agrees with our findings in the homogeneity of variance test.

Figure 6.35b Compressive Strength Residuals vs. Predicted

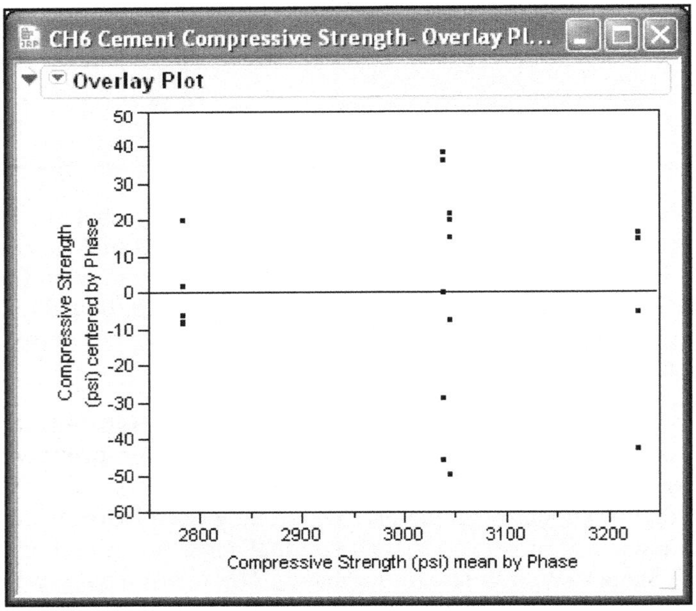

Are the residuals related to time order?

Another useful residual plot is a plot of the residuals vs. time order. This plot helps us to determine whether there is some type of time dependency in our observations. One of the things that we were careful to do was to keep track of the Test Order of the briquettes used in the study. If we plot the residuals vs. the test order, we can detect systematic drifts that might have occurred during the study. Figure 6.35c shows a plot of the residuals vs. the Test Order, generated using **Graph > Overlay Plot**. In Figure 6.35c the points have been connected, and a Y-axis reference line at 0 has been added. No apparent trend or drift can be seen in this plot.

Figure 6.35c Compressive Strength Residuals vs. Time Order

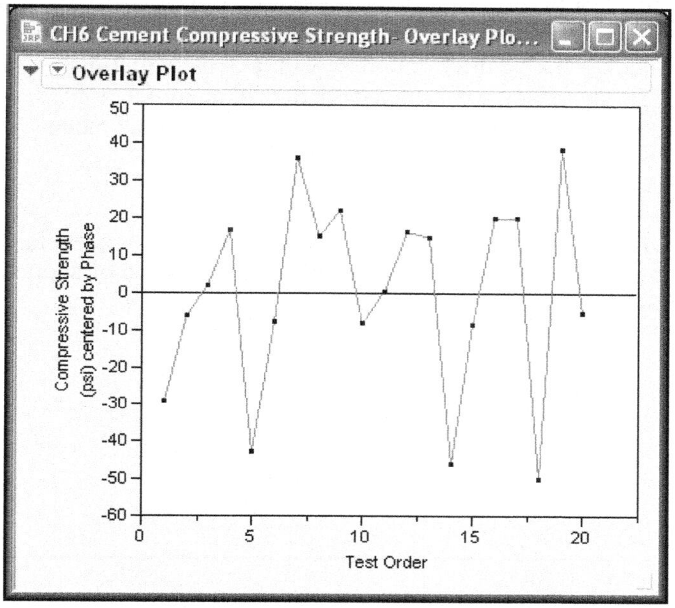

Step 6: Summarize your results with key graphs and summary statistics

It is important to be able to convey our results in a clear and concise way, especially if the audience is not familiar with the statistical techniques which have been used. To quote Walter Shewhart, the father of statistical quality control, "Original data should be presented in a way that will preserve the evidence in the original data for all the predictions assumed to be useful." Below we show a very useful plot to display the results of an ANOVA analysis, the graphical ANOVA, in the latest edition of Box, Hunter, and Hunter (2005).

> *"Original data should be presented in a way that will preserve the evidence in the original data for all the predictions assumed to be useful."*
>
> *W. Shewhart*

Graphical ANOVA

The graphical ANOVA is a really clear way of showing the effects of the different factor levels (signal) as compared to the residuals (noise) distribution. Figure 6.36 shows the graphical ANOVA for the cement data. The bottom part of the graph shows the distribution of the residuals around zero. The top part of the plot shows scaled mean phase deviations from the overall mean. The scaling makes it possible to compare the scaled deviations to the distribution of the residuals. As Box, Hunter, and Hunter (2005) point out, "The graphical ANOVA brings to the attention of the experimenter how big in relation to noise the differences really are." Recall that the residuals represent the variation not accounted for by the different phases. In fact, the variation of the residuals is estimated by the MSE. Therefore, any scaled deviations that are far from the center of the distribution are deemed to be significantly different from the overall mean.

Statistics Note 6.10: The *F*-statistic (shown in equation 6.4) can be written as the ratio of sum-of-squares scaled by the ratio of degrees of freedom:

$$F - statistic = \frac{Signal}{Noise} = \frac{\frac{SS_{Factor}}{df_{Factor}}}{\frac{SS_{Error}}{df_{Error}}} = \frac{SS_{Factor}}{SS_{Error}} \times \frac{df_{Error}}{df_{Factor}}$$

Similarly, in the graphical ANOVA in Figure 6.36 each phase deviation from the overall mean (Phase Mean–Overall Mean) has been scaled by the factor $\sqrt{(df_{Residual(Error)}/df_{Phase})}$. The scaling factor represents the ratio of degrees of freedom between the Residuals (Error) and the factor phase, making the phase deviations comparable to the residuals.

- For Phase 1 the scaled deviation is equal to $(3037.20 - 3023.34) \times \sqrt{16/3} = 13.86 \times 2.31 = 32.02$.
- For Phase 2 the scaled deviation is equal to $(3227.96 - 3023.34) \times \sqrt{16/3} = 204.62 \times 2.31 = 472.63$.
- For Phase 3 the scaled deviation is equal to $(2783.08 - 3023.34) \times \sqrt{16/3} = -240.26 \times 2.31 = -554.86$.
- For Phase 4 the scaled deviation is equal to $(3045.12 - 3023.34) \times \sqrt{16/3} = 21.78 \times 2.31 = 50.31$.

The top plot reveals quite clearly that the compressive strength of the Phase 3 cement is lower than the rest, as it is shown by a negative and large-scaled deviation (–554.86) that is far away from the residual distribution. We can also see that the compressive strengths of the cements used in Phases 1 and 4 are similar, and that the cement used in Phase 2 has a compressive strength higher than the rest. The graphical ANOVA gives us a sense for how large the phases' means are, and this in turn helps us determine whether those means are practically significant.

Figure 6.36 Graphical ANOVA for the Compressive Strength Data

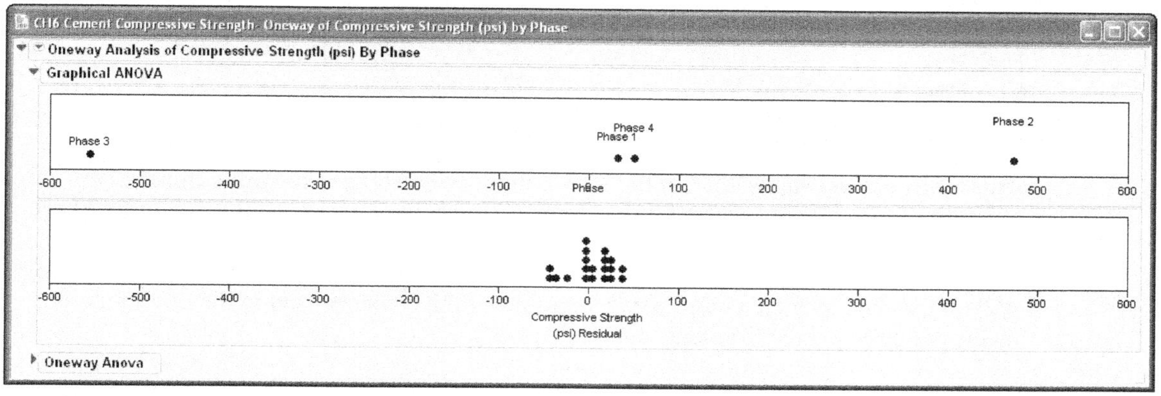

Statistics Note 6.11: The graphical ANOVA is the perfect complement to the ANOVA table. In fact, they should be used together because they combine the statistical and the practical significance. The ANOVA table helps detect that there is a signal, via the *F*-test, while the graphical ANOVA shows us how large the detected differences are. We should resist using the *p*-values from an ANOVA in a mechanical way. We need to understand the magnitude of the effects in conjunction with their statistical significance in order to maximize the discoveries we can make with the use of statistics.

JMP Note 6.7: The graphical ANOVA is not currently implemented in JMP. A script to generate the graphical ANOVA can be downloaded from http://support.sas.com/authors.

Confidence Intervals

As we mentioned before the results of a significance test just indicate that there is a discrepancy between the observed data and our null hypothesis. Confidence intervals, on the other hand, convey this information plus the magnitudes of the observed discrepancies. In order to demonstrate that the average compressive strength for the cement used in a given phase meets the lower bound of 2900 psi, we need to compare its lower confidence bound to 2900. Figure 6.37 shows the 98% two-sided confidence bounds, or equivalently the 99% one-sided confidence lower or upper bounds, for the average compressive strengths of the cements used in the four phases.

Figure 6.37 98% Confidence Bounds for Compressive Strengths

Means for Oneway Anova

Level	Number	Mean	Std Error	Lower 98%	Upper 98%
Phase 1	5	3037.20	12.539	3004.8	3069.6
Phase 2	5	3227.96	12.539	3195.6	3260.4
Phase 3	5	2783.08	12.539	2750.7	2815.5
Phase 4	5	3045.12	12.539	3012.7	3077.5

Std Error uses a pooled estimate of error variance

This table quickly shows that, at the 99% confidence level, the average compressive strength of the cement used in Phase 3 does not meet the lower specification of 2900 psi because even in the best-case scenario, the 99% upper bound, the average compressive strength is only 2815.5 psi. In other words, the 99% confidence upper bound = 2815.5 psi < 2900 psi. The average compressive strength for the cements used in the other phases meets the 2900 psi requirement.

Statistics Note 6.12: The above claims are for the average compressive strength and not a single batch of cement. As we show in Chapter 2, claims for individual values are made using either a prediction interval or a tolerance interval. A sample of five briquettes per phase is not enough to compute these intervals with enough precision.

Having established that the average compressive strength of cement used in Phase 3 was substandard, we can show how different it was from the average compressive strength of the cements used in the other three phases. Figure 6.38 shows the 98% confidence intervals for the differences in average compressive strength with respect to Phase 3.

Figure 6.38 98% Confidence Bounds for Comparisons with Phase 3 Average Compressive Strength

Level	- Level	Difference	Lower CL	Upper CL
Phase 2	Phase 3	444.8862	385.879	503.8931
Phase 4	Phase 3	262.0459	203.039	321.0529
Phase 1	Phase 3	254.1194	195.112	313.1263

By looking at the **Lower CL** and **Upper CL** we can see the following is true:

- At the 98% confidence level, the data shows that the Phase 3 cement's average compressive strength was at least 195 psi (vs. Phase 1) lower than the other three cements.
- The Phase 2 vs. Phase 3 interval shows that the difference in average compressive strength between these two cements can be as large as ~504 psi. For the other two phases this difference can be as large as ~320 psi.
- Since the average compressive strengths of all the cements is about 3023 psi, the percent decrease in average compressive strength of the Phase 3 cement with respect to the Phase 2 cement is between (385.8790/3023.339 ; 503.8931/3023.339) = (13% ; 17%). This information can give valuable insight in terms of what the best type of cement is for minimizing the occurrence of cracks.
- The width of a confidence interval gives us a sense of the noise in our estimated differences. For Phase 2 – Phase 3 the interval width is 503.8931 – 385.879 = 118.014 psi (the width of all these intervals is the same = 118 psi). This value, 118 psi, is about one fourth, or 26.5%, of the Phase 2 – Phase 3 difference of 444 psi, and about one half of Phases 4 – Phase 3 (262 psi) and Phase 1 – Phase 3 (254 psi) differences (45% and 46% respectively). This indicates a good level of precision in the estimation of these differences.

Individuals Control Chart with Phase Variable

The graphical ANOVA, shown in Figure 6.36, is a simple and powerful visual summary of the outcomes from our study. There is another graphical representation that enables us to compare, side by side, the average performance and variation of the four phases. An Individual measurements and moving Range (IR) chart is a trend chart with control limits based on a local measure of variation. The control limits are computed as Average ± 3 [Local Measure of Variation], where the local measure of variation is based on the moving range of two successive observations (see Wheeler 2005 for details).

Wheeler (2005) suggests using an Individual measurements and moving Range (IR) chart "…if there are at least six or more values for each treatment." Although the cement data does not meet this requirement (we have only five briquettes per phase), we will use this data to demonstrate how this chart can be generated in JMP.

In order to generate an IR chart for the compressive strength select **Graph > Control Chart > IR** (Individual measurements and moving Range) to bring up the **IR Control Chart** interface window. Figure 6.39 shows our choices.

JMP Note 6.8: The data needs to be sorted by the **phase** variable before generating a phase control chart.

Figure 6.39 Accessing Individuals Control Charts for Compressive Strength

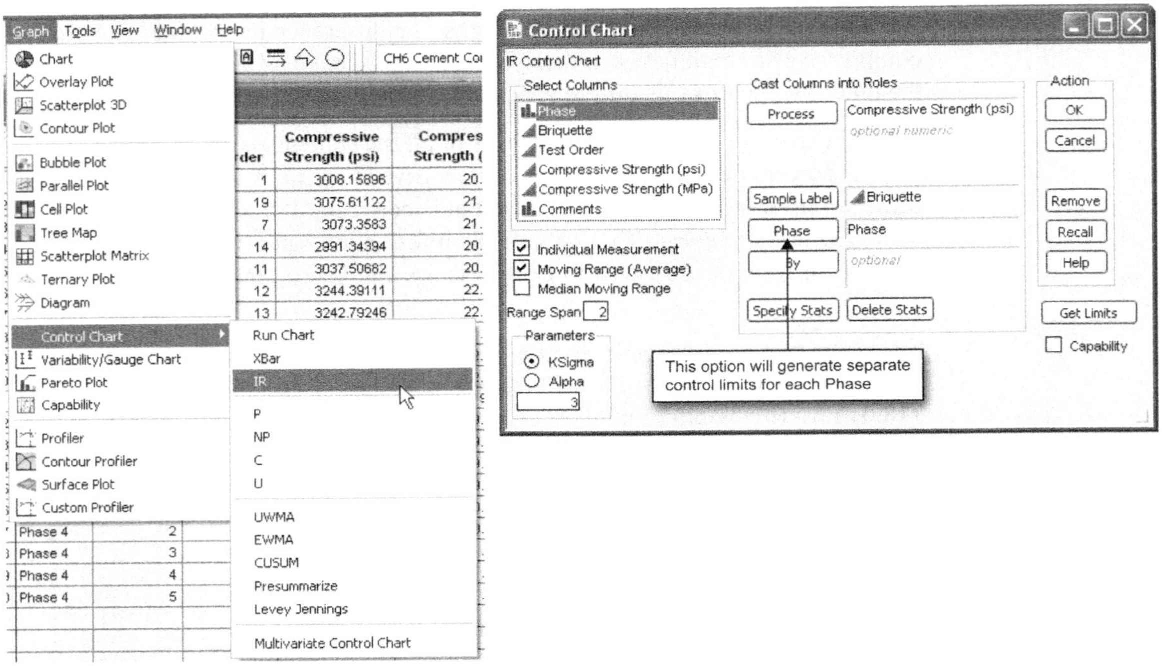

We entered `Compressive Strength (psi)` into the **Process** variable, `Briquette` as the **Sample Label**, and `Phase` as the **Phase** variable (it is just a coincidence that they have the same name). By entering a variable in the **Phase** field, separate control limits will be calculated for each value of the variable phase. Note: The JMP table must be sorted by the phase variable.

The Individual measurements (top) chart is shown in Figure 6.40. From this chart it is clear that the compressive strength of the briquettes made with the cement used in Phase 3 are generally lower than the other phases, and that the compressive strength of the five Phase 3 briquettes is pretty consistent. The compressive strength of the briquettes from the Phase 2 cement is generally higher than the rest. Note that these are visual comparisons and not a significance test.

Figure 6.40 IR Phase Chart for Compressive Strength by Phase

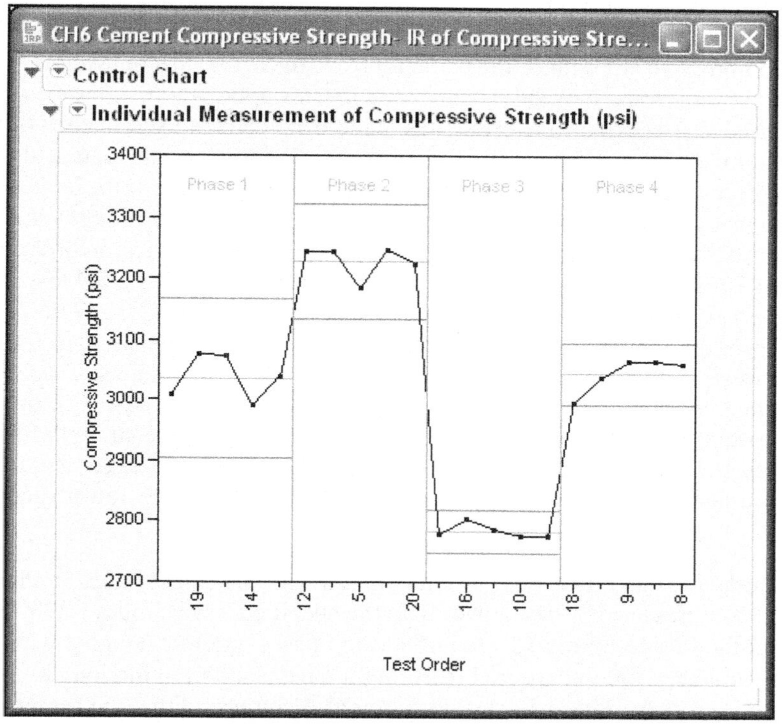

Graphical Tools

In this section we introduced the graphical ANOVA and the Individual measurements and moving Range chart to help us visually support our findings. The graphical ANOVA allows us to see if there are differences between the average compressive strengths, and to visually determine which ones are significantly different from the overall average. It is the perfect complement to the ANOVA table. The IR chart facilitates "...comparisons between treatments while also checking for homogeneity within each treatment..." (Wheeler 2005). Even though the IR is not a significance test, it can indicate where the possible differences are.

Step 7: Interpret the results and make recommendations

We must translate our statistical findings in terms of the context of the situation at hand, so we can make decisions based on the outcomes of the analysis. Therefore, we should always go back to the problem statement in Step 1 and our scientific hypothesis to make sure that we answered the questions that we set out to answer with our study.

One of the main objectives for our study was to determine whether there was something unique about the cement used for the sidewalks that exhibited premature cracking in the houses built in the third phase of the development. We also wanted to see if the average compressive strength of the cement provided by the different suppliers meets the minimum requirement of 2900 psi needed for this application. After measuring the compressive strength of five briquettes made from cement used in each of the four development phases, we carried out a one-way ANOVA on the data and found a statistically significant difference in the average compressive strength. In fact the signal to noise ratio was rather large and yielded an F-statistic of 212.56 with a p-value < 0.0001. This test result substantiated the need for a multiple comparison test in order to determine which phases had different average compressive strength. Tukey's HSD test showed us that the average compressive strength for Phase 2 was the highest (3227.96 psi), and different from the other phases; and that the average compressive strength for Phase 3 was the lowest (2783.08 psi), and also different from the other phases. There is also a 445 psi difference between the average compressive strength of the Phase 2 cement, and the Phase 3 cement. We also discovered that the average compressive strength for Phases 1, 2, and 4, meets, or exceeds, the lower specified value of 2900 psi. The cement used in Phase 3 has an average compressive strength, which is 117 psi below the lower specification limit.

In conclusion, you can report to your client that in fact the cement used in Phase 3 has an average compressive strength lower than the specified lower limit of 2900 psi. You believe that the low compressive strength of the Phase 3 cement coupled with the unusual colder temperatures and unexpected frost might have resulted in the long longitudinal cracks at the top of the Phase 3 sidewalks. You also confirmed that the cement used in the other three phases of the housing development meets or exceeds the minimum requirement, and that the cement used in Phase 2 proved to be the one with the highest compressive strength, about 300 psi higher than the 2900 psi limit. Future investigations might look at the effect of water leaking through small cracks, which might get bigger as the ice expands when the water freezes in cold temperatures. In order to improve the compressive strength of the cement being used one can also look at issues such as, impurities in the cement mix, poor curing of the mix, or laying the cement at too cold temperatures.

We have discovered some relevant information in this study relating to the average cement properties used in the different phases of construction in this housing development. As with most engineering and scientific endeavors, statistics plays an important role in helping develop, frame, and answer the hypothesis of interest, as well as providing important clues to uncover the problem at hand.

6.5 Testing Equivalence of Three or More Populations

The notion of "proving equivalence" rather than significance was presented in Chapters 4 and 5. This approach can also be extended to studies involving three or more population means. For three or more means, JMP conducts equivalence tests on all pairwise combinations of factor levels. We will provide an overview of this JMP feature for the one-way ANOVA platform.

What if we want to show that the average compressive strengths of the cements used in different phases are equivalent? In JMP, we carry out this analysis from within the one-way ANOVA output by using the **Equivalence Test** option from the drop-down menu as is shown in Figure 6.41.

When this option is selected, a dialog box is displayed where we need to enter a value that represents the smallest practical difference between any two means. That is, if the difference in any two means does not exceed this value, then we would consider them to be equivalent. In our case, if two average compressive strengths differ by 200 psi or less then, for all practical purposes, we will consider them to be *equivalent*. We enter `200` in this dialog box.

Figure 6.41 Test of Equivalence to Compare Three or More Means

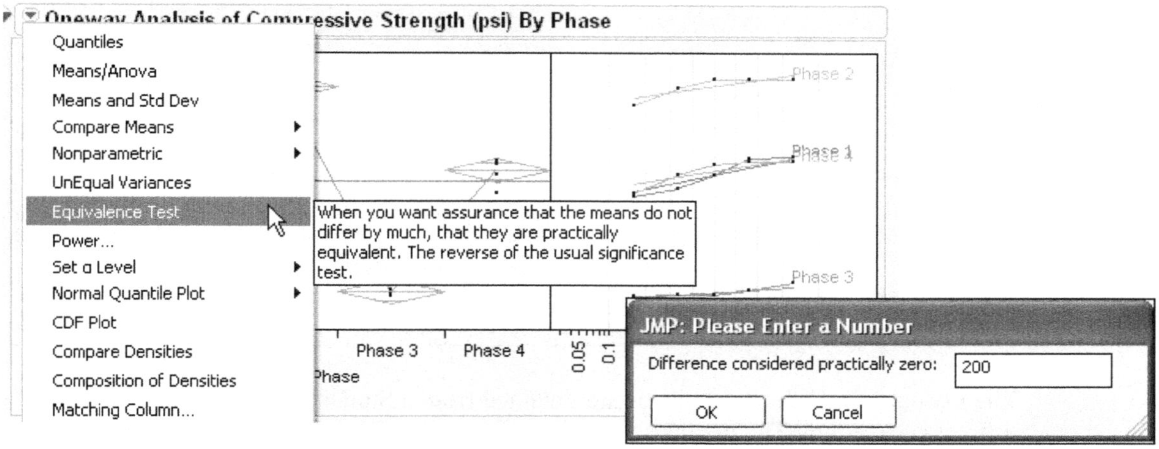

An equivalence test is performed using two one-sided *t*-tests (TOST), as we discussed in Chapter 5, Section 5.5. For our example, we want to know if $|\mu_{\text{phase i}} - \mu_{\text{phase j}}| < 200$, or equivalently, if $-200 < (\mu_{\text{phase i}} - \mu_{\text{phase j}}) < 200$, for any two Phases i and j. For comparing Phase 2 with Phase 1, for example, these two one-sided hypotheses can be written as follows:

a) H_0: $\mu_{\text{Phase 2}} - \mu_{\text{Phase 1}} \leq -200$
H_1: $\mu_{\text{Phase 2}} - \mu_{\text{Phase 1}} > -200$

b) H_0: $\mu_{\text{Phase 2}} - \mu_{\text{Phase 1}} \geq 200$ (Assume)
H_1: $\mu_{\text{Phase 2}} - \mu_{\text{Phase 1}} < 200$ (Prove)

There are six pairwise comparisons between the average compressive strengths of the four phases. The JMP output for the six equivalence tests for the cement example is shown in Figure 6.42.

For comparing Phase 1 and Phase 2, the test statistic for the hypothesis (a) above can be found in JMP with the label **Lower Threshold**, and is equal to 22.036 = (3227.96 – 3037.2 + 200) / 17.73318, while the test for hypothesis (b) above is labeled **Upper Threshold**, and is equal to -0.52 = (3227.96 – 3037.2 – 200) / 17.73318. The other five tests are similarly derived using the estimated difference in the two means, the stated equivalence value of 200 psi, and the standard error of the difference.

Figure 6.42 Dialog Box and Output for Test of Equivalence

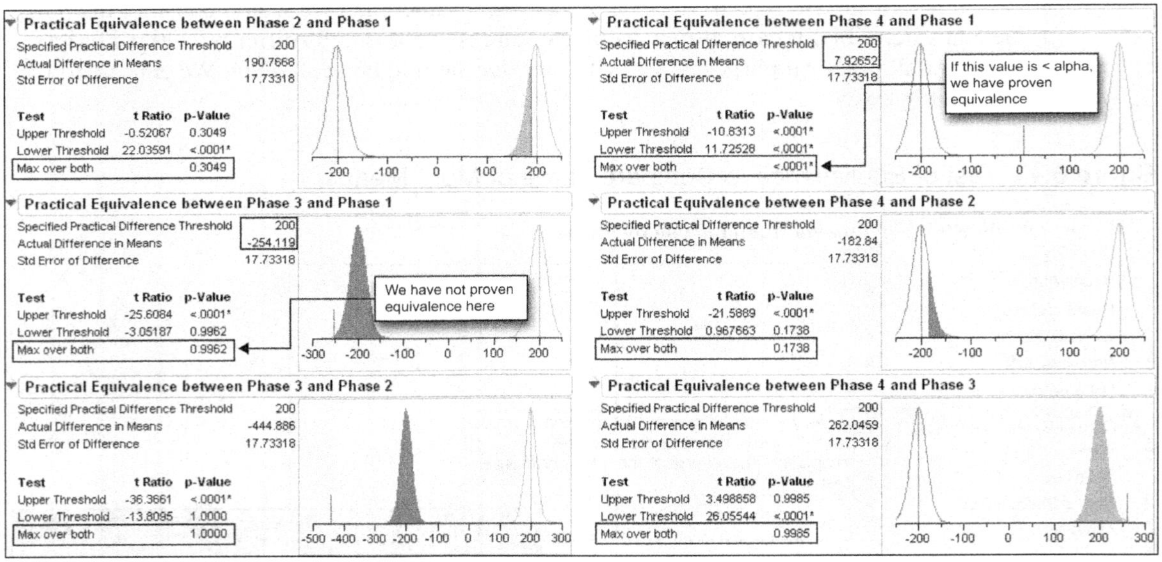

The *p*-values for both test statistics are obtained from a Student's *t*-distribution with 8 degrees of freedom. The *p*-values for (a) and (b) are < 0.0001 and 0.3049, respectively. The JMP output also selects the maximum *p*-value and displays it next to the label **Max over both**. If the maximum *p*-value is < 0.05, our significance level, we reject the null hypothesis in favor of the hypothesis of equivalence.

The six pairwise phase comparisons indicate that only Phase 4 and Phase 1 are equivalent within 200 psi, since the **Max over both** is < 0.0001. From the practical point of view this means that, if we are willing to accept average differences in the compressive strength of 200 psi or less, then the cement from the suppliers used in Phases 1 and 4 can be used interchangeably.

6.6 Summary

In this chapter we learned how to conduct a one-way ANOVA to study the effect of a factor with three or more levels on a continuous response. We discussed why it is called analysis of variance even though it is a comparison of the means, by showing how the data gets partitioned into variation due to the factor of interest, or signal, and the noise in the response. The practice of partitioning the degrees of freedom associated with the signal-to-noise shows us how the data will be allocated to the estimation of the signal and the noise, before the data collection activity. This is a quick check to see if we will have enough degrees of freedom for estimating the noise.

We pointed out that a statistically significant *F*-statistic just tells us that a discrepancy between our data and our null hypothesis exists, but it does not point to the mean, or means, responsible for the discrepancy. In order to determine which factor levels differ, and by how much, we need to perform multiple comparisons, such as **All Pairs, Tukey HSD (Honestly Significant Differences)**.

We provided step-by-step instructions for setting up a comparison of the measured performance of several materials, processes, or products, including how to use JMP to help us to conduct the corresponding statistical analysis, and to make data-driven decisions. We also illustrated how to prove equivalence of the average performance of two populations, in the context of a one-way ANOVA.

Some of the key concepts and practices that we should remember from this and previous chapters include:

- Write out the ANOVA table before running the study in order to identify all terms in the model, and to account for the degrees of freedom corresponding to the factor of interest, signal, and the noise.
- Conduct tests of significance to compare the variances of multiple populations to check the homogeneous variance assumption for the *F*-test *before* the ANOVA analysis is conducted.
- Although for comparing means the ANOVA *F*-test is robust to departures from normality, the tests for unequal variances may not be as robust. Therefore, it is a good practice to check for normality.

- Remember to clearly define the population of interest, the experimental unit, the observational unit, and how the study will be executed in order to conduct the appropriate analysis.
- Use the scientific method to make sure that the statistics are helping us support or refute our hypothesis.
- Select key graphs and analysis output to facilitate the presentation of results.
- *Always translate the statistical results back into the context of the original question.*

6.7 References

Box, G.E.P., J.S. Hunter, and W.G. Hunter. 2005. *Statistics for Experimenters: Design, Innovation, and Discovery.* 2d ed. New York, NY: John Wiley & Sons.

Conover, W.J., M.E. Johnson., and M.M. Johnson. 1981. "A Comparative Study of Tests for Homogeneity of Variances, with Applications to the Outer Continental Shelf Bidding Data." Technometrics, 23:351–361.

Milliken, G.A., and D.E. Johnson. 1984. *Analysis of Messy Data: Volume I: Designed Experiments.* New York, NY: Van Nostrand Reinhold Company.

Montgomery, D.C., and G.C. Runger. 2003. *Applied Statistics and Probability for Engineers.* 3rd ed. New York, NY: John Wiley & Sons.

Sall, J., and B. Jones. 2008. "Leptokurtosiphobia: Irrational Fear of Non-Normality." SPES/Q&P Newsletter, 16:9–12.

Wheeler, D.J. 2005. *The Six Sigma Practitioner's Guide to Data Analysis.* Knoxville, TN: SPC Press, Inc.

Chapter 7

Characterizing Linear Relationships between Two Variables

Chapter Goals

- Become familiar with the terminology and concepts needed to characterize a linear relationship between two materials, product, or process variables.
- Use the **Fit Y by X** and **Fit Model** platforms in JMP to study the linear relationship between two variables and interpret the JMP output accordingly.
- Use residuals analysis to determine whether a linear relationship does an adequate job of characterizing the relationship between two variables.
- Translate JMP statistical output in the context of a problem statement.
- Present the findings with the appropriate graphs and summary statistics.

7.1 Problem Description

What do we know about it?

You are a recent mechanical engineer college graduate who began working for a company that manufactures measuring sensors for different types of applications. The department you have joined specializes in providing weighing technology for different customers including bag filling systems, tanks and silos, trucks, bridges, and so on. Your department's weighing systems cover the range from small scales, capable of weighing up to 200 lb, to high-end systems capable of weighing up to 500,000 lb.

What is the question or uncertainty to be answered?

One key product in your department is a canister-style load cell design that is used in cargo truck weighing systems. In preparation for the certification by a nationally recognized standards bureau, you need to verify several calibration standards for a new high-capacity, canister-style load cell entering this market. Since you are new to this department, your boss wants you come up to speed by reproducing an existing calibration standard for equipment and vehicles up to 3,500 lb, based on a historical study circa 1975 by a National Institute of Standards and Technology (NIST) scientist named Paul Pontius. Your deliverable is a calibration curve, which depicts deflection (inches) as a function of the load (lb). Your boss wants you to go through all of the necessary steps needed to produce this standard, including collecting the measurements, specifying the model, and proving its merit.

How do we get started?

Calibration curves are widespread in scientific and engineering applications, and the methods used to create them involve developing an equation that explains the relationship between two different variables. In your case the calibration curve is used to compare the load cell output, deflection, against standard test loads, measurement systems, or measurement scales. In order to get started, you need to do the following:

1. Describe the requirements in terms of a statistical framework that enables you to determine, with a certain level of confidence, a calibration curve for heavy equipment using a load cell.
2. Translate the statistical findings in a way that clearly conveys the message to your boss without her getting lost in the statistical jargon.

What is an appropriate statistical technique to use?

As was mentioned previously, the calibration curve is based upon an older study that was done to compare deflection as a function of the actual load applied to the load cell. We will use *linear regression* to find a suitable model to represent this phenomenon, and various diagnostics to check the validity of the model. In Section 7.3 we review the key concepts that are relevant for model building using linear regression, and in Section 7.4 we introduce a step-by-step approach for performing linear regression using JMP.

7.2 Key Questions, Concepts, and Tools

Many of the statistical concepts and tools that were introduced in Chapter 2 and shown in previous chapters are also relevant when characterizing the linear relationship between two variables representing key attributes of products, processes, or materials. Table 7.1 outlines these concepts and tools, and the "Problem Applicability" column helps to relate these to the development of the calibration curve, which relates the load of an object to its deflection using a load cell.

Table 7.1 Key Questions, Concepts, and Tools

Key Questions	Key Concepts	Problem Applicability	Key Tools
Is our sample meaningful?	Inference from a random and representative sample to a population	Making sure that we sample the population in a meaningful way is also important for characterizing the relationship between two variables. When determining how to set up the study to evaluate the impact of the load on the deflection, we need to adequately identify the range of loads that enables us to see the possible linear effect. We must also consider how many different loads to use between the minimum and maximum loads and which ones we want to get replicate measures of.	Model specification, replication, sampling schemes, and error degrees of freedom
Are we using the right techniques?	Probability model using the normal distribution	Inferences from tests of significance using the linear regression model depend on the errors being independent and normally distributed with equal variance. For the calibration curve problem, this assumption is checked using the residuals from the fit of the linear model of deflection as a function of load.	Residual plots to check distributional assumptions
What risk can we live with?	Decision theory	The accuracy of the output from the load cell is very important in weighing applications and the level of tolerated risk is low. Therefore, more conservative values are chosen for significance level (α) in this problem.	Type I and Type II errors, statistical power, and confidence
Are we answering the right question?	Modeling linear relationships	For this study, we are using a range of loads with known weights, and measuring their deflection with the load cell. To this data, we will fit a model of the form $Y = \beta_0 + \beta_1 X + \varepsilon_i$ and evaluate its performance and applicability to the data. If the model is appropriate for the data, then we can use the model in reverse to determine the load for a measured deflection for values not included in our original study.	Linear regression analysis, residuals analysis, lack-of-fit, inverse prediction, and R^2
How do we best communicate the results?	Visualizing the results	We study the relationship between two variables, load and deflection, and evaluate the strength of their relationship. This can be visualized using an X-Y scatter plot, overlaid with the prediction equation and prediction bands about the equation.	Scatter plots overlaid with prediction intervals for the mean and individual measurements

7.3 Overview of Simple Linear Regression

7.3.1 Description and Some Applications

Simple linear regression (SLR) is a technique that is used to study the linear relationship between two characteristics of a material, process, or product, in which one variable is the regressor, or independent variable, and the other variable is the response, or dependent variable. In contrast to the one-way ANOVA described in Chapter 6, the dependent and independent variables are both continuous (the case when the dependent variable is discrete is not considered in this book). When using this technique, we usually want to determine whether increasing or decreasing the values of our independent variable, X, results in a linear increase or decrease in our dependent variable, Y. If this holds, we say that there is a *linear correlation* between the two variables. It is important to remember that *correlation is not causation*. In order to establish a cause-and-effect relationship we need to look for the underlying causal relationship, or the physical phenomenon relating the two variables.

> *Correlation is not causation.*

A common representation of a simple linear relationship between two variables, Y and X, can be written as follows:

$$Y = \beta_0 + \beta_1 X + \varepsilon \qquad (7.1)$$

where β_0 is a constant term in the model, the intercept, β_1 is the slope representing the effect on the response as the independent variable changes by one unit, and ε represents the amount by which the model fails to fit the data Y. This is referred to as the simple linear regression model because it includes only one independent variable. However, this model can be easily extended to include more than one independent variable, or additional terms such as quadratics or cubic terms, to make it more useful for real-life applications. This simple equation is described in more detail in Section 7.3.2. Some examples from science and engineering where we could use a simple linear regression model to explain relationships among variables include the following:

- Determining the influence of air temperature on the ozone concentration in the air.
- Predicting systolic blood pressure as a function of age.
- Modeling CPU time as a function of the number of disk I/O.
- Studying the effect of a given stress on fiber breaking strength.

Linear regression can also be a stepping-stone to establishing a more complex model, such as a cubic model or an approximation to a non-linear model by working with a transformation, which more adequately represents the physical mechanisms between the variables.

It is important to consider the type of linear relationship that is present in our system even before we collect any data. In her book, *Experimental Statistics* (1963), Natrella classifies linear relationships into functional, F type, and statistical, S type, depending on the existence of an exact mathematical relationship between the dependent and independent variable. A further classification is given by considering the effect of the measurement error on the measured variables, classifications F1 and F2, and on how the units are sampled and measured, classifications S1 and S2. These classifications help us clarify the underlying mechanism in the linear relationship, as well as the type of analysis that can be performed with the data. These relationships are summarized in Table 7.2.

Table 7.2 Four Cases of Linear Relationships between X and Y

	Functional Relationship (F)		**Statistical Relationship (S)**	
	F1	**F2**	**S1**	**S2**
Distinctive Features	X and Y are linearly related by an exact mathematical formula, which is not observed exactly due to measurement error in one or both variables.		There is no exact (or known) mathematical relationship between X and Y, but a linear statistical relationship can be established based on the characteristics, X and Y, of samples from a population.	
Example	Determination of the spring constant using Hooke's law, where the elongation, Y, is directly proportional to the applied load, X.		X= Height Y= Weight We have a random sample of pairs (X, Y) measured on individuals from a population. The range of X is not restricted.	X= Height (preselected values) Y= Weight (for preselected height) X is measured or known beforehand. Selected values of X are then used to choose the individuals at which to measure the variable Y. The range of X is restricted.

(continued)

Table 7.2 *(continued)*

	Functional Relationship (F)		Statistical Relationship (S)	
	F1	**F2**	**S1**	**S2**
Measurement Error	Affects Y only. The applied load X is accurately known; we only measure the elongation Y.	Affects both X and Y. The true applied load is not known; we measure both the applied load, X, and the elongation, Y.	Usually negligible compared to variation between sample units.	
Applicable Methods	Can fit $Y = \beta_0 + \beta_1 X$ For a given X=*x*, Y values are normally distributed with mean equal to $Y = \beta_0 + \beta_1 x$, and variance independent of the value x.	Can fit $Y = \beta_0 + \beta_1 X$	Can fit $\bar{Y}_X = \beta_0 + \beta_1 X$, or $\bar{X}_Y = \beta'_0 + \beta'_1 Y$. We assume that X and Y are jointly distributed as a bivariate normal.	Can fit only $\bar{Y}_X = \beta_0 + \beta_1 X$, For a given X=*x*, Y values are normally distributed with mean equal to $\bar{Y}_X = \beta_0 + \beta_1 X$, and variance independent of the value of *x*.

Adapted with permission from Table 5-1 of *Experimental Statistics* (Natrella 1963).

Most of the linear relationships we encounter in practice are of the type S1 or S2. In an S1 relationship both the Y and X variables, because they are sampled together, can play an interchangeable role. In other words, we can fit a model $Y = \beta_0 + \beta_1 X$ or $X = \beta_0 + \beta_1 Y$, and the correlation coefficient is meaningful because the ranges of X and Y are not restricted. In an S2 relationship, however, the values of X are preselected before the Y's are measured and the only model we can fit is $Y = \beta_0 + \beta_1 X$. In this scenario, the correlation coefficient could be misleading because the full range of X may not be observed.

Note that in the statistical relationships of Table 7.2 we use the notation $\bar{Y}_X = \beta_0 + \beta_1 X$ to indicate that for a given value of X, the average $\bar{Y}_X$ depends linearly on X. As we continue, we will follow the standard convention of dropping the average and subscript and just write the statistical relationship as $Y = \beta_0 + \beta_1 X$, but please keep in mind the "average" nature of the fit.

In Section 7.3.2 we describe the simple linear regression model, its terms and their interpretation, as well as the assumptions required. In Section 7.3.3, the concept of partitioning the variation due to the model (correlation) and the noise (unexplained or missing terms) is discussed, along with how to determine whether the linear effect of the independent variable is producing a large signal relative to the noise. In Section 7.3.4, we review methods for evaluating how well the model fits the data and its predictive abilities, while in Section 7.3.5 we discuss sample size considerations for SLR studies. Finally, in Section 7.4, we show how to carry out this type of analysis for establishing a calibration curve for deflection (in) as a function of load (lb) using JMP. Although this chapter focuses on the simple form of the linear regression model, at times we extend the concepts to include linear regression models with higher order terms (such as quadratic or cubic terms) in order to improve the models fit.

7.3.2 Simple Linear Regression Model

A simple linear regression model is helpful to describe the relationship between two variables. Figure 7.1 shows scatter plots for variables X and Y illustrating three possible relationships. The cloud of points with no specific pattern in scatter plot (a) suggests no relationship between X and Y; that is, no correlation. The scatter plot (b) shows that Y is increasing in a linear manner as X increases, a *positive linear correlation*, while the scatter plot in (c) shows the opposite, Y decreasing as X increases, a *negative linear correlation.*

Figure 7.1 Some Relationships between Two Variables, Y and X

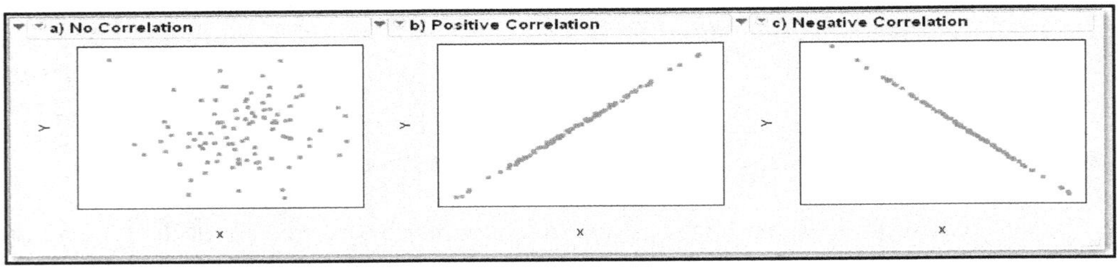

Statistics Note 7.1: We use lowercase italic *x* and *y* to represent observed values of the explanatory variable X and the independent variable Y.

The range of X and Y values used to study a relationship plays a critical role when making claims about the nature of the correlation among the variables. For example, the data shown in plot (a) in Figure 7.1 indicates that there is no correlation between X and Y. Does this mean that X and Y are not related? Let us assume that Y represents Young's Modulus (a measure of stiffness) for sapphire rods and X represents temperature (°C). It would be incorrect to conclude that there is no relationship given that a known functional relationship exists between the two variables. Upon further investigation of the data, we discover that the range of temperatures in the plot is between 300.1 °C and 301.8 °C, a range of only 0.7 °C, and that the measurement technique used to obtain the modulus readings contained large measurement error. Thus, for this scenario, the reason that we do not observe a linear relationship in this data is because we are in a very narrow range of the temperature variable X and there is excessive measurement error.

As we mentioned in Section 7.3.1, the simple linear regression has the form: $Y = \beta_0 + \beta_1 X + \varepsilon$, where Y is the response, β_0 is an intercept term, β_1 is the slope, X is the independent variable, and ε is the model's error. Table 7.3 provides a description of these terms and the formulas used to estimate the parameters β_0 and β_1. The magnitude and direction of the linear relationship between the Y and X is provided by the estimate of the slope parameter, β_1. If there is little or no correlation among the variables, then we expect the slope to be close to 0. In other words, changes in X do not change the response Y, in which case Y is represented solely by the intercept as shown in Figure 7.1. On the other hand, if the slope is significantly different from 0, then as X changes we expect the response Y to change proportional to the slope. The direction of the change in our response is related to the sign of the slope. If the estimate of the slope is positive, then as X increases we expect Y to also increase as shown in b) Positive Correlation (Figure 7.1). If the slope is negative, then as X increases we expect Y to decrease as shown in c) Negative Correlation (Figure 7.1).

Table 7.3 Definitions and Terms for the Simple Linear Regression Model

Model Term	Name	Description	Least Squares Estimator
Y	Response or dependent variable	This is the material, process, or product characteristic that we want to model as a function of the independent variable X.	$\hat{y}_i = b_0 + b_1 x_i$
X	Regressor, or independent variable	We assume that the independent variable is observed with a negligible measurement error. Methods to handle situations when the measurement error in X is not negligible are beyond the scope of this book.	Not applicable.
β_0	Intercept	This represents the predicted response at X = 0.	$b_0 = \bar{y} - b_1 \bar{x}$

(continued)

Table 7.3 *(continued)*

Model Term	Name	Description	Least Squares Estimator
β_1	Slope	The slope is of primary interest because it represents the impact that the factor has on the response—the rate of change. For every unit increase in X, Y increases (or decreases) by this many units.	$b_1 = \frac{\sum_{i=1}^{n}(x_i - \bar{x})y_i}{\sum_{i=1}^{n}(x_i - \bar{x})^2}$
ε	Model errors	Considered to be a random variable representing the degree of error between the predicted and actual values for a given value of X. Assumed to be independent and normally distributed with mean = 0 and constant variance independent of the values of X.	$e_i = y_i - \hat{y}_i$

The last column in Table 7.3 provides the *least squares* equations for estimating the parameters in a simple linear regression model.

What Is Least Squares?

Least squares is an estimation method that seeks to minimize the sum of squares of the residuals (fitting errors), namely $\Sigma(y_i - \hat{y}_i)^2$. The best-fitting least squares line is the one that produces fitted values ($\hat{y}_i$) that minimize this sum of squares. Figure 7.2 shows a scatter plot of Y versus X for seven data points. In this plot the residuals ($y_i - \hat{y}_i$) are the lines projecting from each data point to the fitted line. The squares, defined by the residuals, represent the residuals squared $(y_i - \hat{y}_i)^2$, or each of the terms in the residuals sum of squares. The least squares line in the plot is the line that minimizes the sum of the areas of the seven squares.

Figure 7.2 Least Squares Line and Residuals Squares

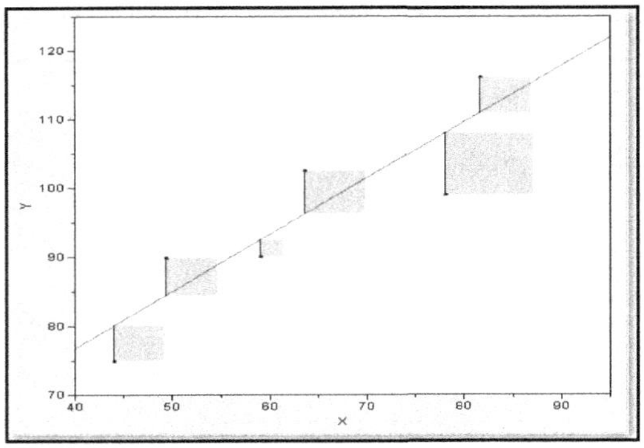

JMP Note 7.1: For a great visual explanation of least squares and minimizing the sum of squares of the residuals, watch the video "Explaining Least Squares" @http://blogs.sas.com/jmp/index.php?/archives/144-Explaining-Least-Squares-with-JMP.html. Figure 7.2 is a display of this demonstration.

Simple Linear Regression in JMP: The Fit Y by X Platform

In 1901, a young Albert Einstein published his first paper "Conclusions Drawn from the Phenomena of Capillarity" in which he investigated the nature of intermolecular forces. In this paper Einstein used least squares to show that the constant c in the equation for the potential energy per unit volume in the interior of a liquid, $P = P_\infty - Kc^2/v^2$, does not depend on the molecular structure but on the atomic composition of the molecule. Here P is the potential energy per unit volume, P_∞ is a constant, K and c are quantities that depend on the molecular composition of the molecule, and v is the molecular volume. In Einstein's words "If the molecule of our liquid consists of several atoms, then it shall be possible to put, in analogy with gravitational forces, $c = \Sigma c_\alpha$ where the c_α's denote the values characteristic for the atoms of the elements."

Table 7.4 shows the data from the Einstein paper, which includes the molecular formula, the value of the constant c obtained from the available literature at the time, and the name of the compound. In addition we have added the atomic composition of the molecule in terms of the number of atoms of Carbon (C), Hydrogen (H), and Oxygen (O). The original paper dealt with the fitting of a model for c as a function of C, H, and O. To illustrate the concepts of simple linear regression, we fit a simple linear regression of the constant c versus C, the number of atomic carbon values.

Table 7.4 Data from Einstein's First Published Paper

Formula	Carbon	Hydrogen	Oxygen	*c*	Name of Compound
C10H16	10	16	0	510	Limonene
CH2O2	1	2	2	140	Formic Acid
C2H4O2	2	4	2	193	Acetic Acid
C3H6O2	3	6	2	250	Propanoic Acid
C4H8O2	4	8	2	309	Butyric Acid and Isobutyric Acid
C5H10O2	5	10	2	365	Valerianic (pentanoic) Acid
C4H6O3	4	6	3	350	Acetic Anhydride
C6H10O4	6	10	4	505	Ethyl Oxalate
C8H8O2	8	8	2	494	Methyl Benzoate
C9H10O2	9	10	2	553	Ethyl Benzoate

(continued)

Table 7.4 *(continued)*

Formula	Carbon	Hydrogen	Oxygen	*c*	Name of Compound
C6H10O3	6	10	3	471	Ethyl Acetoacetate (diacetic ether)
C7H8O	7	8	1	422	Anisole
C8H10O	8	10	1	479	Phenetole and Methyl Cresolate
C8H10O2	8	10	2	519	Dimethyl Resorcinol
C5H4O2	5	4	2	345	Furfural
C5H10O	5	10	1	348	Valeraldehyde
CH14O	10	14	1	587	d-carvone

Data adapted from Einstein, A. (1901), "Conclusions Drawn from the Phenomena of Capillarity," Annalen der Physik, 4, 513–523, as it appears in The Collected Papers of Albert Einstein, 2, The Swiss Years: 1900–1909 (English translation supplement), by Beck, A., and Havas, C., Princeton: Princeton University Press. Reprinted by permission of the publisher.

Figure 7.3 shows a scatter plot of *c* versus Carbon, which was produced in JMP using the **Analyze > Fit Y by X** platform. There appears to be a positive linear relationship between the two quantities. In other words, as the number of carbon atoms increases so does the value of the constant *c*. The number of Carbon atoms ranges from 1 to 10, while the data for the constant *c* ranges from approximately100 to 600. Why is it important to note the range of values represented by the data? The interpretation of the linear relationship might be valid only for the range of the observed data. For Einstein's molecular data, we see a positive linear relationship within the ranges stated previously, but we cannot be certain, based on the data alone, that the relationship remains linear outside of these ranges and if the same rate of change (slope) applies.

Statistics Note 7.2: The variable Carbon (the number of carbon atoms in the molecule) is an integer variable rather than continuous. We can treat this variable as "continuous" because the slope still represents the change in *c* when the number of Carbon atoms increase by one. In other words, the linear regression equation enables us to predict *c* as a function of Carbon. As we discussed in Section 2.2.2, the scale type is not an inherent attribute of the data, but rather depends upon the questions we intend to ask of the data.

Figure 7.3 Einstein's Data: *c* Versus Carbon

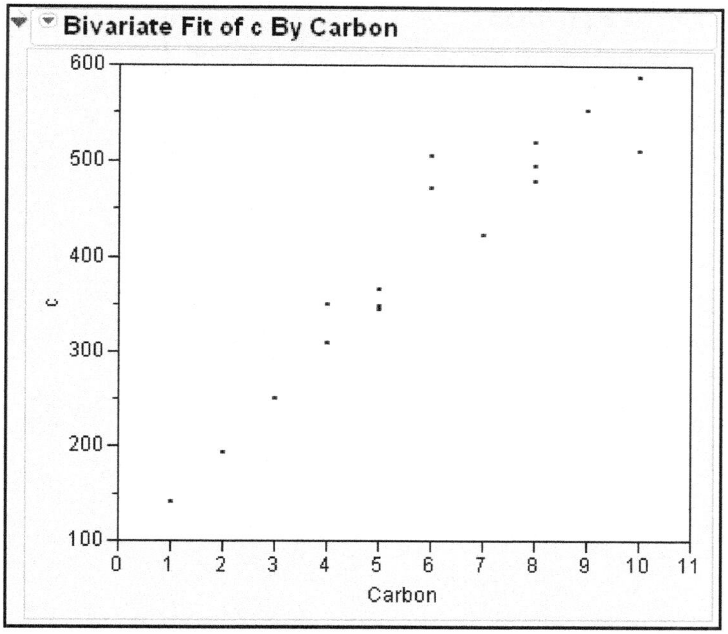

We can use the equations shown in Table 7.3 to obtain estimates for the intercept and slope terms in the simple linear regression model, $c = \beta_0 + \beta_1\text{Carbon} + \varepsilon$. This model can be easily fit using JMP by selecting the **Fit Line** option when clicking the red triangle next to **Bivariate Fit of c By Carbon** shown at the top of the output window in Figure 7.3. The estimated linear regression model for this data is $c = 130.61 + 45.74\mathit{Carbon}$, and a graphical representation of the fitted line is shown in Figure 7.4.

In Figure 7.4, the actual data is shown by the dots, the solid line represents the simple linear regression model, and, as it was shown in Figure 7.2, the distance between each dot and the line represents the residual, or model error. The least squares estimate for the slope is 45.74, which means that the constant *c* increases by 45.74 units when the number of Carbon atoms increases by 1. The estimate of the intercept is 130.61, which means that for a carbon content of 0 the value of $c = 130.61$. The intercept not being 0 indicates that the constant *c* depends on the other components, Hydrogen and Oxygen, which were not included in the model (in Section 7.5 we show this is the case). We are just using Carbon to illustrate the basics of the simple linear regression model.

Figure 7.4 Einstein's Data: *c* Versus Carbon with Line Fit

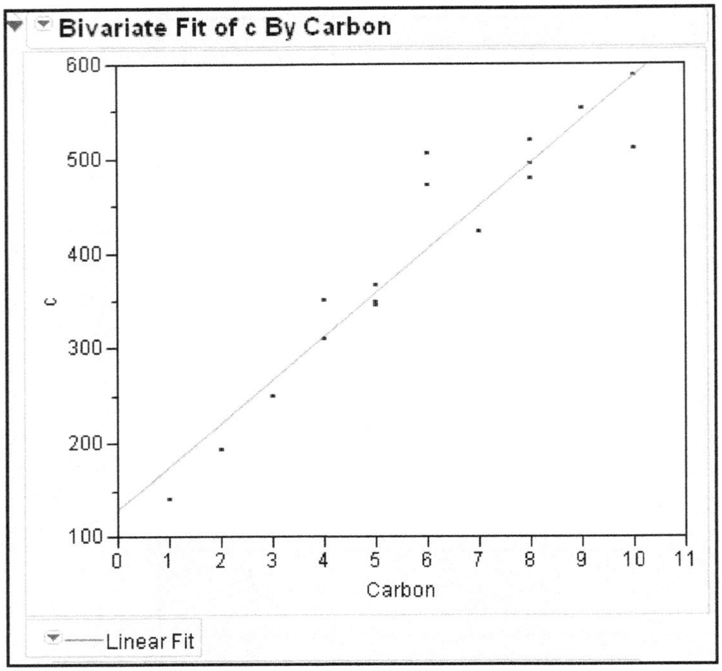

The equation $c = 130.61 + 45.74Carbon$ serves as a prediction equation for the constant c as a function of the number of Carbon atoms. For example, for a number of 10 Carbon atoms the predicted value of the constant is $\hat{c} = 130.61 + 45.74 \times 10 = 588.01$. For a given number of Carbon atoms the residual values are calculated as the observed value minus the predicted value. For a number of 10 Carbon atoms the Residual = 510 – 588.0015 = –78.0015.

Table 7.5 Einstein's Data: Data, Predicted Values, and Residuals

Formula	Carbon	*c*	Predicted *c*	Residual *c*
C10H16	10	510	588.001535	–78.001535
CH2O2	1	140	176.345957	–36.345957
C2H4O2	2	193	222.085466	–29.085466
C3H6O2	3	250	267.824974	–17.824974
C4H8O2	4	309	313.564483	–4.5644831
C5H10O2	5	365	359.303992	5.69600819
C4H6O3	4	350	313.564483	36.4355169
C6H10O4	6	505	405.043501	99.9564995
C8H8O2	8	494	496.522518	–2.5225179
C9H10O2	9	553	542.262027	10.7379734
C6H10O3	6	471	405.043501	65.9564995
C7H8O	7	422	450.783009	–28.783009
C8H10O	8	479	496.522518	–17.522518
C8H10O2	8	519	496.522518	22.4774821
C5H4O2	5	345	359.303992	–14.303992
C5H10O	5	348	359.303992	–11.303992
CH14O	10	587	588.001535	–1.0015353

Table 7.5 shows the predictions and residuals for Einstein's data. The residuals range from –78.0015 to 99.9565, including a value of –1.0015. This indicates that the predicted values from the model were as close as one unit from the actual values, and as far off as 100 units from the actual values. Based solely on the appearance of the plot in Figure 7.4, the model seems like a good fit to the data and a reasonable representation of the relationship between the constant *c* and the number of Carbon atoms. However, a few questions still remain regarding this relationship:

- Is the linear relationship statistically significant?
- Are there any outliers or influential observations in the data?
- Are the model assumptions met?
- How much error is associated with the predictions?
- How much variation in the response is explained by our factor?

- How much variation is unexplained?
- Do we need to consider other models for our data?
- How should we collect data for future studies?

These questions are answered with the help of the statistical tools and methods associated with linear regression analysis.

7.3.3 Partitioning Variation and Testing for Significance

It might seem odd that concepts associated with *analysis of variance* (ANOVA) also have a place in simple linear regression analysis. Remember, however, that one of the goals of a statistical analysis is to be able to sort the signals from the noise. In the simple linear regression case the signal is the presence of the linear relationship. In this section we show the connection between ANOVA and how we can use it to test for the linear relationship (signal) in our simple linear regression. In Chapter 6 we introduced the *sums of squares* concept that was used to partition the total variation in the measured response into two components representing the factor (signal) and the error (noise):

$$SS_{Total} = SS_{Factor} + SS_{Error} \quad (7.2)$$

In Chapter 6 the factor sum of squares represented the variation *between* the different levels, or treatments, of a factor, while the error sum of squares represents the variation *within* the different levels. How is this relevant for simple linear regression?

The partitioning of the sums of squares provides us with a relatively simple way to form tests of significance for the parameters, intercept (β_0) and slope (β_1), of the simple linear regression model. The mathematical expression for the *total sums of squares* decomposition shown in equation 7.1 is as follows:

$$\sum_{i=1}^{n}(y_i - \bar{y})^2 = \sum_{i=1}^{n}(\hat{y}_i - \bar{y})^2 + \sum_{i=1}^{n}(y_i - \hat{y}_i)^2 \quad (7.3)$$

The total variation in the response is represented by the left-hand side of equations 7.2 and 7.3, SS_{Total}. Our job is to figure out how much of that variation is due to the explanatory variable (factor) under study and how much is due to noise or unaccounted sources of variation. The variation contributed by our factor (SS_{Factor} or SS_{Model}) is represented by the first term on the right-hand side in equations 7.2 and 7.3, and will be used to measure the signal in our data. The second term on the right-hand side of equations 7.2 and 7.3 is the "residual" sum of squares that is minimized using the least squares method, and which serves as the signal yardstick, and is a measure of the experimental error in our data. We will form a signal-to-noise ratio for our simple linear regression model using this partitioning of variation.

Table 7.6 shows the ANOVA table for the simple linear regression model, $Y = \beta_0 + \beta_1 X + \varepsilon$. The Model test is a test for the slope parameter, β_1 being 0, which is given by the ratio of the model mean squares to the error mean squares (F Ratio column). As mentioned in Chapter 6, this ratio follows an *F*-distribution. If the *p*-value for this F ratio is less than our stated significance level (α), then we conclude that the linear relationship between the two variables is statistically significant; otherwise, we conclude that it is not. Recall that statistical inferences using ANOVA are valid only if the following three assumptions are met:

1. Independent observations between and within the populations are defined by the factor levels.
2. Observations follow a normal distribution.
3. Homogeneous Variances of the populations defined by the factor levels.

These assumptions are further explored in the context of simple linear regression in Section 7.3.4 and Table 7.8.

Table 7.6 Analysis of Variance for Simple Linear Regression

Source	DF	Sum of Squares	Mean Square	F Ratio	Prob > F (P-value)
Model	1	SS_{Model}	$MS_{Model} = SS_{Model} / 1$	MS_{Model} / MSE	Prob > F Ratio
Error	n – 2	SS_{Error}	$MSE = SS_{Error} / (n-2)$		
Total	n – 1	SS_{Total}			

What exactly do we mean by a significance test for the slope parameter, β_1? We need to write out the formal tests of hypothesis statements for checking the significance of the slope, which is shown below. The null hypothesis H_0 assumes that the slope is 0, which means that the response does not change with the explanatory variable, the rate of change is 0 (a "flat-liner"), while the alternative hypothesis (H_1) attempts to prove with statistical rigor that the relationship is linear; that is, the rate of change is different from 0.

Null Hypothesis	H_0: $\beta_1 = 0$	(Assume)
Alternative Hypothesis	H_1: $\beta_1 \neq 0$	(Prove)

Statistics Note 7.3: If we do not reject the null hypothesis for the slope, it does not necessarily mean that there is no correlation among the variables, as is shown in plot (a) in Figure 7.1. Instead, it may mean that the relationship is not linear, but takes on some other form, such as quadratic, or the range of the data is too narrow to observe the linear relationship. We discuss this later in the chapter.

Statistics Note 7.4: For more complicated models involving higher order terms, such as quadratic effects or additional factors, the Model Sums of Squares (SS_{Model}) in the ANOVA table provides a more general test that at least one of the model terms is significant, although it does not indicate which ones are significant. The degrees of freedom associated with the model would represent all of the terms in the model, excluding the intercept. We will see an example of this in Section 7.4.

The ANOVA table for Einstein's published data is presented in Table 7.7. The Model F-test is testing the hypothesis that the slope, β_1, in the equation $c = \beta_0 + \beta_1 Carbon + \varepsilon$ is 0. Using the significance level a = 0.05, we reject the null hypothesis in favor of the alternative hypothesis. If the slope for Carbon is not 0, then what is it?

Table 7.7 Einstein's Data: Analysis of Variance

Source	DF	Sum of Squares	Mean Square	F Ratio	Prob > F (P-value)
Model	*1*	240468.74	240468.74	136.70 = (240468.74/1759.14)	< 0.0001
Error	*15=(17–2)*	26387.14	1759.14= (26387.14/15)		
Corrected Total	*16=(17–1)*	266855.88			

In Section 7.3.2 we estimated the slope as $\beta_1 = 45.74$. The ANOVA results indicate, what may seem obvious, that 45.74 is statistically different from 0; that is, larger than the noise. In Chapter 2, we introduced the concept of using a confidence interval around a descriptive statistic in order to see how well it is estimated. For example, we have shown confidence intervals for the mean and standard deviation many times throughout this book and, at times, used them to conduct a test of significance. We can apply the same concept to our slope parameter, β_1, and get a $(1-\alpha)\%$ confidence interval for its estimate, b_1, using the Student's *t*-distribution:

$$b_1 \pm t_{\alpha/2,n-2}\sqrt{\frac{MSE}{\sum_{i=1}^{n}(x_i-\overline{x})^2}} \tag{7.4}$$

JMP can output the 95% (default) confidence interval for the slope as well as for the intercept. When the **Fit Line** option is selected from within the **Analyze > Fit Y by X** platform, the ANOVA table and parameter estimates are automatically produced. If you right-click the parameter estimates table, additional columns to be displayed can be selected from a drop-down list, including the confidence intervals for the slope and intercept model terms. The output is shown in Figure 7.5. We see that the 95% confidence interval for the slope estimate of 45.74 is (37.40; 54.08). Since this confidence interval for the slope does not include 0, we conclude that the slope is statistically different from 0 (we are testing H_0: $\beta_1 = 0$ versus H_1: $\beta_1 \neq 0$). The confidence interval also gives us a sense for how different the slope is from 0. While the estimated slope is 45.74, it could be as low as 37.40 or as high as 54.08.

Figure 7.5 also shows a *t*-ratio and *p*-value for the slope parameter that can also be used to conduct the test of significance for β_1 instead of the ANOVA table presented earlier. For the simple linear regression model, these two tests result in the same outcome. In fact, the F ratio = 136.70 in the ANOVA table, Table 7.7, is equal to the square of the *t*-ratio = 11.692 in Figure 7.4, that is, $136.70 = 11.692^2$. For more complex models involving more than one factor or higher order terms, this relationship is not true and you need to use the *t*-ratios for each model term to determine which ones are statistically significant.

Figure 7.5 Einstein's Data: Confidence Intervals for Parameter Estimates

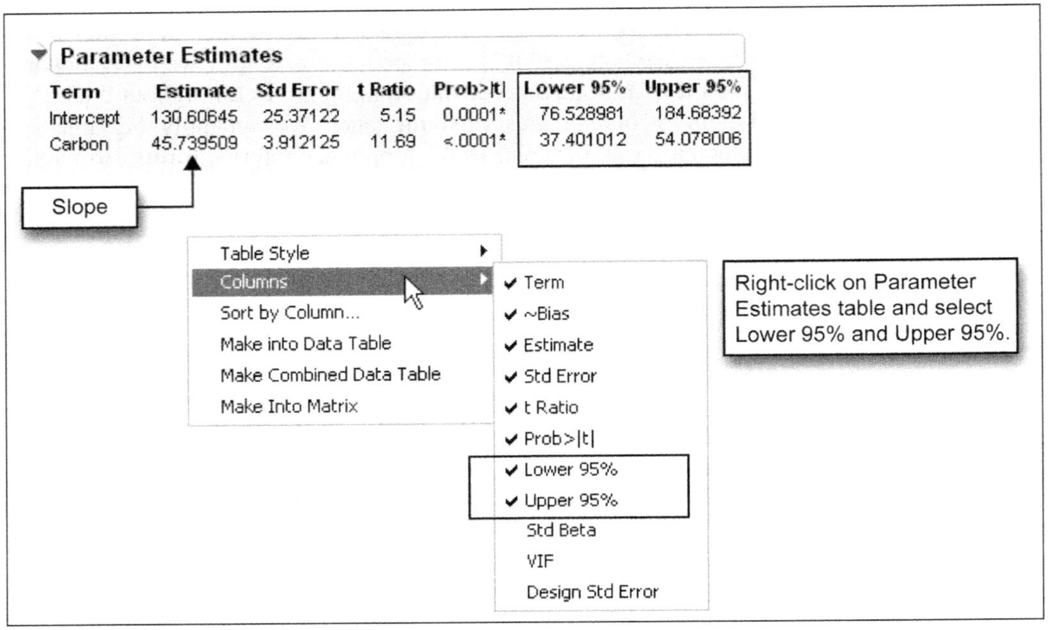

Although there is a *t*-ratio and *p*-value associated with the intercept term, β_0, we typically ignore it and always keep the intercept, which represents the baseline, term in the final prediction equation, unless the regression equation passes through the origin, in which case the estimated intercept should be close to 0. In the next section we show a useful application for accounting for the uncertainty in the intercept.

7.3.4 Checking the Model Fit

When fitting the simple linear regression model, $Y = \beta_0 + \beta_1 X + \varepsilon$, by least squares, the errors ε_i are assumed to be independent, have a mean equal to 0, have a constant variance, and follow a normal distribution. If these assumptions are met we have "good" (in the statistical sense) estimates of the slope and intercept, and we can perform tests of significance using an *F*- or *t*-distribution. As we discussed in Chapters 6, it is important to check these assumptions before drawing any final conclusions about the linear relationship between the two variables.

In general, if the model does a good job at explaining the behavior of the data then the estimated errors or residuals, being the observed minus the predicted values $(y_i - \hat{y}_i)$, should behave like "white noise" in the sense that there is no predominant signal left after the model is fit. In other words, we can think of a statistical model as a recipe for transforming data into white noise as follows:

$$\text{Data} - \text{Model} = \text{White Noise (Residuals)} \qquad (7.5)$$

The residuals are surrogates of the model errors in equation 7.1, and can be used to check the assumptions and adequacy of the linear regression model. Table 7.8 shows several model checks, along with diagnostic tools, that enable us to evaluate the quality of the simple linear regression model fit. As Draper and Smith (1998) point out, the question we should ask when looking at the residuals diagnostics is: "Do the residuals make it appear that our assumptions are wrong?" As it is the case with tests of significance, we are looking either for evidence to reject the claim that the assumptions hold, or to say that we do not have enough evidence to say that the assumptions are violated.

Table 7.8 Model Checks and Diagnostic Tools for Simple Linear Regression (SLR)

Model Check	Relevance	Diagnostic Tool
Errors follow a normal distribution	This is a key assumption for the validity of the tests of significance, F- or *t*-tests, and confidence intervals for the slope and intercept. Slight departures from normality are tolerated, but more extreme departures can result in unreliable *p*-values and non-sensible model predictions.	Normal quantile plot of studentized residuals
Errors have constant variance	The homogeneous variance assumption is important for the least squares estimators for the slope and intercept and their standard errors. If the variance is increasing with the mean response, this should be taken into account in the model fit, and may result in unreliable *p*-values.	Plot of residuals versus predicted values
Errors have zero mean	If the errors do not have zero mean, it is an indication of model misspecification. Least squares guarantees that the mean of the residuals is zero (up to rounding error).	Distribution or summary of the residuals
Errors are independent	Similar to the *t*-test and ANOVA, the assumption of independent errors is key for the reliability of the estimate of the error variance. This in turn affects the standard error of the slope and intercept, as well as tests of significance and confidence intervals. Violation could be due to data collected over time resulting in auto-correlated errors and model misspecification.	– Plot of residuals versus time order – Durbin-Watson statistic – Autocorrelation plot of residuals

(*continued*)

Table 7.8 *(continued)*

Model Check	Relevance	Diagnostic Tool
Linearity	For SLR, this means that only one explanatory variable may not be enough. Therefore, the MSE is inflated by the ignored terms and the model does not predict as well as it could.	– Plot of Y versus X – Plot of studentized residuals versus predicted values – Lack-of-fit test
Outliers	Points that fall significantly away from the linear trend highlight a potential breakdown in the model at that point. This might raise suspicion about the validity of the linear relationship between the two variables. Outliers tend to have large residual values.	– Check data – Plot of studentized residuals versus predicted values
High leverage observations	A high leverage observation is one with an unusual value for the independent variable X. A leverage observation is "far" from the independent variable average.	– Plot of studentized residuals versus hat values
Influential observations	An observation is influential if the estimates of the slope and intercept changes substantially when it is removed. You should study influential observations further in order to make sure that they are valid and should be included in the study.	– Plot of Cook's D influence versus observation number – Bubble plot of Y versus X by Cook's D influence
Model adequacy	When prediction is the primary objective for the SLR model, we must make sure that we have a model that best predicts our response.	– Coefficient of determination, R2 – Lack-of-fit test – Prediction intervals

Many of the model checks shown in Table 7.8 rely upon the residuals from the model. There are several types of residuals that can be used to validate our model; however, the two most popular ones are as follows:

- **(Raw) Residuals:** the actual response minus the predicted response, $e_i = (y_i - \hat{y}_i)$, from the SLR model. For these residuals mean = 0 and variance = MSE. It is important to note that the raw residuals are estimates of the true errors and do not have equal standard errors.
- **Studentized Residuals:** residuals standardized by their standard errors, or $e_i/\sqrt{Var(e_i)}$. We use these for checking model assumptions. JMP studentized residuals are also known in the literature as internally studentized residuals because the estimate of σ used in the standard errors includes all the data. These

residuals are preferred because they have a common variance, which is equal to 1 if the model is correct.

> **JMP Note 7.2:** In JMP raw residuals can be saved to the data table from the **Analyze > Fit Y by X** platform, but the studentized residuals must be saved from the **Analyze > Fit Model** platform.

Let's conduct some of the model checks for Einstein's first published paper data using some of the diagnostic tools shown in Table 7.8. The raw residuals were already presented in Table 7.5, along with the predicted values from the estimated model. We need to get the studentized residuals by going into the **Analyze > Fit Model** platform, as is shown in Figures 7.6 and 7.7.

Figure 7.6 Einstein's Data: Fit Model Platform

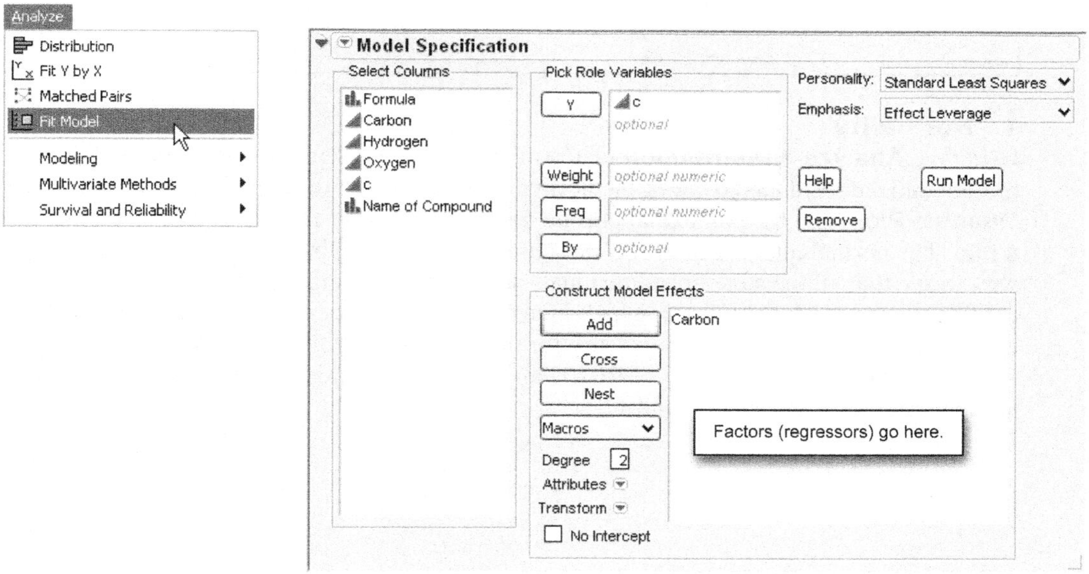

Figure 7.7 Saving Studentized Residuals from Fit Model Platform

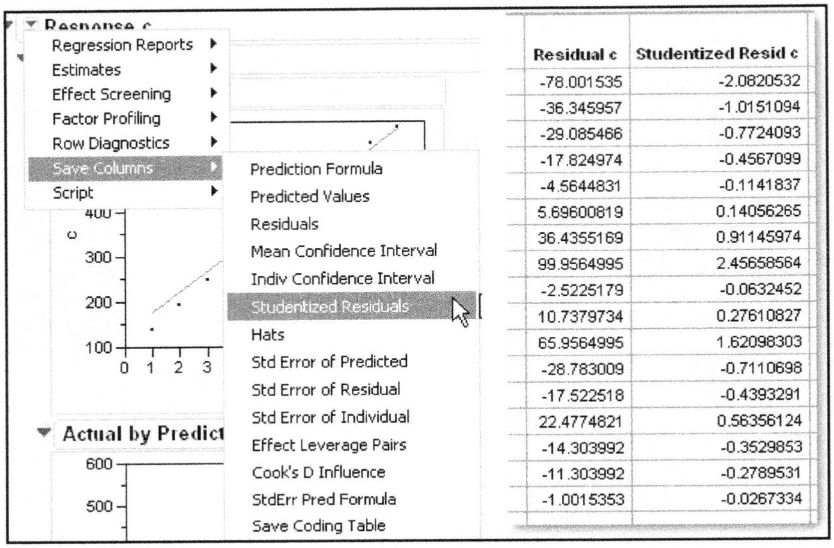

1. Normality

Using the **Analyze > Distribution** platform, we can generate a normal quantile plot of the studentized residuals to check the normality assumption. To do this we select **Normal Quantile Plot** from the contextual pop-up menu to the left of the **Studentized Resid c** title. The normal quantile is shown in Figure 7.8, above the box plot. From this plot we can see that all the studentized residuals are close to a straight line and are within the confidence bands (dashed lines), which indicates that the studentized residuals follow a normal distribution. It is not necessary for the point to fall right on the line but to be "close." The estimated mean = –0.020207 and standard deviation = 1.03 are close enough to their theoretical values of 0 and 1.

Figure 7.8 Einstein's Data: Normality Check for Studentized Residuals

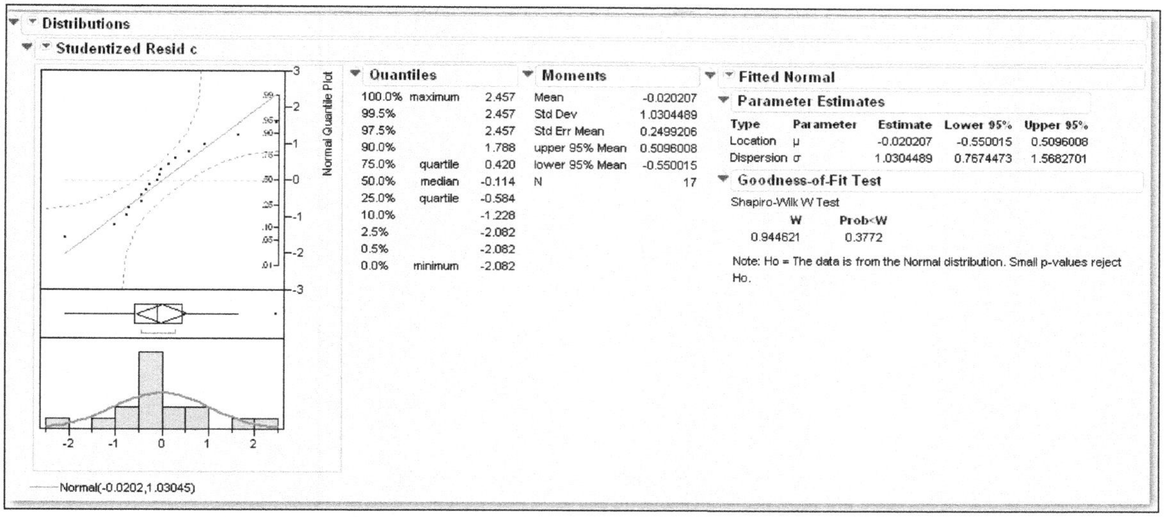

2. Constant Variance (Homogeneous)

The assumption of constant variance can be checked by creating a scatter plot of the studentized residuals versus the predicted values using the **Graph > Overlay** platform in JMP. Select Studentized Resid c for **Y** and Predicted c for **X**. This plot is shown in Figure 7.9. In order to look for patterns in the residuals, it is helpful to enhance this plot with a reference line on the y-axis at 0. An ideal shape for this plot is a random scattering of the residuals around 0, with no particular pattern or shape across the range of the predicted values for Y. This would suggest that the variance is indeed constant across the levels of Y. For Einstein's data, the studentized residuals seem to increase for *c* values < 400, and decrease after that. This is an indication that the constant variance assumption may be violated.

Figure 7.9 Einstein's Data: Constant Variance Check

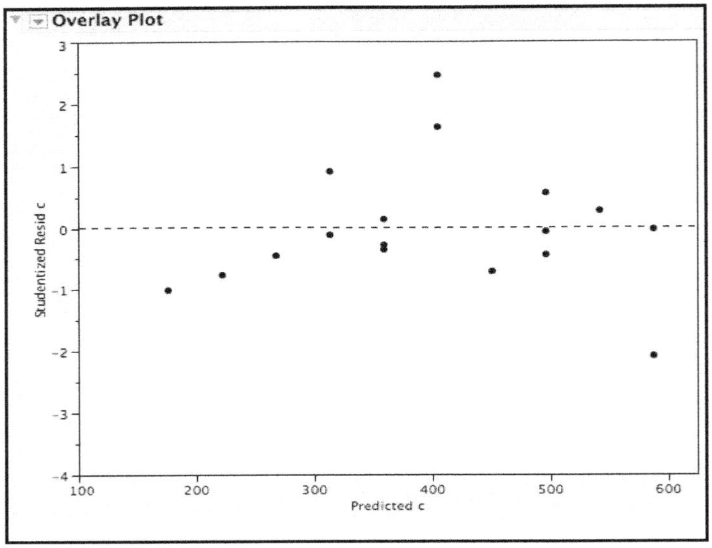

When the data exhibit variation that is not constant, we might see a funnel shape with the opening at either end of the x-axis. For illustrative purposes, a residuals plot showing heterogeneous variance is provided in Figure 7.10. In this plot we see more spread in the studentized residuals for smaller values of the predicted response Y than we do for larger values of the predicted response Y. In other words, the variation is decreasing as Y increases. This could happen if, for example, it is harder to measure the response at lower values. There are methods, such as weighted least squares estimation, that enable us to deal with non-constant variance but they are beyond the scope of this book (for examples see Sections 9.3 and 9.4 of Draper and Smith [1998]).

Figure 7.10 Example of Heterogeneous Variance

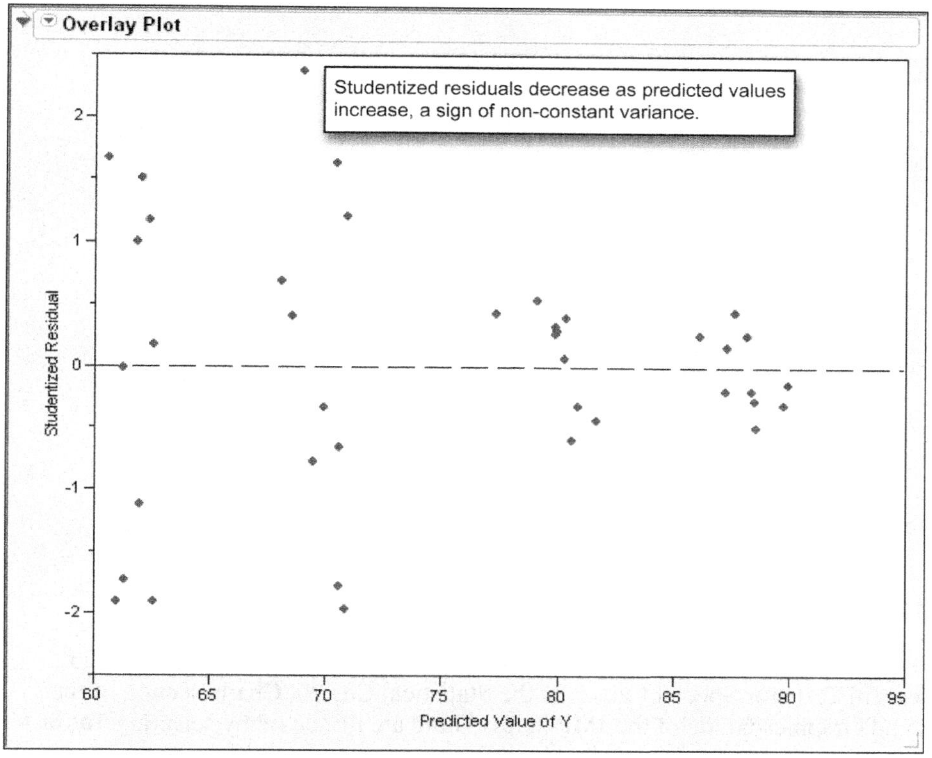

3. Independence

Although the assumption of independence is better guaranteed by the way the data is collected, there are a few checks that we can perform with the residuals. One of them is to plot the residuals in the time order that the data was collected. We do not have any time order information for Einstein's data but, for argument's sake, let us assume that the data were collected in the order presented in Table 7.4. We can use a time-ordered plot of the regular or studentized residuals to get a picture of any type of time-ordered effect. Since we are looking for unusual patterns or trends, a control chart is probably the best way to check this assumption. An Individual Measurements Moving Range chart for the studentized residuals was created in JMP using the **Graph > Control Chart > IR** platform and is shown in Figure 7.11.

Figure 7.11 Einstein's Data: Individual Measurements Chart for Studentized Residuals

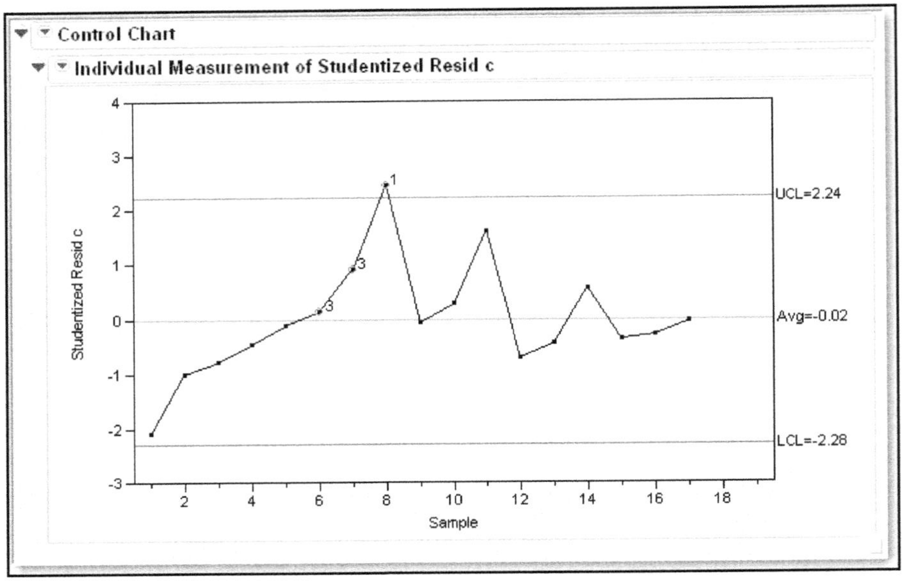

The numbers in the chart correspond to the different tests for special causes (you can find the tests in Tests for Special Causes in the Statistical Control Charts section of the Statistics and Graphics Guide of the JMP help). These are turned on by selecting **Tests > All Tests** in the contextual pop-up menu (red triangle) next to **Individual Measurement of Studentized Resid c**. We see a run of increasing values from observations 1 to 7 (indicated by the number 3), we see that observation 8 has a value that seems larger than the rest (indicated by the number 1), and we see a possible cycle after observation 8. If this were the actual time order in which the data was collected we would conclude that there is a lack of independence in Einstein's data.

We can also a use a more quantitative check for first-order autocorrelation in the residuals and we can output the Durbin-Watson statistic from within the **Analyze > Fit Model** platform, as is shown in Figure 7.12.

Figure 7.12 Einstein's Data: Durbin-Watson Test for First-Order Autocorrelation

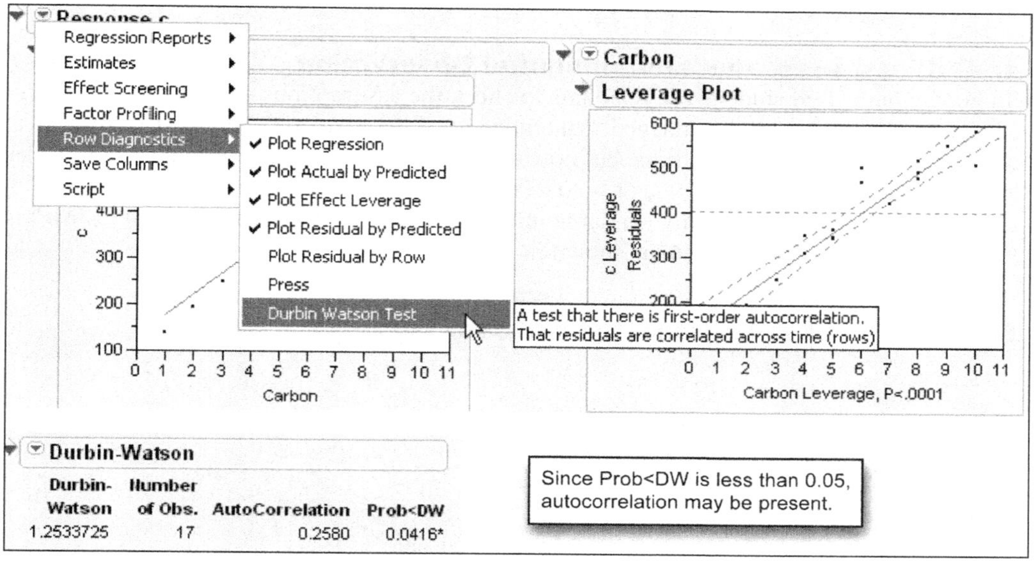

This statistic is calculated using the raw residuals, e_i:

$$DW = \frac{\sum_{t=2}^{n} (e_t - e_{t-1})^2}{\sum_{t=1}^{n} e_i^2} . \qquad (7.6)$$

In equation 7.6 the summation index *t* represents the time ordering. If the residuals exhibit positive first-order autocorrelation, then the difference between consecutive residuals becomes small, resulting in a small numerator and a small value of *DW* with limiting value equal to 0. On the other hand, if the residuals exhibit negative first-order autocorrelation the difference between consecutive residuals becomes large, resulting in a large numerator and a large value of *DW* with limiting value equal to 4. The Durbin-Watson test is a significance test for the null hypothesis of first-order autocorrelation equal to 0:

$$H_0: \rho = 0 \text{ versus } H_1: \rho \neq 0. \qquad (7.7)$$

The JMP output for Einstein's data shows that the Durbin-Watson statistic = 1.2534, with Prob < DW = 0.0416 < 0.05, indicating that there is some positive autocorrelation present in the data. The first-order positive autocorrelation is estimated to be = 0.258, in agreement with the results in Figure 7.11. Once again, if this were the actual time order in which the data was collected we would conclude that there is a lack of independence in Einstein's data.

4. Outliers, Leverage, and Influential Observation

So far, we have used studentized residuals to check the assumptions of normality, constant variance, and independence. Studentized residuals can also be used to identify observations that might be outliers, but how do we know whether the outliers are leverage, influential observations, or both? Two more diagnostics measures that can be used for this purpose are readily available in the **Analyze > Fit Model** platform in JMP: **Hats** and **Cook's D Influence**, as shown in Figure 7.13.

Figure 7.13 Outlier, Leverage, and Influence Diagnostics in JMP

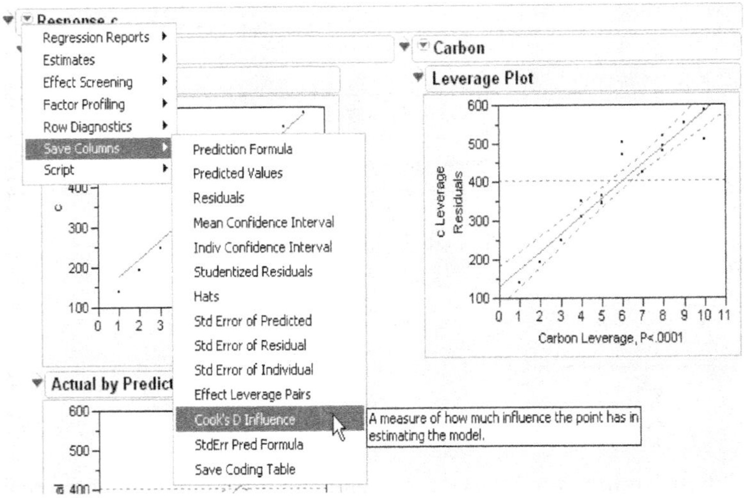

Studentized Resid c	h c	Cook's D Influence c
-2.0820532	0.20214944	0.54916726
-1.0151094	0.27123849	0.19176157
-0.7724093	0.19396111	0.07178333
-0.4567099	0.13408393	0.01614923
-0.1141837	0.09160696	0.0006574
0.14056265	0.06653019	0.00070409
0.91145974	0.09160696	0.04188897
2.45658564	0.05885363	0.18869045
-0.0632452	0.09570113	0.00021166
0.27610827	0.14022518	0.00621685
1.62098303	0.05885363	0.08215671
-0.7110698	0.06857728	0.01861349
-0.4393291	0.09570113	0.01021304
0.56356124	0.09570113	0.01680573
-0.3529853	0.06653019	0.00444019
-0.2789531	0.06653019	0.00277301
-0.0267334	0.20214944	9.0538e-5

Potential outliers have large studentized residuals, hats and Cook's D.

The hats are the diagonal values of the Hat matrix $H=\mathbf{X}(\mathbf{X'X})^{-1}\mathbf{X}$, with one for each observation, denoted h_i. The magnitude of each h_i reflects the influence of each observation on each fitted value. They are related to the variance of each residual by the formula $Var(e_i) = \sigma^2(1-h_i)$. As you can see, as h_i approaches 1, the variance of the residual goes to 0. The h_i value is also called the leverage of the *i*th observation.

For the *i*th observation, Cook's D_i measures the change in the estimated intercept and slope caused by deleting the *i*th observation. For a simple linear regression model the D_i are given by the following:

$$D_i = \frac{\left(\hat{\beta}_{(i)} - \hat{\beta}\right)'(\mathbf{X'X})\left(\hat{\beta}_{(i)} - \hat{\beta}\right)}{2s^2} \qquad (7.8)$$

Where $\hat{\beta}_{(i)}$ is the matrix of regression coefficients estimated.

From equation 7.8 we can see that the D_i's measure the level of influence that an observation has on the estimated parameters. For more details on these two regression diagnostics, as well as other useful ones, see Section 3.2 of *Regression Using JMP* by Freund et al. (2003).

Residual plots using studentized residuals, and the diagnostics h_i and D_i, can shed light on which observations are possible outliers and which might have an influence on the regression fit. Figure 7.14a shows needle plots of hat values and Cook's D_i versus the observation number. We can see that observation 2, corresponding to the compound Formic Acid (CH_2O_2), has a high hat value, h_2.= 0.271. A recommended cutoff for the hat values is $2p/n$, where p is the number of parameters in the model, and n the number of observations. For a simple linear regression p=2 and the cutoff becomes $4/n$, or 2(2)/17 = 0.2353 in this example. We also notice that observation 1, corresponding to the compound Limonene ($C_{10}H_{16}$), has a high Cook's D value, 0.549. For SLR models, a similar rule-of-thumb applies to the D_i values; that is, cutoff = 4/n, or 4/17 = 0.2353 in our case. This suggests that observations 2 and 1 could be influential on the regression. However, this is not the complete picture because these diagnostics cannot be used in isolation.

Figure 7.14a Einstein's Data: Needle Plots of Hat Values and Cook's D

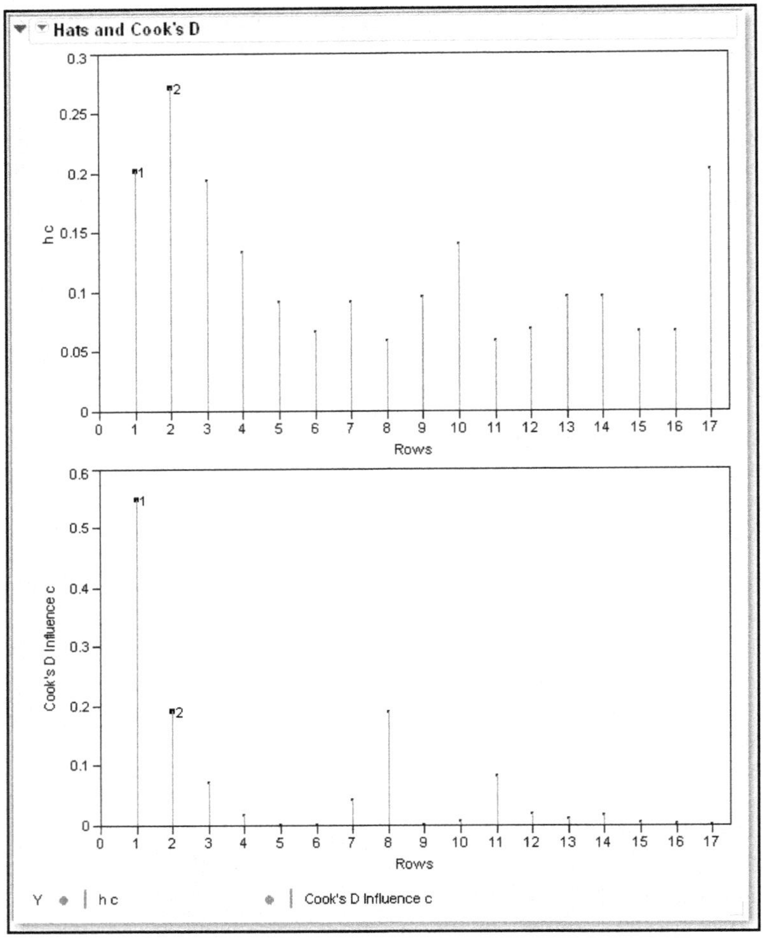

A very useful plot for combining the information of all three of the major regression diagnostics (studentized residuals, hat values, and Cook's D values) is the *bubble plot*, which is generated using **Graph > Bubble Plot**. We place Studentized Resid c in the **Y**, h c on the **X**, and Cook's D Influence c on the **Sizes** (the size of the circles). Figure 7.14b shows a bubble plot of the studentized residuals versus hat values with the size of the circles proportional to Cook's D influence. This plot enables us to identify outliers, high leverage, and influential observations in the same plot.

Figure 7.14b Einstein's Data: Diagnostic Bubble Plot

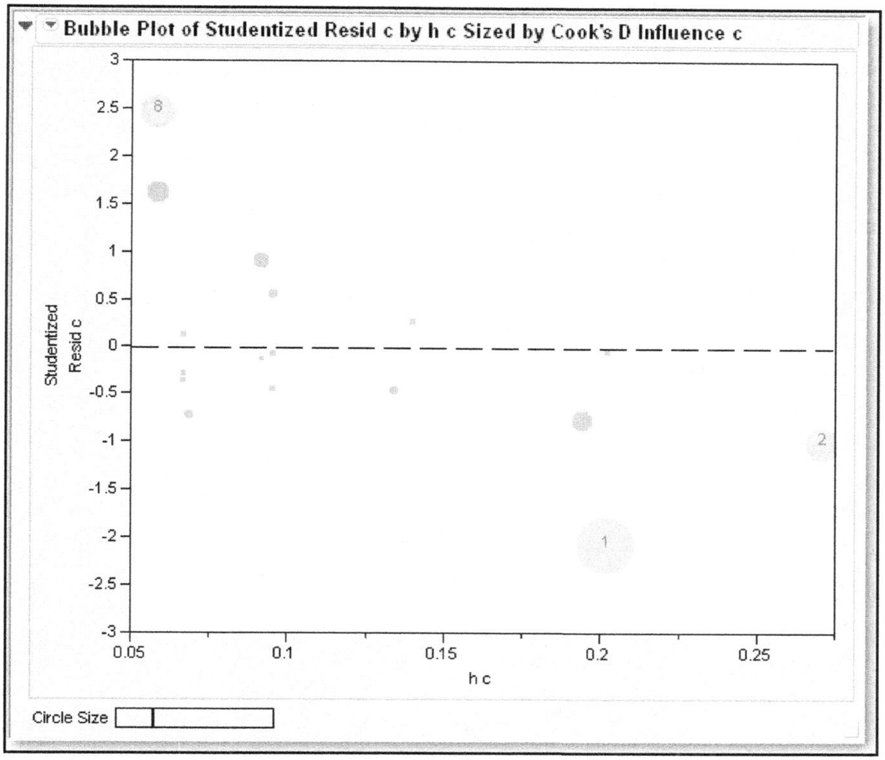

From Figure 7.14b we can quickly see that observation 1 has a large Cook's D influence value, followed by observations 8 and 2. Observation 1 also has the second largest studentized residual value (–2.082), and the second largest Hat value (0.202). This suggests that observation 1 could be influential. Observation 2 has the largest hat value (0.271) but a small studentized residual value (–1.015), so it might not be influential. Observation 8 has the largest studentized residual value (2.457) indicating a possible outlier, or an observation that the simple linear regression does not fit well.

In order to check the possible influence of observation 1 on the simple linear regression fit, we fit a model without observation 1, and then compare the results with the model using all the data. This can be easily done in JMP by following these steps:

1. Fit the model with all the data.
2. Select the first row in the data table row.
3. Select **Rows > Exclude/Unexclude** to exclude observation number 1.

4. Select **Script > Redo Analysis** from the contextual menu next to **Response c**. The model fit without observation number 1 appears in a new window.

The equation using the complete data is $c = 130.61 + 45.74Carbon$, while the equation without observation 1 is $c = 115.85 + 49.19Carbon$. The exclusion of observation 1 has resulted in a decrease in the intercept and an increase in the slope. Is this difference something to be concerned about? Before jumping to conclusions it is helpful to visualize these two models by overlaying the two fitted lines in the same plot, as shown in Figure 7.14c where the solid line corresponds to the fit using all the data, and the dashed line corresponds to the fit without observation 1.

Figure 7.14c Einstein's Data: Fits with and without Observation 1

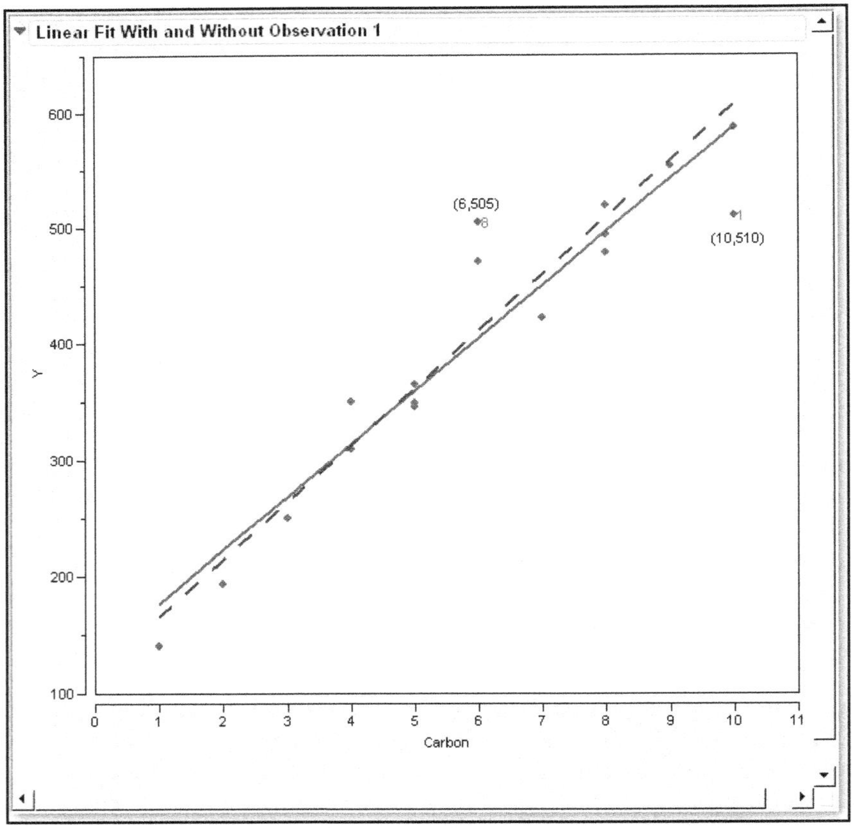

We can see that observation 1, with values (Carbon=10, *c*=510), pulls the regression line slightly downwards resulting in a smaller slope and larger intercept. However, with the two lines in the same plot we now see that, within the range of the data, the two simple linear regression models do not appear all that different, and that observation 1 is not that influential. In Figure 7.14a observation 8 was identified as a possible outlier. We can see in Figure 7.14c that the regression line does not fit that point well. According to the regression line, for a Carbon value of 6 we expect a *c* value of about 405, but the actual value is 505.

5. Model Adequacy

For Einstein's data, is the regression equation $c = 130.61 + 45.74Carbon$ adequate to describe the constant *c* as a function of Carbon? In other words, how well can we predict the value of *c* from a given value of Carbon? The partitioning of variation, shown in Table 7.7, can be used to determine how much of the variation in the response, Y, is explained by the linear regression model through a statistic called the *coefficient of determination* or R^2.

Figure 7.15a shows the R^2 (RSquare) = 0.9011 produced by either the **Fit Line** or **Fit Model** platforms. The *coefficient of determination* $R^2 = SS_{Model} / SS_{Total} = 240468.74 / 266855.88$, and since $SS_{Model} \leq SS_{Total}$, it is bounded by 0 and 1. R^2 is normally expressed as a percentage, and it is interpreted as "90.11% of the variation in the observed values of the constant *c* can be explained by the Carbon linear model." On the flip side, we can also say that 9.89% of the variation in the observed values of the constant *c* remains unexplained or unaccounted for.

Figure 7.15a Einstein's Data: Coefficient of Determination (R^2)

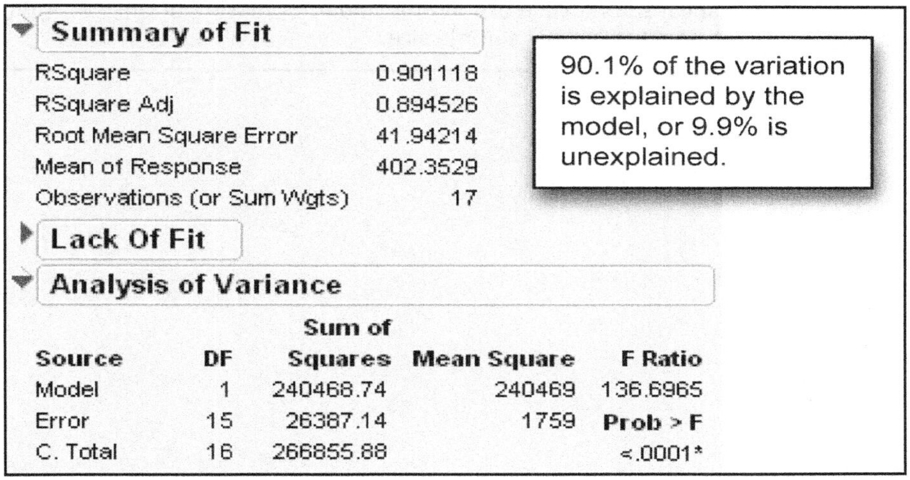

Summary of Fit

RSquare	0.901118
RSquare Adj	0.894526
Root Mean Square Error	41.94214
Mean of Response	402.3529
Observations (or Sum Wgts)	17

Lack Of Fit

Analysis of Variance

Source	DF	Sum of Squares	Mean Square	F Ratio
Model	1	240468.74	240469	136.6965
Error	15	26387.14	1759	Prob > F
C. Total	16	266855.88		<.0001*

For an SLR model, the R^2 statistic also represents the degree of linear correlation between the two variables. The linear correlation between X and Y can be estimated using this equation:

$$r = \frac{\sum_{i=1}^{n}(x_i - \bar{x})y_i}{\sqrt{\sum_{i=1}^{n}(x_i - \bar{x})^2 \sum_{i=1}^{n}(y_i - \bar{y})^2}}. \qquad (7.9)$$

With a little bit of algebraic manipulation, it can be shown that Equation 7.9 reduces to $\sqrt{R^2}$, with the sign of the correlation coefficient *r* the same as the sign for the estimate for the slope parameter. For Einstein's data, since $R^2 = 0.9011$ and the slope is positive, the linear correlation between the constant *c* and Carbon is then $\sqrt{0.9011} = 0.9493$. With an $R^2 = 90.11\%$, or a correlation of 0.9493, how do we feel about the adequacy of the simple linear regression model, $c = 130.61 + 45.74Carbon$, to represent the relationship between the constant *c* and Carbon?

Statistics Note 7.5: While the coefficient of determination, R^2, is often the first number considered to judge the quality of the model fit, it has its limitations. A high value of R^2 does not necessarily assure a statistically significant regression equation. In fact, Hahn (1973) showed that as the sample size increases, a statistically significant simple linear regression requires smaller and smaller R^2 values. In other words, a high R^2 value is not an indication of a good model fit. Figure 7.15b shows the R^2 values required to establish statistical significance of a simple linear regression at the 5% significance level, as a function of sample size.

Figure 7.15b R^2 Values Required to Establish Statistical Significance of an SLR at the 5% Significance Level

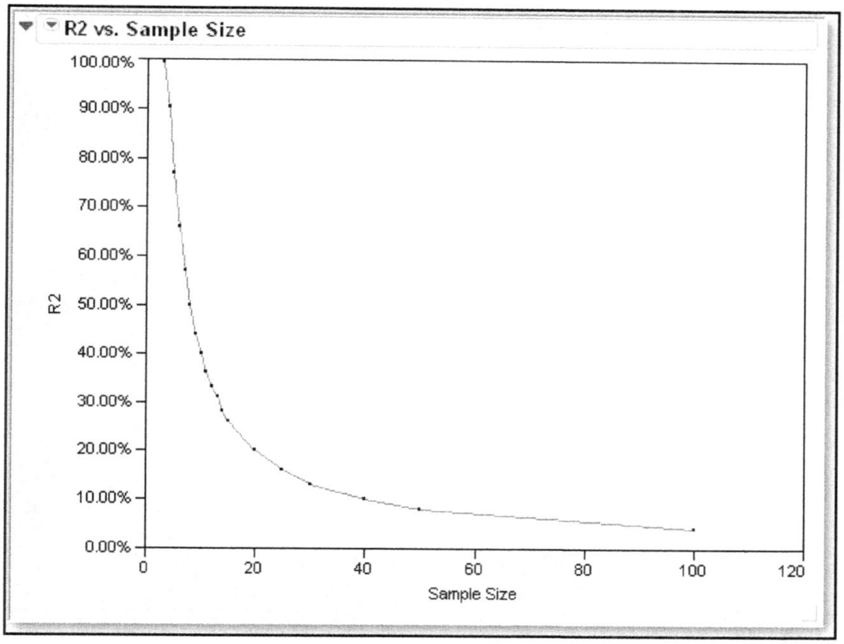

If we are going to use the regression equation to predict future values of the response, we need to make sure that the equation is up to the task and, as we saw above, a high R^2 value can be misleading. It would be unwise to use the model for predictions before performing the model checks presented in Table 7.8.

For a given value of the independent variable $X=x_0$, the estimated regression equation provides a prediction of the estimated response Y at that value. The predicted value at $X=x_0$, denoted as $\hat{y}(x_0)$, is an estimator of the average value of the population of response values at that given value of x_0. The distribution of constant values *c* for Carbon=6, with mean $\hat{c}(6) = 405.04$, is illustrated in Figure 7.16. This was generated by using a normal distribution centered at the predicted value of 405, with standard deviation equal to the standard error of an individual prediction = 43.

Figure 7.16 Distribution of c Values for Carbon=6 ~ N(405,43)

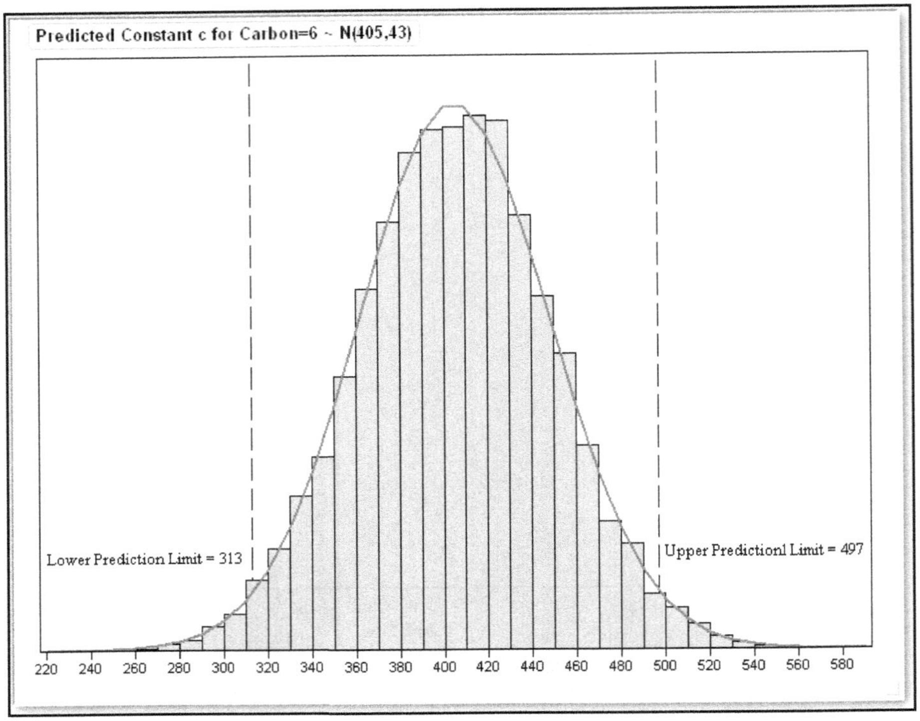

These predictions are statistical quantities derived from data and as such are subject to uncertainty. As we have seen in previous chapters, statistical intervals are a great way of quantifying the uncertainty around statistical estimates. Two types of intervals are commonly used in linear regression to accompany a predicted response value for a given value of the independent variable (see Weisberg (1985), 2nd Edition p. 22):

1. A confidence interval for the average value of $\hat{y}(x_0)$ at a given value of X=x_0:

$$\hat{y}(x_0) \pm t_{1-\frac{\alpha}{2},n-1}\,\hat{\sigma}\left[\frac{1}{n}+\frac{(x_0-\overline{x})^2}{\sum_{i=1}^{n}(x_i-\overline{x})^2}\right]^{2} \qquad (7.10a)$$

2. A prediction interval for a future individual value of $\tilde{y}(x_0)$ for a given value of X=x_0:

$$\hat{y}(x_0) \pm t_{1-\frac{\alpha}{2},n-1}\,\hat{\sigma}\left[1+\frac{1}{n}+\frac{(x_0-\overline{x})^2}{\sum_{i=1}^{n}(x_i-\overline{x})^2}\right]^{2} \qquad (7.10b)$$

The first interval applies to those situations where we might be interested in quantifying the uncertainty around the average value of the predicted response at a given value of x_0. The second type of interval is more relevant since most of the time we use a regression equation to "predict the future." This interval quantifies the uncertainty of the predicted value of an *unobserved* value of the response $\tilde{y}$ for a given value of x_0. In other words, it is the prediction of a value of the response that was not used to estimate the regression equation.

Using Einstein's data as an example, the predicted constant *c* for a Carbon = 6, along with the 95% confidence interval for the mean response, and the 95% prediction interval for a future observation are as follows:

- predicted value $\hat{c}$ (6) = 405.04
- 95% confidence interval for the mean response = (383.36; 426.73)
- 95% prediction interval for a future individual value = (313.05; 497.03)

What do you notice about the two intervals? Because of the added uncertainty of predicting an unobserved value of Y, the confidence interval for a mean is always narrower than the prediction interval for an individual value. The confidence interval for the mean tells us, with 95% confidence, that a Carbon value of 6 gives an estimated constant *c* between 383.36 and 426.73, on average. Similarly, the prediction interval tells us, with 95% confidence, that a future value of the constant *c* for a Carbon = 6 should fall between 313.05 and 497.03, as illustrated in Figure 7.16.

We could carry out a similar analysis for the remaining 16 observations in Einstein's data. Rather than generating an interval for each individual value of *x*, a confidence band or a prediction band can be generated to cover the range of *x* values in the linear regression. This is done in JMP using the **Analyze > Fit Y by X** platform as shown in Figure 7.17. The narrower bands represent the 95% (the default value for JMP) confidence bands for the average value of the predicted response, while the wider bands represent the prediction bands for a future individual observation.

The width of both intervals is narrowest at the average value for X and then gets wider as you move away from the center of X in both directions (easier to see in the confidence bands for the average response). This means that there is less uncertainty in the predictions at the center of X, but as we move away from the center the uncertainty in the predictions goes up. This is one of the reasons why extrapolating beyond the range of X in our study is not recommended. As we go outside the range of X values in our study the uncertainty gets too large. The other reason is that we do not know whether the linear relationship is the same outside the observed range.

For Einstein's data, Figure 7.17 shows that most of the points are within the bands for individual values, except for observation 8 (Ethyl Oxalate). Recall that we flagged this observation as having the largest studentized residual, and as one for which the linear regression model did not fit very well.

Figure 7.17 Einstein's Data: SLR Confidence and Prediction Bands

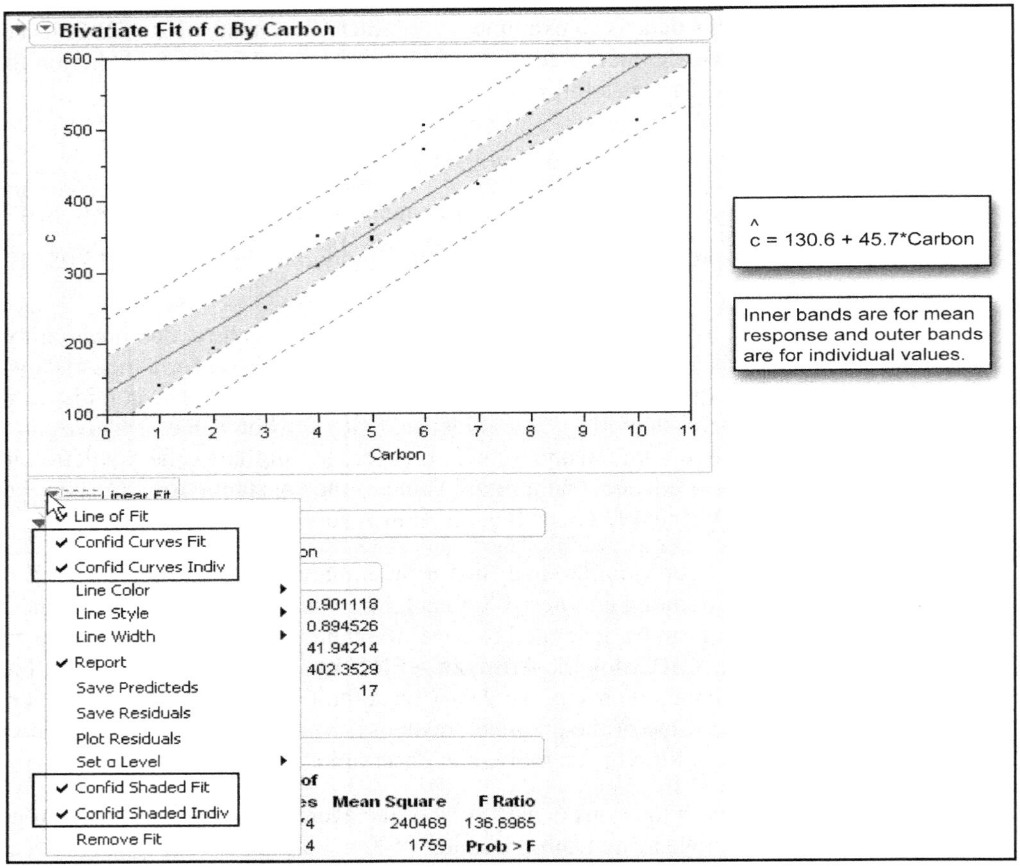

We also look at the width of the individual prediction intervals, shown in Figure 7.18, and see that they range from a minimum width of 116 units at the center of Carbon values 5 and 6, to a maximum width of 128 units at the Carbon value of 1. Once again, you can see that there is less uncertainty in the predictions at the center of Carbon, but as we move away from the 5 and 6 Carbon values, the uncertainty in the predictions goes up.

Figure 7.18 Einstein's Data: Prediction Interval Width versus Carbon

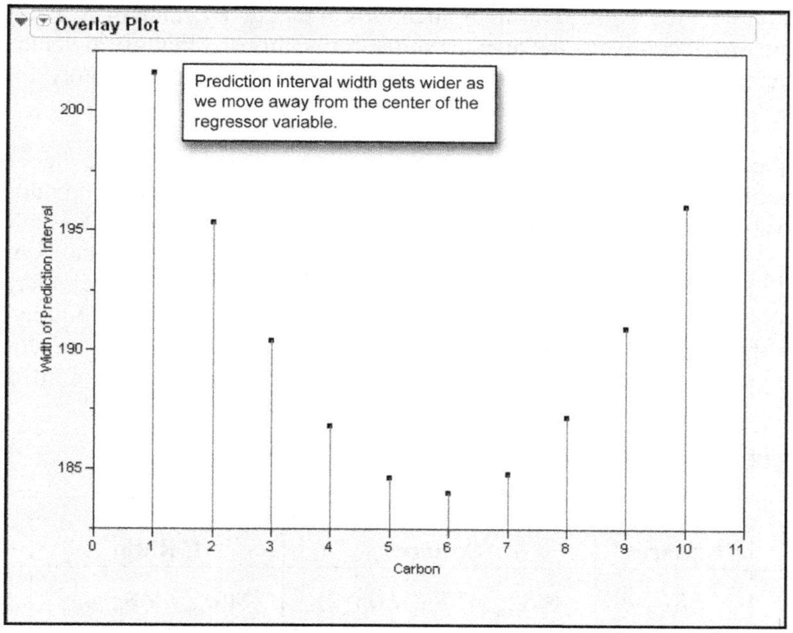

Lack-of-Fit (LOF) Test

There is a formal check for the model adequacy called the *lack of fit* (LOF) test, which is available when our data has replications of the regressor variable. The LOF test is based on a further partitioning of variation of the ANOVA error sum of squares. When replicates exist (that is, different experimental units—see Section 7.3.5), we have an estimate of noise based on the replications that is called *pure error*. The error sum of squares can be partitioned into a component due to the pure error and a component due to unexplained variation. We then compare the pure error with the estimate of noise coming from the unexplained error, which could include unspecified terms in the model, called *lack of fit*. The partitioning of the error sums of squares term for the LOF test is given by these equations:

$$SS_{\text{Error}} = SS_{\text{LOF}} + SS_{\text{Pure error}} \tag{7.11a}$$

$$\sum_{i=1}^{m}\sum_{j=1}^{n_i}\left(y_{ij}-\hat{y}_i\right)^2 = \sum_{i=1}^{m}\sum_{j=1}^{n_i}\left(y_{ij}-\bar{y}_i\right)^2 + \sum_{i=1}^{m} n_i\left(\bar{y}_i-\hat{y}_i\right)^2 \tag{7.11b}$$

Here *m* is the number of unique levels of the regressor variable and n_i is the number of replicates at each unique level of the regressor variable. The first term (LOF) compares an observation with the average of the replications at that given value of X, while the second term (Pure error) compares the average of the replications at that given value of X with the predicted value at that value of X. If the model fits the data adequately, then these two estimates of noise should be similar and their ratio should be close to 1.

The simple linear regression ANOVA table can be expanded to include the terms corresponding to the LOF test as shown in Table 7.9, where *n* is the total number of observations, and *m* is the number of unique levels of the regressor variable. The hypothesis statement for the LOF test is loosely given as H_0: SLR Model is Adequate versus H_1: SLR Model is Not Adequate. Therefore, if the LOF produces a large F ratio, whose *p*-value is less than the significance level, α, then we reject the null hypothesis and conclude that our model is "missing something." If, on the other hand, we do not reject the null hypothesis, we then conclude that our model does a good job of fitting the data.

Table 7.9 Analysis of Variance for Lack-of-fit Test for an SLR Model

Source	DF	Sum of Squares	Mean Square	F Ratio	Prob > F (P-value)
Lack of fit	m–2	SS_{LOF}	$MS_{LOF} = SS_{LOF} / (m-2)$	$MS_{LOF} / MS_{Pure\ error}$	Prob > F Ratio
Pure error	n–m	$SS_{Pure\ error}$	$MS_{Pure\ error} = SS_{Pure\ error} / (n-m)$		
Total error	n–2	SS_{Error}			

Statistics Note 7.6: The pure error estimate is not related to the regression model but depends solely on the replications. It also depends on the assumptions of constant variance and independent observations (Weisberg 1985).

The LOF test can be accessed in the **Analyze > Fit Model** and **Analyze > Fit y by X** platforms in JMP and is automatically included in the output if replicates exist in the data. Figure 7.19 shows the lack-of-fit test for Einstein's data. Recall that the error sum of squares has 17–2 = 15 degrees of freedom, corresponding to the 17 observations minus the 2 terms in the model (intercept and slope). These 15 degrees of freedom can be further partitioned, according to Table 7.9, into 7 degrees of freedom coming from the replications, and 8 degrees of freedom due to unexplained variation. Equations 7.11a and 7.11b yield

$$SS_{Error} = SS_{LOF} + SS_{Pure\ error}$$

$$26387.14 = 20954.81 + 5432.33$$

The LOF test is done using the MS_{LOF} versus the $MS_{Pure\ error}$ as shown in Figure 7.19. The *p*-value for the LOF test is $0.0632 > 0.05$, suggesting that the model does a good job at fitting the data. However, as was the case with the coefficient of determination R^2, this is not the complete picture. The LOF test is just another diagnostic to help us determine whether the model adequately fits the observed data.

Figure 7.19 Einstein's Data: Lack-of-fit Test

Total Error is split into Lack-of-fit and Pure Error.

Analysis of Variance

Source	DF	Sum of Squares	Mean Square	F Ratio
Model	1	240468.74	240469	136.6965
Error	15	26387.14	1759	Prob > F
C. Total	16	266855.88		<.0001*

Lack Of Fit

Source	DF	Sum of Squares	Mean Square	F Ratio
				3.3752
Lack Of Fit	8	20954.808	2619.35	Prob > F
Pure Error	7	5432.333	776.05	0.0632
Total Error	15	26387.142		Max RSq
				0.9796

Statistics Note 7. 7: A significant lack-of-fit result tells us there is something wrong with the model but it doesn't indicate what is wrong. For Einstein's data we have noted that the model does not fit observation 8 very well, and that we have included only one of the three explanatory variables.

A significant LOF just indicates that the model does not fit well, but why? For an SLR model, a potential cause for LOF could be the omission of higher order terms like $Carbon^2$. This, however, will be evident in a plot of the residuals versus predicted values. It is up to us to further explore the data and determine the root cause and, if possible, adjust the model.

7.3.5 Sampling Plans

Many of the concepts for sampling plans presented in previous chapters apply to data collection methods needed for determining the linear relationship between two variables. For example, we must still consider the concept of an experimental and observational unit, which is crucial to carrying out a meaningful significance test. Recall, the *experimental unit (EU)* is the smallest unit that is affected by the "treatment" or condition we want to investigate, while the *observational unit (OU)* is the smallest unit on which a measurement is taken. As we discussed in Section 2.3.2, OUs are not replications but repeated measurements on the *same* EU. Therefore, if more than one observational unit is taken per experimental unit, the total number of measurements increases but the total sample size for our experiment will not. For example, if for Einstein's data we took two measurements (OUs) for each of the 17 compounds (EUs) we will end up with 34 values of the constant *c* but no true replication for those original 17 EUs or compounds.

In order to conduct a study to establish the linear relationship between two variables, we must define our *sampling plan*, consisting of the *sample size* (that is, how many), as well as the *sampling scheme*. A *random sampling* is a scheme in which each pair (X,Y) has an equal chance of being selected is still preferred (Situation S1, Table 7.2). In this section we discuss general concepts associated with how many samples to take and how to take them. However, we are not using any sample size calculations in JMP, as we did for determining sample sizes for one-sample, two-sample, and one-way ANOVA techniques.

Sample Size

A simple linear regression requires only two paired observations, (x_1, y_1) and (x_2, y_2), to fit a straight line with slope $\beta_1 = (y_2 - y_1) / (x_2 - x_1)$, the change in y over the change in *x*, and intercept $\beta_0 = (y_1 \times x_2 - y_2 \times x_1)/(x_2 - x_1)$. With two pairs of observations, n = 2, we are able to estimate the slope and the intercept (two degrees of freedom), but not the experimental error because there is no data left (error DF = n–2 = 2–2 = 0). Therefore, we will not be able to carry out any tests of significance but, we will get a perfect fit (that is, an $R^2 = 1$, which is the limiting case in Figure 7.15b). So, what is wrong with using only two paired observations for establishing a linear relationship between two variables? Maybe nothing! Recall the discussion by Natrella regarding the functional and statistical relationships between X and Y (Table 7.2). If we truly have a known linear functional relationship between two variables (situations F1 or F2), with little measurement error in Y, then sampling two points along the curve is probably a very efficient way to estimate the SLR model. However, this is usually not the case and we either have non-negligible measurement errors in the Ys, or we do not have a known functional relationship but a statistical relationship that can be described by a linear model.

If we are setting up the data collection study with the intent of fitting an SLR model, then we want to consider the following items:

- Do we have enough samples to adequately estimate the experimental error, σ^2_{error}, using MSE in the ANOVA table?
- Have we sampled a wide enough range of the regressor variable in order to observe the relationship between the response, Y, and the regressor, X?
- Have we sampled in several locations across the range of the regressor variable in order to determine whether higher order terms are also needed to adequately describe the relationship between X and Y?
- Did we include replications in the sampling plan so we can partition the experimental error into *lack-of-fit* and *pure error* in order to conduct a LOF test?

Estimating Experimental Error

Let's address these questions in more detail, beginning with the first one. Estimating σ^2 reliably generally requires enough degrees of freedom for the MSE estimate. How many degrees of freedom do we need? We can get a good idea of what enough degrees of freedom are by looking at the relationship between the degrees of freedom and the coefficient of variation of estimates of the standard deviation. In general, for any statistic computed from the data, the CV is the ratio of the standard deviation of the statistic to the mean of the statistic, and the larger the CV, the more uncertainty we expect in a given statistic (in Chapter 2, Table 2.3, we introduced the coefficient of variation for the estimate of the mean). It can be shown (Wheeler (2005)) that the coefficient of variation for the standard deviation statistic is proportional to the degrees of freedom:

$$Coefficient\ of\ Variation \approx \frac{1}{\sqrt{2df}} \qquad (7.12)$$

Equation 7.12 clearly shows that in order to cut the uncertainty in the estimate in half we must quadruple the number of degrees of freedom of the estimate. Figure 7.20 shows a plot of the coefficient of variation versus the number of degrees of freedom of the estimate of σ. We can see that having less than 5df produces an estimate of σ with a CV in excess of 35%, estimates of σ with $8 < df \leq 12$ have a CV between 20% and 32%, and estimates with > 50df have CVs < 10%. Although intuition tells us that that the more data the better, Figure 7.20 shows that there is a limit in terms of diminishing returns; that is, starting around 50 degrees of freedom, adding more samples decreases the CV by only ~ 0.057%.

Figure 7.20 Coefficient of Variation versus Degrees of Freedom for the Estimate of σ

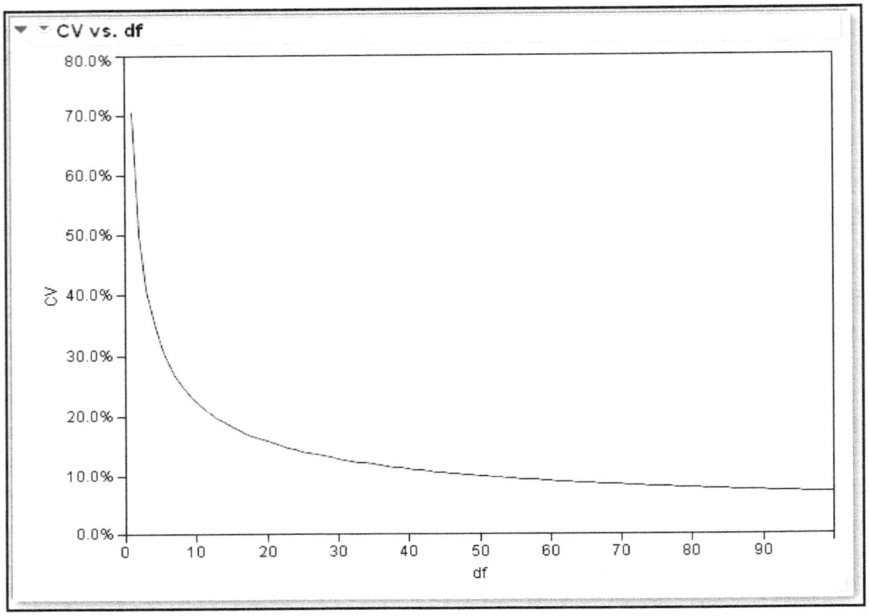

Table 7.10 provides general guidelines for determining sample sizes based on the number of degrees of freedom of the MSE estimate (Figure 7.20). Remember that for an SLR model, the total sample size is two more than the degrees of freedom associated with the error term in an ANOVA table.

Table 7.10 Sample Size Guidelines for a Simple Linear Regression Model

Degrees of Freedom for MSE Estimate	Quality of Estimate	Total Sample Size, n
< 5	Expected coefficient of variation > 35%. "Soft" estimate of σ.	< 7
5 < df ≤ 12	Coefficient of variation between 20% and 32%. Better estimate of σ but can be improved.	7 < n ≤ 14
12 < df ≤ 30	Coefficient of variation between 13% and 20%. Good estimate of σ.	14 < n ≤ 32
> 50	Coefficient of variation less than 10%. Having this much data might be overkill unless the system exhibits a lot of noise.	n > 52

Using the guidelines presented in Table 7.10, we should include enough samples to get a reliable estimate of the experimental error, σ^2_{error}.

Caution: In defining sample sizes, you must take into account the specific application for the model, the risks associated with being wrong, and the costs associated with collecting the sample.

Sample Levels

Once we know how many samples we need to collect, we still need to figure out how and where to take the samples from the larger population. Relevant questions are:

1. How many levels of the independent variable X?

   As we mentioned before, we could just sample two values of X since that enables us to get an estimate of the slope or linear effect. However, using only two X values without enough knowledge of the relationship between X and Y could lead to the wrong model. In Figure 7.21, the underlying relationship between X and Y is that of a cubic model. If we sample only at the two levels shown as the squares in the graph, and mistakenly fit a simple linear regression model, we completely miss the cubic behavior of Y for X values in the interval [–2; 1]. Using only two levels does not give us information about departures from linearity.

   If you are sure that the relationship is linear, then using only two levels is acceptable. However, three levels enable you to check for departures from linearity. More than three levels gives you a good indication of whether the linear relationship between X and Y holds across the range of X values, or whether another type of functional relationship is required.

Figure 7.21 Linear Fit for Cubic Model Sampled at Only Two Levels

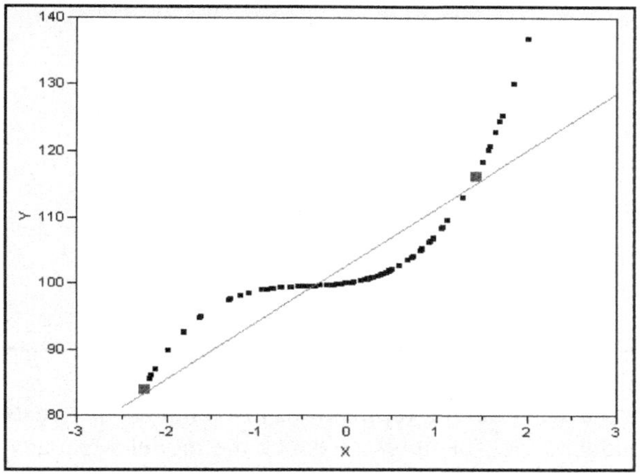

2. What about the extremes x_{min} and x_{max}?

   Try to select a wide enough range for the independent variable X in order to be confident about the linear effect in Y. If this range is too narrow the slope might not be significant because the variance of the slope, being a function of the sum-of-squares of the Xs, is maximized when the range, $x_{max} - x_{min}$, is minimized. If this range is too large then the relationship might take on nonlinear forms or produce a response that cannot be measured. Also, predictions outside the x_{min} and x_{max} range (extrapolation) always have very large prediction errors. Therefore, go wide enough to include any possible predictions that you might want to make in the future.

3. How many replications per X level?

   If given the choice, we should always include replicates in order to get a sense for the experimental error in the system. Figure 7.22 illustrates two different replication scenarios where the linear relationship between X and Y is similar, as X increases so does Y. However, we have more conficence in the SLR model in Figure 7.22 (a) because the replications are close together; that is, they have less prediction error, which is reflected in a small $MSE_{pure\ error} = 0.72$, and $R^2 = 91.74\%$. The replications in Figure 7.22(b) have more spread, which results in a large $MSE_{Pure\ error} = 21.29$, and $R^2 = 44.16\%$. In this case, the replications help us evaluate the quality of the fit between the two data sets, which we might have never uncovered because the regression lines look very similar.

Figure 7.22 SLR with Replicates

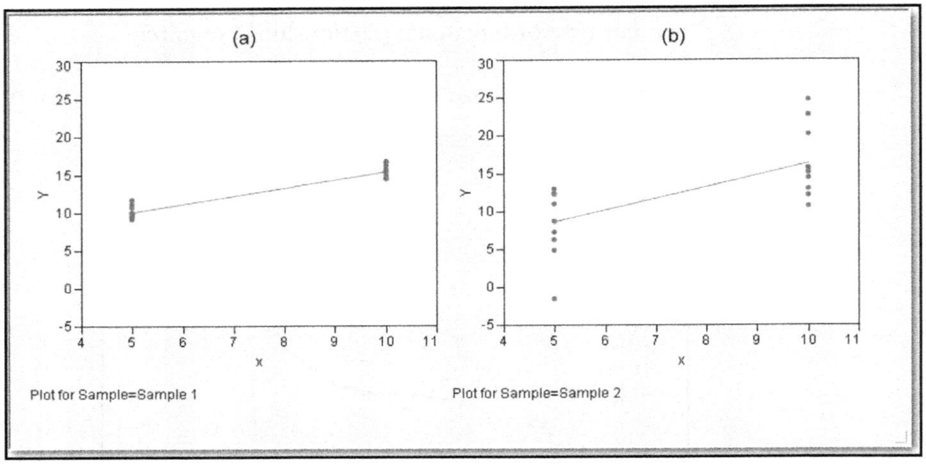

As we saw in Section 7.3.4 having replications also enables us to estimate pure error that can be used to conduct a lack-of-fit test to check the model adequacy.

In conclusion, use the suggested guidelines along with common sense and good judgment when deriving a sampling plan for a study aimed at establishing a linear relationship between a response and a factor of interest.

7.4 Step-by-Step JMP Analysis Instructions

For the load cell calibration curve described in Section 7.1, we will walk through the steps necessary to establish a linear relationship between load and deflection. As we have done in other chapters, we follow the 7 steps shown in Table 7.11. The first two steps should be stated before any data is collected or analyzed, and they do not necessarily require the use of JMP. However, Steps 3 through 6 are performed using JMP and the output from these steps helps us complete Step 7.

Table 7.11 Step-by-Step JMP Analysis for Load Cell Calibration Curve

Step	Objectives	JMP Platform
1. Clearly state the question or uncertainty.	Make sure that we are attempting to answer the right question with the right data.	Not applicable.
2. Specify the hypotheses of interest.	Is there evidence to indicate that a linear relationship between the deflection and load is appropriate? Should we include higher order terms in the model?	Not applicable.
3. Determine the appropriate sampling plan and collect the data.	Identify how many experimental units are needed, how many levels of load we will use, and where we will get replicate readings. State the significance level (α); that is, the level of risk (of saying that there is a linear dependency, when in reality there is none) that we are willing to take.	• **Formula Editor**
4. Prepare the data for analysis and conduct exploratory data analysis.	Plot deflection (Y) versus load (X) and look for a linear relationship. Check for any outliers or influential observations in the data.	• **Analyze > Fit Y by X**

(continued)

Table 7.11 *(continued)*

Step	Objectives	JMP Platform
5. Perform the analysis to verify your hypotheses, and answer the questions of interest.	Fit a simple linear regression model to the data. Is there a significant linear correlation? How much variation is explained by it? Is the model adequate to describe the relationship? Check the analysis assumptions to make sure your results are valid.	• **Analyze > Fit Model** • **Graph > Overlay Plot** • **Graph > Bubble Plot**
6. Summarize the results with key graphs and summary statistics.	Find the best graphical representations of the results that give us insight into our uncertainty. Select key output for reports and presentations.	• **Analyze > Fit Y by X** • **Graph > Overlay Plot**
7. Interpret the results and make recommendations.	Translate the statistical jargon into the problem context. Assess the practical significance of your findings.	Not applicable.

Step 1: Clearly state the question or uncertainty

In preparation for the certification by a nationally recognized standards bureau, you need to verify several calibration standards for a new high-capacity canister style load cell entering this market. Since you are new to this department, your boss wants you come up to speed by reproducing an existing calibration standard for equipment and vehicles up to 3,000 lb based on a historical study circa 1975 by a National Institute of Standards and Technology (NIST) scientist named Paul Pontius. This standard is based on establishing the relationship between a specified load and the corresponding deflection measured by the load cell. You must derive a calibration curve, which depicts deflection (inches) as a function of the load (lb) and determine its merits.

Step 2: Specify the hypotheses of interest

We must develop a calibration curve that shows the relationship between the load (lb) of an object and its measured deflection (in). A simple linear regression model is often appropriate for establishing calibration curves and can be used here to specify a model for our calibration problem. The calibration curve between deflection and load can be written as follows:

$$Deflection = \beta_0 + \beta_1 Load + \varepsilon, \tag{7.13}$$

where β_1 represents the slope or change in deflection as load changes by one unit. The hypothesis statement for determining whether the slope is statistically significant is shown below:

H_0: $\beta_1 = 0$ (Assume)
H_1: $\beta_1 \neq 0$ (Prove)

In addition to determining the significance of the effect of load on deflection, we also want to make sure that a simple linear regression model provides a good fit to the data and, therefore, can be used to predict a deflection for a load that was not included in the study, but that is within the range of loads used. The lack-of-fit (LOF) test (discussed in Section 7.3.4) is used to determine the adequacy of the model and is shown in the hypothesis statement below:

H_0: SLR Model is Adequate (Assume)
H_1: SLR Model is Inadequate (Prove)

In the event that the model fit is inadequate, we should be ready to modify our model to include higher order terms; i.e., *Deflection* = $\beta_0 + \beta_1 Load + \beta_2 Load^2 + \varepsilon$, where β_2 represents a quadratic effect of load on deflection.

Step 3: Determine the appropriate sampling plan and collect the data

We will use the general guidelines presented in Section 7.3.5 to determine a sampling plan that will include how many different load values we need to include in the study, and which loads should be replicated.

Determine the Experimental Unit (EU) and Observational Unit (OU)

The first question that needs to be answered is what are the experimental unit (EU) and observational unit (OU) for our study? In order to answer this question, we need to understand how the study physically will be carried out. Different certified standard weights are placed upon a scale containing a load cell and the amount of the deflection of the weight is measured using a device. Therefore, the EU for this study is a weight (load) standard. What about the OU?

Each time a given weight is placed upon the load cell a *single* deflection measurement is produced. In this case the observational unit is also the weight standard. However, in Step 2 we indicated that we want to have some replicated measurements for some of the weight standards. Do we need to use two different standards with the same weight, or can we simply measure the exact same weight standard multiple times on the load cell? If we use the same weight standard and measure it multiple times, then the error coming from the replications represents only the measurement error of the load cell. However, if we use two different weight standards, then the error coming from the replications includes variation due to the standards, as well as variation due to the measurement error. Since these are controlled standards with guaranteed accuracy, we expected the variation from standard-to-standard to be negligible. Based on this we should be fine using one weight standard and measuring it multiple times and use the OUs as "replications."

Statistics Note 7.8: Note the distinction between replications and repetitions: leaving the same weight standard on the load cell and measuring it repeatedly provides repeated measurements that reflect the load cell's repeatability. However, if the same weight is measured at different times, perhaps with measurements of different weight standards in between, these measurements are replications that reflect the load cell's reproducibility.

Select the Extremes and Intermediate Levels of the Explanatory Variable

Selecting the appropriate levels for our factor is probably one of the most critical decisions to establish a useful calibration curve for a load cell, since we are interested in using the model for prediction of future values. As was discussed in previous sections, using a model to predict factor levels outside of the range included in the study is risky, and perhaps even inaccurate and imprecise. Therefore, we should identify the range of load values for which we would like to obtain future predictions. We know that the load cell can be used for heavy equipment and vehicles in the range from 150 lb to 3,000 lb. This provides the minimum, 150 lb, and maximum, 3000 lb, levels for the load factor.

How many load levels should be sampled in between these two extremes? In Step 2, we identified a potential need to include higher order terms in the model, so we need to include several loads in between 150 lb and 3,000 lb. The company has load standards available in 150 lb increments and, since it is not too difficult to obtain the deflection measurements, we decide to include all load standards in the range of interest, such as 150, 300, 450, …, 2,850, and 3000 lb. Our levels will include 20 load standards (EUs). The determination of the levels should also take into account the limiting factors of time and cost of measurement.

Since the levels of the independent variable, Load, are preselected and in the 150-lb to 3000-lb range, this is an S2 type of statistical relationship (see Table 7.2).

Determine Levels to Replicate

In order to check for model adequacy using the lack-of-fit test we must replicate some or all of the chosen levels. As was discussed, the replications can be obtained using the same 20 load standards, and represent the measurement error in the load cell device. One issue that we want to investigate is if the measurement error in the load cell is similar for different load standards. Once again, since this is a fairly easy study to run, we decide to replicate all 20 EUs giving a total of 40 = 2 x 20 paired observations of (load, deflection).

Once the number of levels and replications has been determined, we need to generate the data collection sheet in which the order of the weights has been randomized. We can create the run sheet in JMP using the following steps:

1. Click the **New Data Table** icon to bring up a new JMP table.
2. From the main title bar, select **Rows > Add Rows** (or double-click the Row area of the table to bring up the **Add Rows** window). Then, enter `40` in the dialog box and click **OK**. This adds 40 rows to the JMP table.
3. Right-click the **Column 1** column heading and select **Formula.** This brings up the formula dialog box. Then, select **Row > Sequence** from the **Functions** drop-down list. In order to create the load standard levels from 150 to 3,000 in increments of 150, fill in the sequence parameters as is shown in Figure 7.23 and click **OK** when you are finished. The JMP table should contain the sequence of numbers starting with 150 and ending with 3000 in increments of 150, repeated once. Double-click the column name to change it to **Load (lb)**.

Figure 7.23 Formula Editor and Results

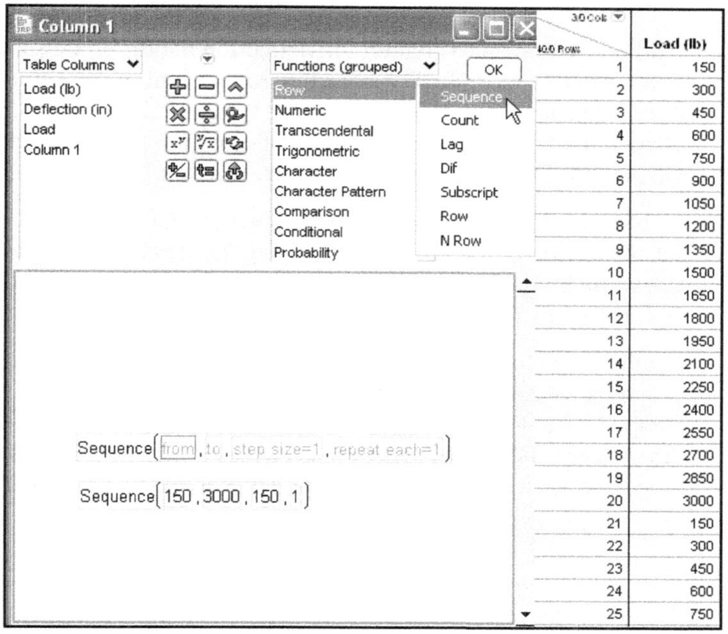

4. To randomize the run order, double-click to the right of the Load (lb) column to add another column and label it `Random Order`. Right-click the label and select **Formula;** the **Formula Editor** window opens. From the **Functions** list, select **Random > Random Uniform,** and then click **OK**. This fills in the column with random numbers generated from the uniform distribution.

5. To sort the table using the Random Order column generated in Step 4, select **Tables > Sort** and put Random Order as the **By** variable. Either select **Replace Table** to replace the current JMP table, or enter the name for the new table containing the sorted rows. Then click **OK.**

   We can also use **Formula > Random > Col Shuffle** to randomize the order of the load values without having to sort by the random number. The procedure described in Steps 4 and 5 is more general and enables us to select different distributions.

6. To add a new column to the table that reflects the randomized run order, double-click at the top of the table and enter `Run Order` as the column label, and then right-click the column label **Run Order** and select **Formula**. From the Functions list in the **Formula Editor** window, select **Row > Row,** and then click **OK**. This column will contain the values of 1 through 40.

7. Add new columns to contain the deflection measurements and any comments taken during the study. Double-click to the right of the column **Random order** once and change the label to `Deflection (in)`. Repeat this but this time change the label to `Comments`. Right-click the label and select **Column Info**. Change the **Data Type** to **Character**, and then click OK.

8. Columns can be rearranged by clicking their label in the **Columns** listing on the left-hand side of the table and dragging the label to the desired placement in the list while holding down the mouse key. Release the mouse key when placement is completed. The run sheet for this study is shown in Figure 7.24.

9. Finally, don't forget to save the JMP table for future use. For data collection purposes, we can print the JMP table directly, or save the table to Excel and print out a copy to use while running our study. From the main title bar, select **File > Save As.** A **Save JMP File** As dialog box appears. Set the **Save as type** to **(*.XLS)**. Select the appropriate file location and type in the filename before clicking **Save**.

Figure 7.24 Run Sheet for Load-Deflection Calibration Curve Study

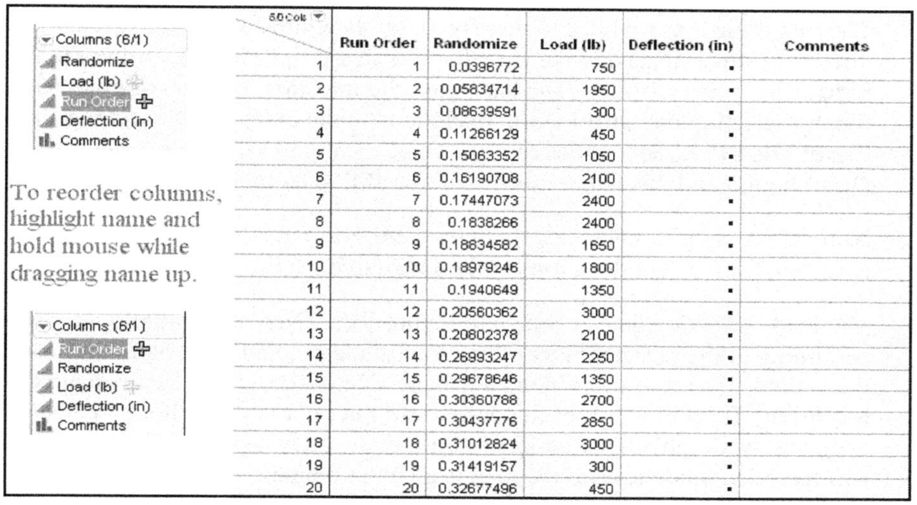

	Run Order	Randomize	Load (lb)	Deflection (in)	Comments
1	1	0.0396772	750	•	
2	2	0.05834714	1950	•	
3	3	0.08639591	300	•	
4	4	0.11266129	450	•	
5	5	0.15063352	1050	•	
6	6	0.16190708	2100	•	
7	7	0.17447073	2400	•	
8	8	0.1838266	2400	•	
9	9	0.18834582	1650	•	
10	10	0.18979246	1800	•	
11	11	0.1940649	1350	•	
12	12	0.20560362	3000	•	
13	13	0.20802378	2100	•	
14	14	0.26993247	2250	•	
15	15	0.29678646	1350	•	
16	16	0.30360788	2700	•	
17	17	0.30437776	2850	•	
18	18	0.31012824	3000	•	
19	19	0.31419157	300	•	
20	20	0.32677496	450	•	

Write the ANOVA Table

We can write out our ANOVA table in order to see how many degrees of freedom (df) are associated with the experimental error term, or MSE, and the pure error coming from replications. The partitioning of the 39 degrees of freedom is shown in Table 7.12. The total error degrees of freedom is 38, which consists of 20 degrees of freedom, from measuring each of the 20 load standards twice, and 18 degrees of freedom for lack-of-fit. Based on the guidelines in Table 7.10, the 38 df for error should provide a solid estimate of error.

Table 7.12 Partial ANOVA Table for Load-Deflection Calibration Curve

Source	DF
Model	1
Error	38
Lack of Fit	18
Pure Error	20
Total	39

Step 4: Prepare the data for analysis and conduct exploratory data analysis

The first thing that we need to do is read our data into JMP, if we saved it in a format different from JMP, and make sure that it is in the right format for the **Fit Y by X** and **Fit Model** analysis platforms. The data was read into JMP from its original source, an Excel file, by selecting the **Open Data Table** icon in the JMP Starter window and selecting the **Excel Files (*.XLS)** for files of type. Once the correct filename is selected, we click the **Open** button and the data is converted to a JMP data table.

Note: If we want to save the data in this JMP format, we need to select **File > Save** and enter a new name and location for the JMP table.

The load-deflection data is shown in Figure 7.25. Each row contains all information pertaining to the measurement of one load standard, and one column for each of the variables in the study. On the left-hand side of the table we can identify how the **Modeling Type** is set for each column variable. For an SLR, both the response and the explanatory variable must be set as continuous. As was previously noted, the deflection was recorded in inches and the load is given in pounds.

Fitting a simple linear regression can be done using the **Fit Y by X** platform. However, the **Fit Model** platform provides more options for plotting and saving diagnostic statistics, and checking assumptions Therefore, we will conduct some parts of the analysis using the **Fit Y by X** platform and other parts using the **Fit Model** platform. We begin with the **Fit Y by X** platform to visualize the data and check for outliers or unusual observations.

Figure 7.25 JMP Table Containing Results for Load-Deflection Calibration Curve Study

NIST Deflection Study

	Run Order	Randomize	Load (lb)	Deflection (in)	Comments
1	1	0.0396772	750	0.54798	
2	2	0.05834714	1950	1.41599	
3	3	0.08639591	300	0.21956	
4	4	0.11266129	450	0.32939	
5	5	0.15063352	1050	0.76562	
6	6	0.16190708	2100	1.52399	
7	7	0.17447073	2400	1.73947	
8	8	0.1838266	2400	1.73965	
9	9	0.18834582	1650	1.20001	
10	10	0.18979246	1800	1.30818	
11	11	0.1940649	1350	0.98292	
12	12	0.20560362	3000	2.16829	
13	13	0.20802378	2100	1.52408	
14	14	0.26993247	2250	1.63159	
15	15	0.29678646	1350	0.983	
16	16	0.30360788	2700	1.95392	
17	17	0.30437776	2850	2.06128	
18	18	0.31012824	3000	2.16844	
19	19	0.31419157	300	0.22018	
20	20	0.32677496	450	0.32949	
21	21	0.34582382	1650	1.20004	
22	22	0.39351379	900	0.65694	
23	23	0.39933288	900	0.65739	
24	24	0.46590315	600	0.43886	
25	25	0.51313469	1050	0.76596	
26	26	0.54119847	1950	1.41613	
27	27	0.57539136	1200	0.87487	
28	28	0.57569975	2250	1.63194	
29	29	0.63063277	1800	1.30822	
30	30	0.70568893	2700	1.95445	
31	31	0.74428619	750	0.54803	
32	32	0.78408929	1200	0.87474	
33	33	0.78738728	600	0.43899	
34	34	0.79045451	2550	1.84646	
35	35	0.82150401	150	0.11052	
36	36	0.87220172	2550	1.84696	
37	37	0.88463301	2850	2.06177	
38	38	0.90797142	1500	1.0915	
39	39	0.92842745	150	0.11019	
40	40	0.95398897	1500	1.09146	

NIST Deflection Stu
Source

Columns (5/0)
Run Order
Randomize
Load (lb)
Deflection (in)
Comments

Rows
All rows 40
Selected 0
Excluded 0
Hidden 0
Labelled 0

Note: This example is based on a load cell calibration study conducted by Paul Pontius, circa 1975, at the National Institute of Standards and Technology (NIST). The data used in this chapter is available from the data sets archives of the Statistical Reference Datasets of NIST (http://www.itl.nist.gov/div898/strd/lls/data/LINKS/DATA/Pontius.dat).

Steps for Launching the Fit Y by X Platform

1. From the primary toolbar, select **Analyze > Fit Y by X**. A dialog box appears that enables us to make the appropriate choices for the analysis.
2. Select Deflection (in) from the Select Columns window, and then click **Y, Response**. This populates the window to the right of this button with the name of our response.
3. Select Load (lb) from the Select Columns window, and then click **X, Factor**. This populates the window to the right of this button with the name of the explanatory variable. See Figure 7.26.

Figure 7.26 Fit Y by X Dialog Box for SLR

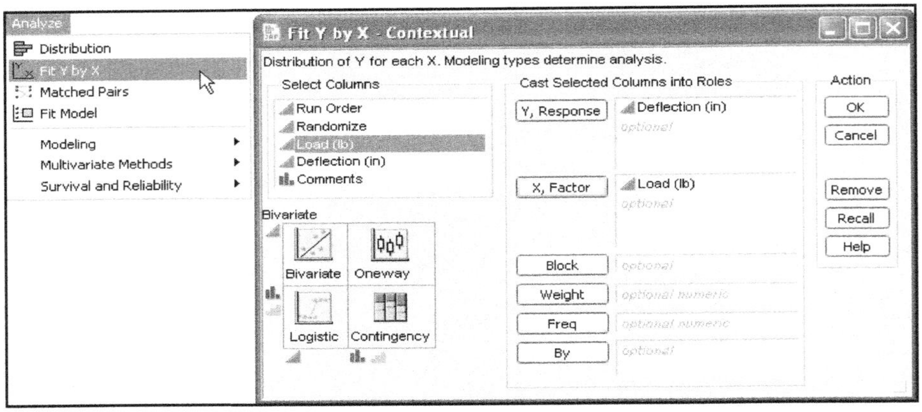

4. Click **OK**.

The default JMP output is shown in Figure 7.27 with deflection plotted on the y-axis and load plotted on the x-axis. Even with this basic plot we can readily see a very strong linear relationship between the two variables. Also, there are no obvious outliers in the data set, since all pairs of points fall along a straight line.

Figure 7.27 Default **Fit Y by X** Platform Output for Load-Deflection Calibration Curve

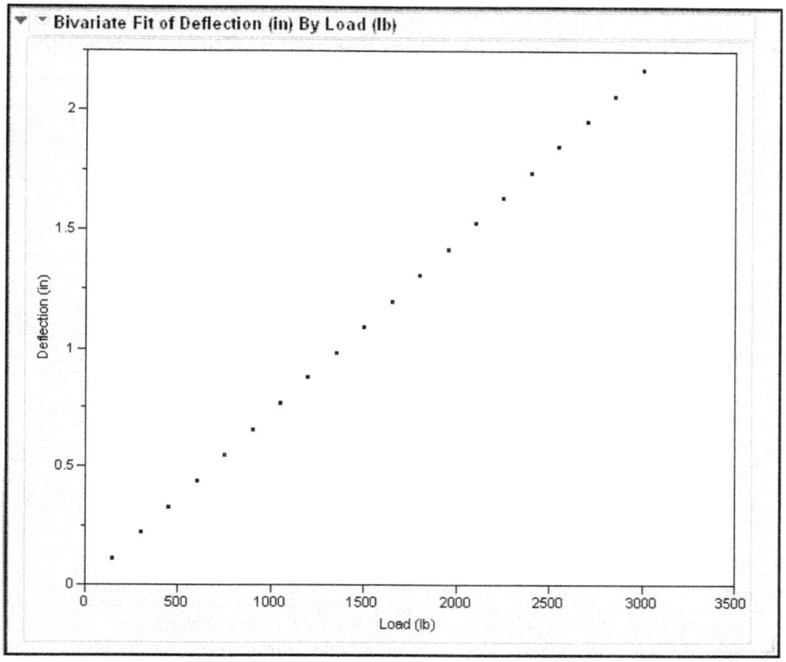

Step 5: Perform the analysis to answer the questions of interest

Our hypothesis is that the model Deflection = $\beta_0 + \beta_1$Load + ε, where β_0 is the intercept and β_1 is the slope, can be used to adequately represent the relationship between load and deflection, and in turn be used to establish our calibration curve. The scatter plot in Figure 7.27 can be enhanced by adding a simple linear regression line through the points by selecting **Fit Line** from the list that appears when clicking the contextual pop-up menu (triangle) at the top of the plot. The results are shown in Figure 7.28. In addition to the fitted line there is some tabular output below the plot.

The simple linear regression equation using least squares estimation is provided right below the plot under the banner **Linear Fit**. For the calibration study, the fitted model is as follows:

$$Deflection = 0.0061497 + 0.0007221Load. \qquad (7.14)$$

The intercept term is the measured deflection for a load = 0, which is 0.0061497 inches. Sometimes the intercept term does not make physical sense, but is merely a constant term, or correction factor, in the equation. For our study, this interpretation might make sense if there is some inherent bias in the load cell that measures the deflection. Otherwise, we would expect a deflection = 0 if no load (load = 0 lb) was placed on the load cell. The slope = 0.0007221 means that for each 1 pound increase in load, the deflection increases by 0.0007221 inches.

Figure 7.28 Enhanced Output for Load-Deflection Calibration Study

Bivariate Fit of Deflection (in) By Load (lb)

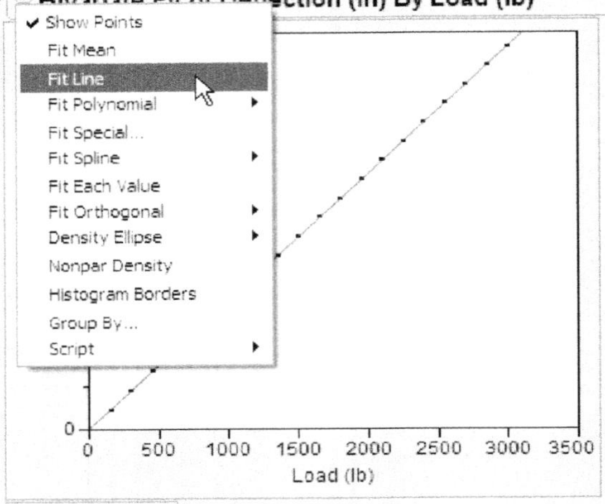

Linear Fit

Linear Fit

Deflection (in) = 0.0061497 + 0.0007221*Load (lb)

Summary of Fit

RSquare	0.999989
RSquare Adj	0.999988
Root Mean Square Error	0.002171
Mean of Response	1.143461
Observations (or Sum Wgts)	40

99.99% of the variation in Deflection can be explained by the linear effect of Load.

Includes experimental error and unspecified terms in the model.

Lack Of Fit

Source	DF	Sum of Squares	Mean Square	F Ratio
				214.7469
Lack Of Fit	18	0.00017823	9.9014e-6	Prob > F
Pure Error	20	0.00000092	4.6107e-8	<.0001*
Total Error	38	0.00017915		Max RSq
				1.0000

Analysis of Variance

Source	DF	Sum of Squares	Mean Square	F Ratio
Model	1	15.603857	15.6039	3309811
Error	38	0.000179	4.714e-6	Prob > F
C. Total	39	15.604036		<.0001*

Parameter Estimates

Term	Estimate	Std Error	t Ratio	Prob>|t|
Intercept	0.0061497	0.000713	8.62	<.0001*
Load (lb)	0.0007221	3.969e-7	1819.3	<.0001*

Used to test the hypothesis that the slope = 0.

The **Summary of Fit** in Figure 7.28 provides information about the coefficient of determination and the root mean square. Recall that the R-square (R^2) statistic is the amount of variation in the response that is explained by the model. For the calibration data, the linear model is explaining 99.9989% of the deflection variation, while 0.0011% is unexplained by the model.

The Root Mean Square Error (RMSE) is an estimate of the error that remains after the linear effect of the Load (lb) has been removed from the data. To get a better understating of what the RMSE means it is helpful to compare it with the standard deviation of the 40 Deflection measurements. This estimate can be obtained by either selecting Deflection (in) in the **Analyze > Distribution** platform, or by using the **Tables > Summary** menu in the load cell calibration table. The standard deviation of the deflection data is 0.632537 inches. After the effect of load is removed (the linear model is fitted), the standard deviation is reduced to RMSE = 0.002171 inches, an almost 300% decrease. The RMSE also represents the experimental error for our study, and includes the replication error of the load standards plus any unspecified terms in our model. The RMSE is used as the yardstick to determine the statistical significance of the slope, β_1, and will also be used to obtain prediction intervals for the predicted values obtained with the fitted line.

Finally, at the bottom of the output, we find the parameter estimates for the slope and intercept, as well as their tests of significance. The least squares estimate for the slope is 0.00072 with a standard error of 3.969E–7. The signal-to-noise ratio for testing the significance of the slope is 1,819.3, with a corresponding p-value < 0.0001. These results lead us to reject the null hypothesis, H_0: $\beta_1 = 0$, in favor of the alternative hypothesis, H_1: $\beta_1 \neq 0$. Since the slope is greater than 0, there is a positive correlation between the load and the deflection.

Checking the Adequacy of the Model

The analysis thus far seems to indicate a very strong linear relationship between load and deflection, as is seen in the scatter plot of the data and the high R^2 value obtained from the SLR fit to the data (Figure 7.28). We have also discussed that the R^2 value might not be a good indicator of the adequacy of the fit. The replications enable us to check for lack-of-fit, or how well a linear model describes the relationship between load and deflection. These lack-of-fit results (Figure 7.29) can be viewed by clicking the arrow next to the **Lack of Fit** label in the output shown in Figure 7.28. The ANOVA output is also included in Figure 7.29, in order to see the breakdown of the **Total Error** into the **Pure Error** (replications) and **Lack-of-Fit** components.

Figure 7.29 Lack-of-fit Test for Load-Deflection Calibration Data

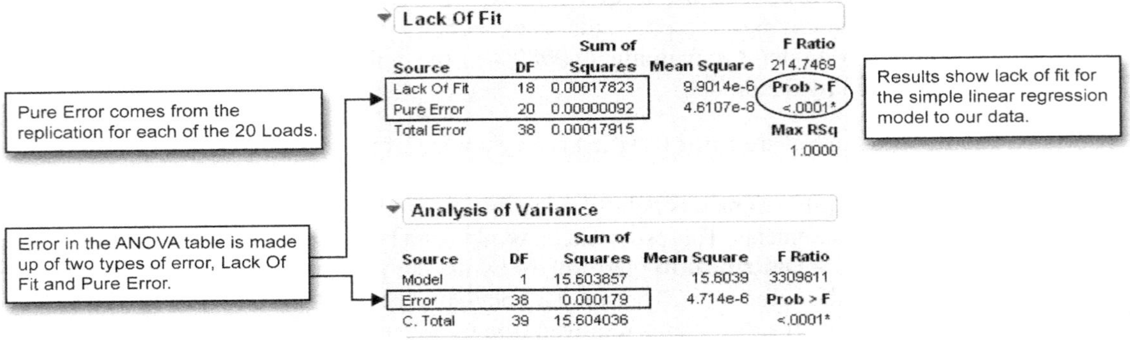

The experimental error of 0.000179 can be split into a component coming from unspecified terms in our model, Lack-of-fit = 0.00017823, and a component from the replication of the 20 load conditions, Pure Error = 0.00000092. If the model is appropriate then these two sources of variation should be similar and their ratio should be close to 1. For our study, the ratio of the normalized sums of squares, or mean squares, is 214.7469 (F Ratio), with a corresponding *p*-value < 0.0001. In other words, per degree of freedom, the error from the possibly unspecified terms in the model is approximately 215 times larger than the error from the replication of the loads. This is another example that, even though the R^2 is very high (that is, the linear model is explaining 99.9989% of the variation in the data), the linear model does not fit the data well. But where is the lack-of-fit coming from?

Residual Diagnostics

The lack-of-fit test in Figure 7.29 revealed that the simple linear regression model, *Deflection = 0.0061497 + 0.0007221Load* does not adequately fit the data. We must look a little deeper to understand why. In addition, we need to verify the model assumptions of normally distributed residuals, independence, and homogeneous variance. For this we switch to the **Fit Model** platform because it provides more diagnostics tools than the **Fit Y by X** platform.

Steps for Launching the Fit Model Platform

1. From the primary toolbar, select **Analyze > Fit Model**. A dialog box appears that enables us to make the appropriate choices for our data.
2. Select Deflection (in) from the **Select Columns** window, and then click **Y**. This populates the window to the right of this button with the name of our response.

3. Select Load (lb) from the **Select Columns** window, and then click **Add** in the right-hand side of the window. This populates the window to the right of this button with the name of the factor under investigation (Figure 7.30).

Figure 7.30 Dialog Box for Fit Model

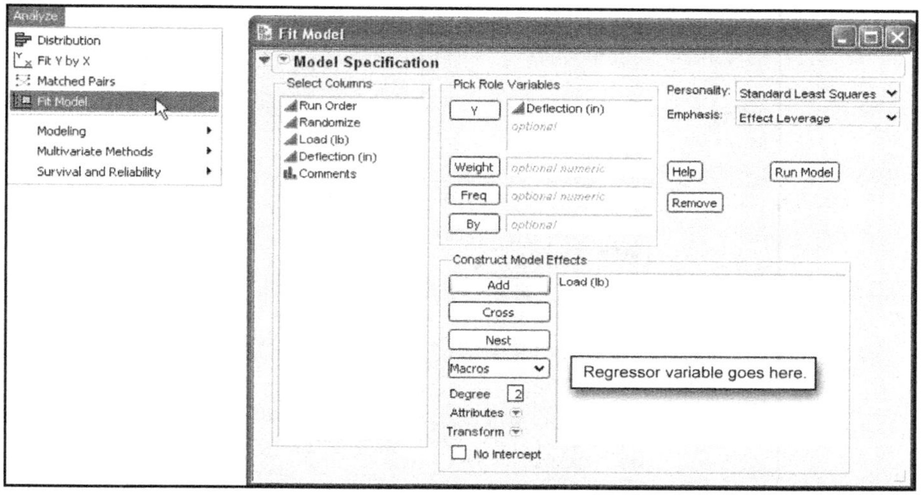

4. Click the **Run Model** button.

The JMP output for the load-deflection calibration study using the **Fit Model** platform, shown in Figure 7.31, is similar to the output that we got using the **Fit Y by X** platform. Some of the differences, however, have to do with the additional features that are available when we select from the contextual pop-up menu (triangle) at the top of the output windows. Two of these options for the Fit Model platform, **Row Diagnostics** and **Save Columns**, are shown in Figure 7.32.

Figure 7.31 Fit Model Output for Load-Deflection Calibration Study

Response Deflection (in)

Load (lb)

Whole Model

Regression Plot

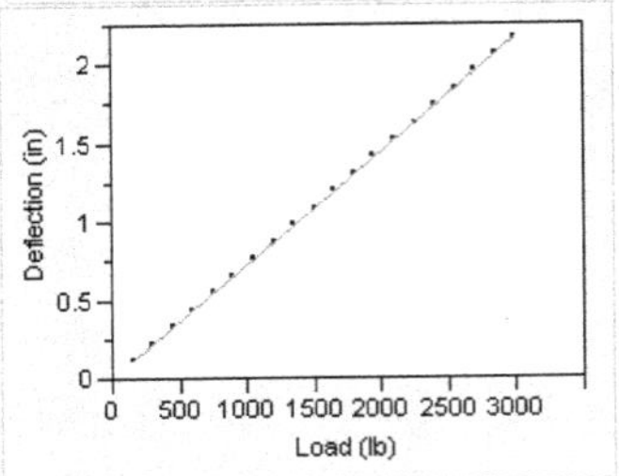

Summary of Fit

RSquare	0.999989
RSquare Adj	0.999988
Root Mean Square Error	0.002171
Mean of Response	1.143461
Observations (or Sum Wgts)	40

Analysis of Variance

Source	DF	Sum of Squares	Mean Square	F Ratio
Model	1	15.603857	15.6039	3309811
Error	38	0.000179	4.714e-6	Prob > F
C. Total	39	15.604036		<.0001*

Lack Of Fit

Source	DF	Sum of Squares	Mean Square	F Ratio
				214.7469
Lack Of Fit	18	0.00017823	9.9014e-6	Prob > F
Pure Error	20	0.00000092	4.6107e-8	<.0001*
Total Error	38	0.00017915		Max RSq
				1.0000

Parameter Estimates

Term	Estimate	Std Error	t Ratio	Prob>|t|
Intercept	0.0061497	0.000713	8.62	<.0001*
Load (lb)	0.0007221	3.969e-7	1819.3	<.0001*

Effect Tests

Source	Nparm	DF	Sum of Squares	F Ratio	Prob > F
Load (lb)	1	1	15.603857	3309811	<.0001*

The **Row Diagnostics** options include plots and statistics that help us check the fit of the SLR model. For example, **Plot Residual by Predicted** produces a very common plot that is used to look for both heterogeneous variance and unspecified terms in our model, such as quadratic terms. If we select this option, a scatter plot of the raw residuals from the SLR model versus the predicted values is shown at the bottom of the output (Figure 7.33).

Figure 7.32 Diagnostics in the Fit Model Platform

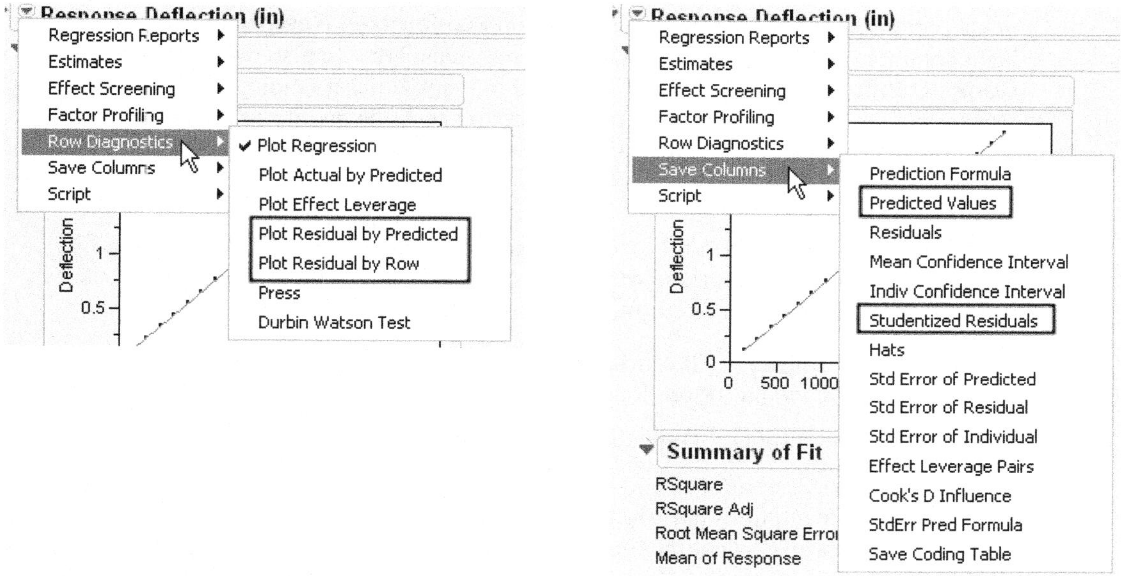

Figure 7.33 shows a very telling pattern: the residuals behave in a quadratic way. When a quadratic pattern shows up in the residuals versus predicted plot it usually means that we need to add a quadratic term to the model, $Y = \beta_0 + \beta_1 X + \beta_2 X^2 + \varepsilon$. We also use this plot to look for a model violation for homogeneous variance; however, because of the quadratic pattern in the residuals, it is difficult to assess this assumption at this time.

Figure 7.33 Residuals versus Predicted Plot for Calibration Study

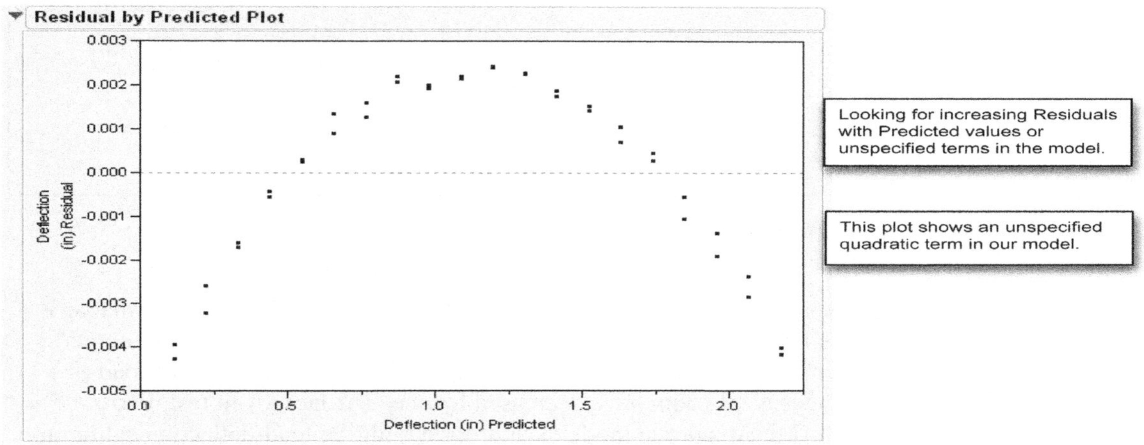

The **Fit Model** platform also offers more options for us to save key diagnostics to the JMP table, as is shown by the **Save Columns** options in Figure 7.32. While there are many options here, the **Predicted Values** and **Studentized Residuals** choices are the most commonly used to conduct analysis of residuals. We will also use the **Hats** and **Cooks D Influence** to look for leverage and influential observations on the model fit. However, there is not much reason to continue to check the quality of the linear model fit using the residuals until we refit the model with an additional quadratic term.

Refitting the Model

We can add a quadratic term for load to our model in either the **Fit Y by X**, or the **Fit Model** platform. However, if we want access to the more extensive diagnostics then it is better to use the Fit Model platform.

1. Select **Analyze > Fit Model** to launch the dialog box needed to specify the quadratic model for our load-deflection calibration curve.
2. Click the response, Deflection (in), and add it to the **Y** field.
3. Next, select Load (lb), and then click **Macros** and a list of options is revealed. One of the options is **Polynomial to Degree**, which is the one that we will use to specify a quadratic model. By default, selecting this option automatically populates the model effects with Load (lb) and Load (lb)*Load (lb) model terms, as shown in Figure 7.34.

Figure 7.34 Specifying a Polynomial Model Using the Fit Model Platform

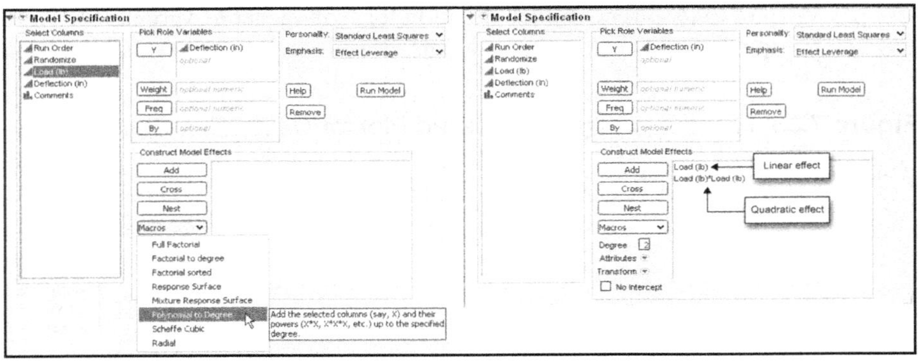

4. Click **Run Model** when finished.

The resulting output is shown in Figure 7.35. The JMP output has the same format as it did for the simple liner regression model shown in Figure 7.31, except that the degrees of freedom for the model have increased by one due to the quadratic term for load (lb) and the error degrees of freedom have decreased by one. The lack-of-fit test, **Prob > F** = 0.662, indicates that the quadratic model is appropriate for the load-deflection calibration study. Further down in the JMP output in the **Effect Tests** section, we see the tests of significance for Load (lb) and Load (lb)*Load(lb). For both of them, the **Prob > F**, or *p*-values, are both < 0.0001 and therefore we conclude that the linear and quadratic terms are statistically significant and should be included in the final model. Note that in Figure 7.35 $R^2 = 1$. By default JMP displays a limited number of decimal places. By double-clicking the number "1" value we can add more decimal places and see that $R^2 = 0.999999900179$. In other words, the quadratic equation explains 99.9999900179% of the variation.

Figure 7.35 JMP Output for Polynomial Model for the Load-Deflection Calibration Study

Response Deflection (in)

Whole Model

Regression Plot

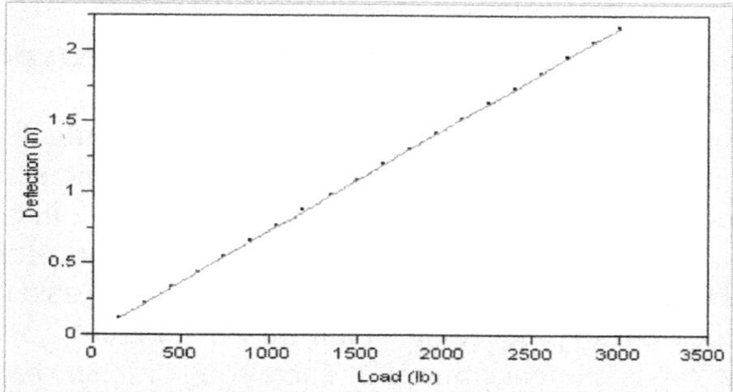

Summary of Fit

RSquare	1
RSquare Adj	1
Root Mean Square Error	0.000205
Mean of Response	1.143461
Observations (or Sum Wgts)	40

Large R-square value.

Reduction in root MSE from 0.002171.

Analysis of Variance

Source	DF	Sum of Squares	Mean Square	F Ratio
Model	2	15.604034	7.80202	1.853e+8
Error	37	1.55762e-6	4.21e-8	Prob > F
C. Total	39	15.604036		<.0001*

Lack Of Fit

Source	DF	Sum of Squares	Mean Square	F Ratio
Lack Of Fit	17	6.35468e-7	3.738e-8	0.8107
Pure Error	20	9.2215e-7	4.6107e-8	Prob > F
Total Error	37	1.55762e-6		0.6662
				Max RSq
				1.0000

Parameter Estimates

Term	Estimate	Std Error	t Ratio	Prob>|t|
Intercept	0.0085144	7.66e-5	111.15	<.0001*
Load (lb)	0.0007221	3.751e-8	19252	<.0001*
(Load (lb)-1575)*(Load (lb)-1575)	-3.161e-9	4.87e-11	-64.95	<.0001*

Effect Tests

Source	Nparm	DF	Sum of Squares	F Ratio	Prob > F
Load (lb)	1	1	15.603857	3.707e+8	<.0001*
Load (lb)*Load (lb)	1	1	0.000178	4218.525	<.0001*

Both liner and quadratic terms are statistically significant.

> **JMP Note 7.3:** By default, JMP centers the interactions and higher polynomial terms in a linear model by subtracting the middle point. Note that in the **Parameter Estimates** section in Figure 7.35 the quadratic term is reported as **(Load (lb) –1575)*(Load (lb)–1575)**, where 1575 is the mean of the load values. This makes the test for the linear term independent of the test for the quadratic term, and can improve the accuracy of the calculations.

Rechecking the Model Fit

Since the quadratic model does not exhibit lack of fit, we can now conduct a more thorough residuals analysis and look for outliers or influential observations. From the top of the window we can click the contextual pop-up menu and select **Save Columns**, and then select **Predicted Values**. We repeat this for **Studentized Residuals**, **Hats**, and **Cooks D Influence,** and then close the JMP output window to return to the JMP table.

Does the spread of the residuals increase (or decrease) as the response increases (or decreases)?

A plot of the studentized residuals versus the predicted values should display a random pattern of points around zero. If the studentized residuals tend to increase or decrease as the predicted values increase, then a "funnel shape" is displayed, indicating that the spread of the residuals is not constant. To create this plot, select **Graph > Overlay Plot**, and assign Predicted Deflection (in) to the **X** role, and Studentized Resid Deflection (in) to the **Y** role, and then click **OK** to produce the graph shown in Figure 7.36a, which has been enhanced by adding a Y-axis reference line at 0. There is no "funnel shape" pattern in this plot so we conclude that the variance is not increasing or decreasing with the mean. We also no longer see the quadratic pattern that we saw in the same plot for the simple linear regression model, or any other pattern for that matter.

Figure 7.36a Deflection Studentized Residuals versus Predicted Deflection

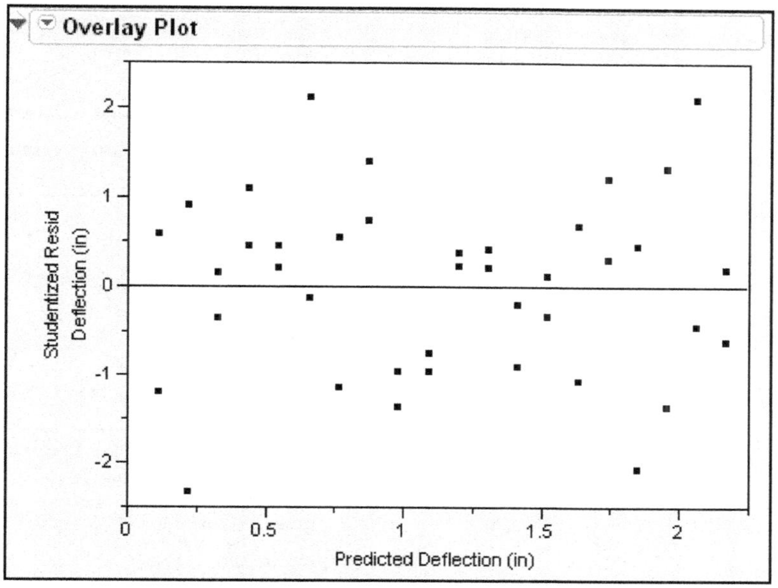

Are the residuals normally distributed?

The studentized residuals should follow a Normal (0, 1). We can verify this using the **Analyze > Distribution** platform in JMP and selecting Studentized Resid Deflection (in) for the **Y, Columns** field. From within the **Distribution** platform, we can select **Fit Distribution > Normal** to superimpose a normal curve on the histogram, and to run a significance test for normality. Click the contextual menu (triangle) next to the **Fitted Normal** label and select **Quantile Plot** and **Goodness of Fit** to get the output shown in Figure 7.36b. As we mentioned before, the normal quantile plot is a quick, visual way of checking for normality. We see that most points follow the straight line, and fall within the confidence bands of the quantile plot. Additionally, the *p*-value for the goodness of fit test is greater than our significance level, $\alpha = 0.05$, so we can conclude that the normal assumption is a reasonable one for this data set.

Figure 7.36b Normal Distribution Fit to Studentized Residuals from the Quadratic Model

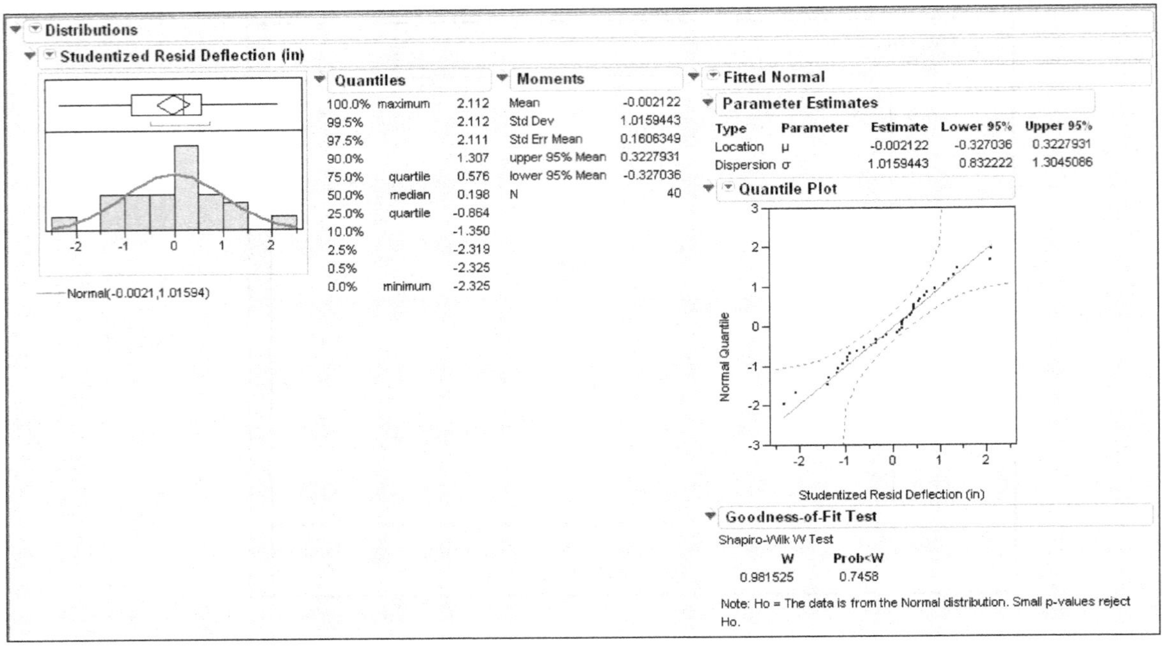

JMP 8 Note 7.1: To get the same output in JMP 8, from within the **Distribution** platform select **Continuous Fit > Normal.**

Are the residuals related to time order?

Another useful residuals plot is a plot of the studentized residuals versus time order, which represents the run order for the experiment. This plot helps us determine whether there is some type of time dependency in our observations. The run sheet shown in Figure 7.25 includes a column for the run order of the study. When plotted in time order, the studentized residuals should randomly move about a centerline of 0 and contain no systematic drifts or patterns. We will use an Individual Measurements and Moving Range control chart to examine the randomness of the studentized residuals by selecting **Graph > Control Chart > IR**.

In the dialog box, select Studentized Resid Deflection (in) for the **Process** role and Run Order for the **Sample Label,** and then click **OK**. Figure 7.36c shows the plot of the studentized residuals versus the Run Order. From within the control chart output, we can turn on some tests for runs violations by clicking the contextual pop-up menu at the top of the IR chart and selecting **Tests > All Tests**. In Figure 7.36c, we do not see any violations in runs rules and the data appears to be random around the centerline of 0. We conclude that no apparent trend or drift is present in our data. The natural process variation of the studentized residuals is in the range of –2.59 to 2.59, in accord with the Empirical Rule for a Normal(0,1) (see Section 2.5.1).

Figure 7.36c Deflection Studentized Residuals from Quadratic Model versus Run Order

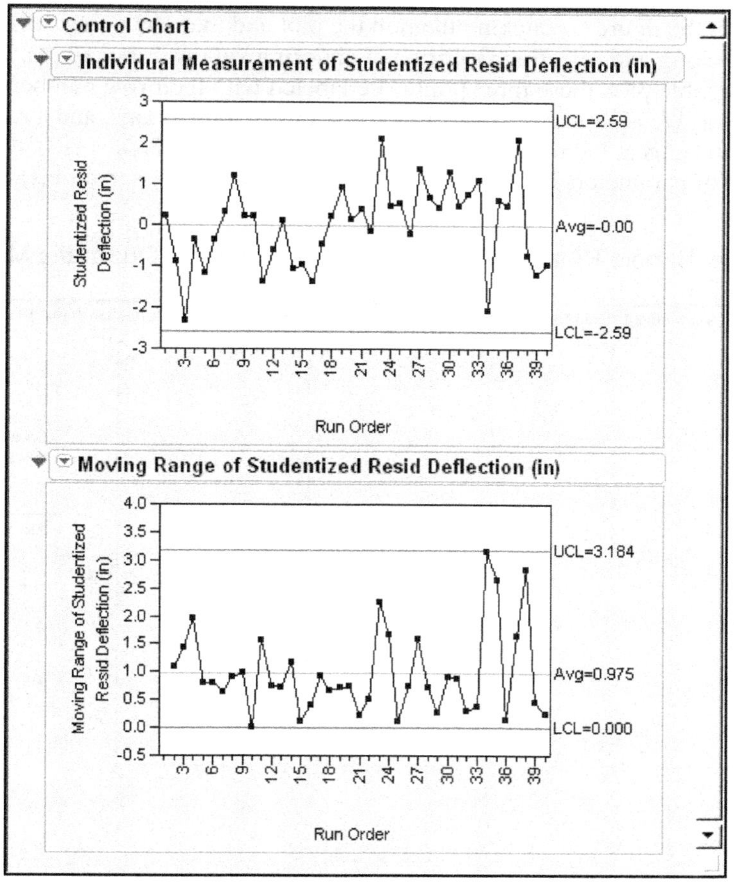

Are there any influential observations in the data?

One final thing we should do to check the fit of our model is to look for observations that have a larger influence on the parameter estimates for our quadratic model. If we do find one or more observations that heavily influence the fit, we want to make sure that they are valid and warrant such influence on our calibration curve. We provided several guidelines for how to identify influential observations in Table 7.8, using a combination of studentized residuals, hats, and Cook's D influence statistics.

The **Graph > Bubble Plot** is a great way to visualize data in three dimensions and enable us to look for influential observations using the three diagnostic measures mentioned above. From within the **Bubble Plot** dialog box, select Studentized Resid Deflection (in) for the **Y** role, h Deflection (in) in **X**, and Cook's D Influence Deflection (in) in the **Sizes** field, and then click **OK**. The bubble plot shown in Figure 7.37a has been enhanced from the default JMP output. For example, observations 37, 3, and 39 have been labeled in the figure by clicking them in the plot and then, by going back to the JMP table, right-clicking and selecting **Colors** and choosing yellow, for example. When we return to the bubble plot, these three points are labeled with their row numbers and have a different color. We also added a 0 reference line to the vertical axis, and a reference line to the horizontal axis at the recommended cutoff for the hats of $2p/n = 2(3)/40 = 0.15$ (p is the number of parameters in the model, and n the number of observations).

Figure 7.37a Bubble Plot for Load-Deflection Calibration Quadratic Model

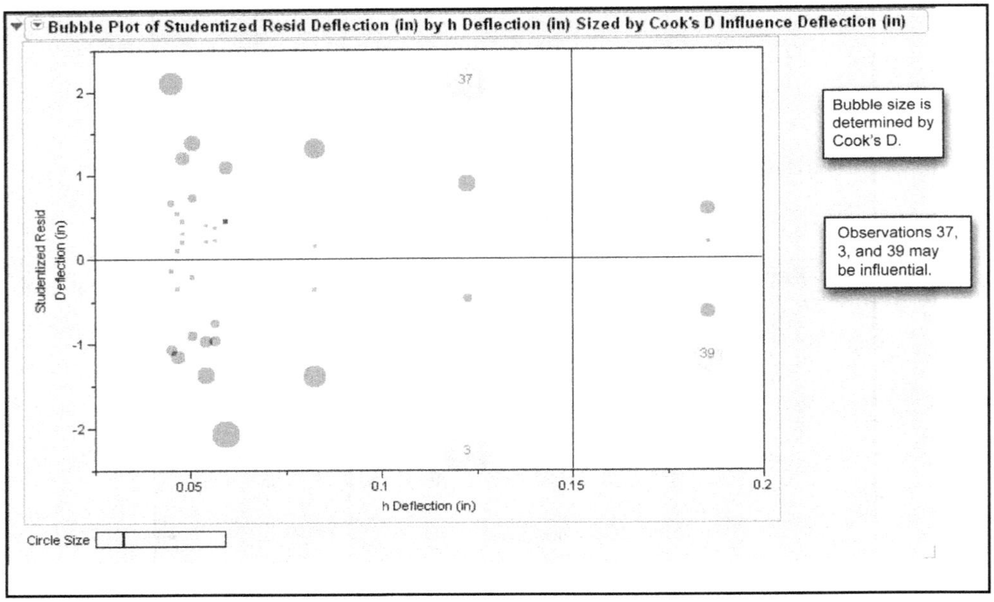

Observations 37 and 3 both have high Cook's D, which gives them a large bubble size and large studentized residuals. Observation 39 has a high leverage value (hat > 0.15) and a slightly larger Cook's D value. While this plot is helpful to identify potentially influential observations, it does not help us to visualize their influence on the fit of the quadratic model. Another helpful bubble plot uses the Deflection (in) for the **Y** role,

Load (lb) in **X**, and Cook's D Influence Deflection (in) in the **Sizes** field. This plot is shown in Figure 7.37b. As long as the rows still have the color options turned on, then these points are automatically labeled in the second bubble plot.

Figure 7.37b Bubble Plot to Visualize Influential Observations

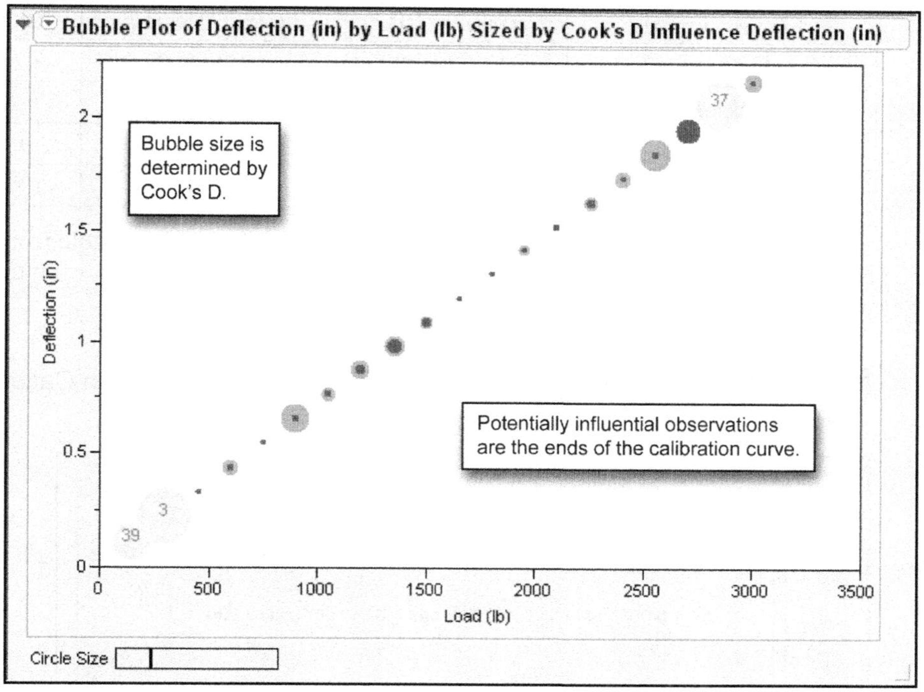

The three influential points, observation 39 = (150, 0.11019), observation 3 = (300, 0.21956), and observation 37 = (2850, 2.06177), are located around the two extreme loads of 150 and 3000 lb that we used in our calibration. In order to see just how influential these observations are on the estimates of our model parameters, we can refit the quadratic model without these three rows of data. The two sets of parameter estimates are shown in Table 7.13 and the two sets of predictions are plotted in Figure 7.38. The parameter estimates are slightly different, but their differences do not appear to be of practical significance. For example, the difference between the two linear term slopes is 0.000722103 – 0.000721802 = 3.0077E-7.

Table 7.13 Model Fit Comparison for Calibration Study

Parameter Estimate or Statistic	Model Using All Data, N = 40	Model Using Reduced Data N = 37	Difference
Intercept, b0	0.008514372	0.008980795	–4.6642E-04
Linear term, b1	0.000722103	0.000721802	3.0077E-07
Quadratic term, b2	–3.1608E-09	–3.1286E-09	–3.2194E-11
Lack-of-fit	None	None	
Root MSE	0.000205177	0.000175778	2.9399E-05
R^2 Adj.	0.999999895	0.999999913	–1.8000E-08

In Figure 7.38 the predicted lines from both models are indistinguishable from each other. Note that JMP provides predictions for the three suspect observations, even though they were not used in the model's fit.

Figure 7.38 Impact of Influential Observations 37, 3, and 39 on Calibration Model

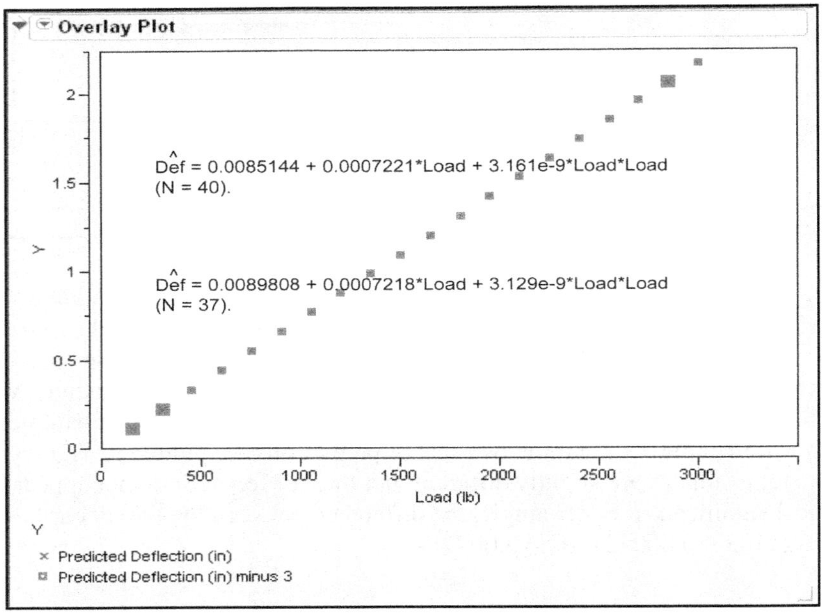

Based on these results, we are not going to investigate these 3 observations further and we are going to use them to estimate the parameters of the quadratic model using all 40 observations.

A window into the future: Prediction Intervals

In addition to performing a thorough check of the model assumptions, including a lack-of-fit test, it is helpful to examine the quality of the predictions from our estimated quadratic model:

$$\text{Deflection} = 0.0085144 + 0.0007221{*}\text{Load} - 3.161\text{E}{-9}{*}(\text{Load}-1575)^2 \quad (7.15)$$

For a linear regression model with one factor, the **Fit Y by X** platform can be used to visualize the prediction error. The JMP steps and output for the Einstein data were previously presented in Section 7.3.4. These are repeated here for the load-deflection calibration study.

1. Select **Analyze > Fit Y by X** and highlight Deflection (in), and click **Y, Response**. Then select Load (lb) and click **X, Factor**. Finally, click **OK**.
2. From within the output window, click the contextual pop-up menu at the top of scatter plot and select **Fit Polynomial > 2, quadratic**. The quadratic line should appear in the scatter plot and a triangle with the label 'Polynomial Fit Degree = 2' should appear underneath the plot (Figure 7.39).
3. Click the triangle located right underneath the plot and select **Confid Shaded Fit** and **Confid Shaded Indiv**. The plot should now have two sets of bands surrounding the prediction line (Figure 7.39).

Figure 7.39 Creating Shaded Confidence Bands in Fit Y by X

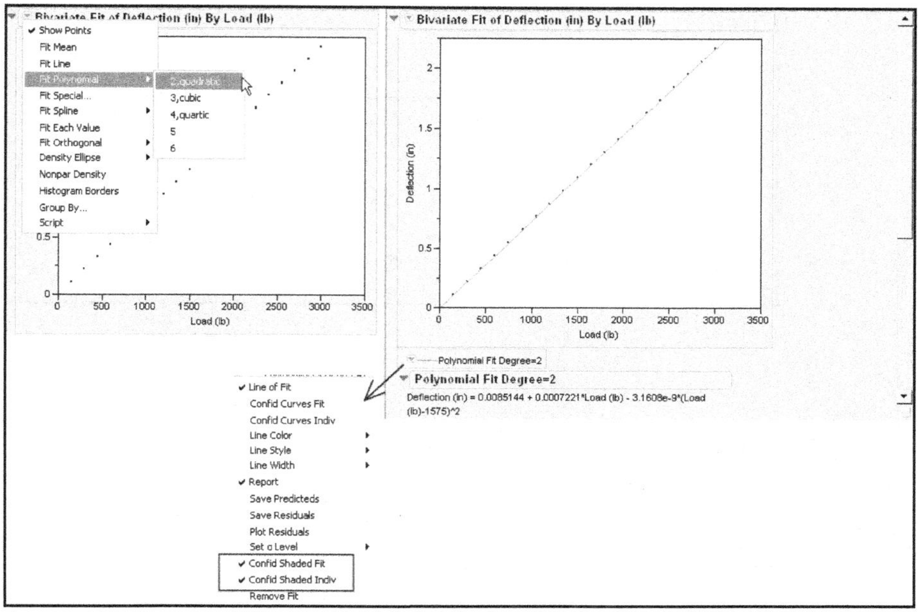

The final plot is shown in Figure 7.40 and it looks almost identical to the plot shown in Figure 7.39 without the shaded confidence intervals turned on. Why? For the calibration study, the fit of the quadratic model to our data has a very small RMSE = 0.000205177, which in turn results in a very small prediction error. That is why we cannot see the bands around the predicted values.

Figure 7.40 Load-Deflection Quadratic Model with Shaded Confidence Bands

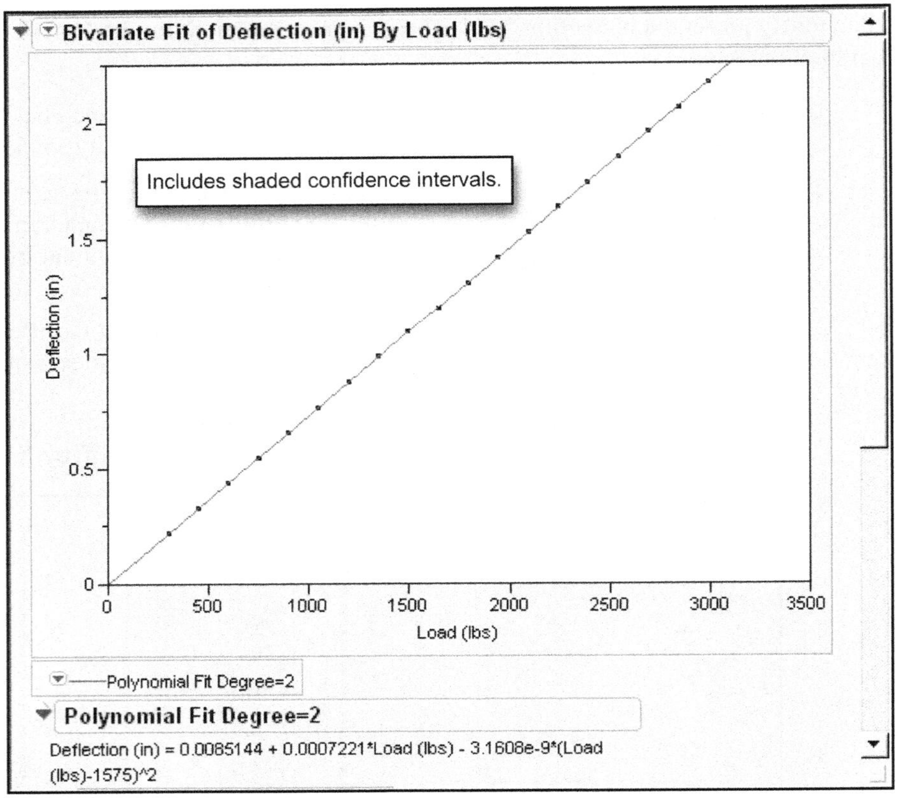

If, for example, we want to know the values for the 95% prediction interval for a load = 150 (lb), then we might need to save the intervals to the JMP table from the **Fit Model** platform, and read them from the tabular output. The mean and individual 95% confidence intervals for a load = 150 (lb) are provided below:

Mean: 0.11041 ± 0.000179 = (0.11023; 0.11059)

Individual: 0.11041 ± 0.000453 = (0.10996; 0.11086)

Remember that the interval for an individual value is always wider than the one for the mean because it allows for the additional variation of a not-yet-observed value. This information can be shown graphically using the **Graph > Overlay Plot** platform. After saving the intervals to the JMP table, we can add two new columns to the table, one for the half width of the mean confidence interval and one for the half width of the individual confidence interval. The formula editor can be used to create these columns. For example, the half width of the interval for the individual confidence limits is (Upper 95% Indiv – Lower 95% Indiv) / 2. This value represents the ± margin-of-error added and subtracted from the predicted value to calculate the upper and lower bounds of the interval. To create this plot, follow the steps below.

1. From within the **Overlay Plot** dialog box, select Predicted Deflection (in), Half Width of Mean, and Half Width of Indiv, and assign them to **Y**.
2. Select Half Width of Mean in the **Cast Selected Columns Roles** field, and then click the **Left/Scale/Right Scale** button. Make sure the arrow is pointing to the right.
3. Repeat this for Half Width of Indiv and make sure the arrow is pointing to the right.
4. Select Load (lb) and assign it to the **X** role.
5. Click **OK**.

The final assignments of the variables should match those shown in Figure 7.41.

Figure 7.41 Overlay Plot with Two Vertical Axes Using Predicted Values

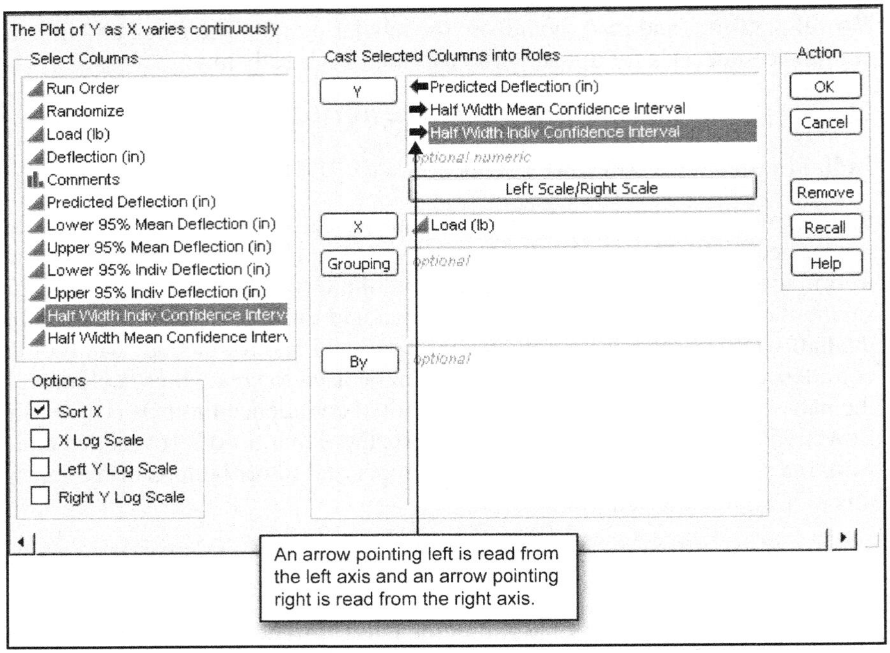

The resulting overlay plot, along with a partial view of the data used to create it, is shown in Figures 7.42a. and 7.42b. The predicted Deflection (in) value for a given Load (lb) is read off of the left-hand axis and the ± margin-of-error is read from the right-hand axis. This is a visual way to illustrate the difference in size between the mean confidence interval and the individual confidence interval. We can also see how the half widths of these intervals change across the different Load (lb) values used in our study. These widths are largest at the two extremes of our load values and then get smaller as we approach the center value for load.

Figure 7.42a Mean and Individual Confidence Interval Half Width (Overlay Plot)

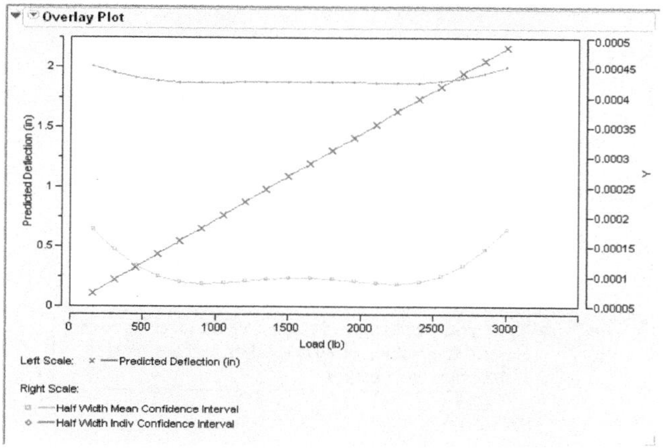

Figure 7.42b Mean and Individual Confidence Interval Half Width (JMP Table)

Load (lb)	Predicted Deflection (in)	Lower 95% Mean Deflection (in)	Upper 95% Mean Deflection (in)	Lower 95% Indiv Deflection (in)	Upper 95% Indiv Deflection (in)	Half Width Mean Confidence Interval	Half Width Indiv Confidence Interval
150	0.11041132	0.11023232	0.11059032	0.10995869	0.11086395	0.000179	0.00045263
300	0.22000684	0.21986125	0.22015243	0.21956636	0.22044732	0.00014559	0.00044048
450	0.32946012	0.32934081	0.32957943	0.32902761	0.32989263	0.00011931	0.00043251
600	0.43877117	0.43867012	0.43887221	0.43834334	0.439199	0.00010104	0.00042783
750	0.54793998	0.54784889	0.54803106	0.54751439	0.54836557	9.10823e-5	0.00042559
900	0.65696655	0.65687832	0.65705477	0.65654156	0.65739153	8.82274e-5	0.00042499
1050	0.76585088	0.76576097	0.76594079	0.76542554	0.76627622	8.99074e-5	0.00042534
1200	0.87459298	0.87449959	0.87468637	0.87416689	0.87501907	9.33877e-5	0.00042609
1350	0.98319284	0.98309618	0.9832895	0.98276602	0.98361966	9.66615e-5	0.00042682
1500	1.09165046	1.09155191	1.09174902	1.09122321	1.09207772	9.85576e-5	0.00042725
1650	1.19996585	1.19986729	1.20006441	1.1995386	1.2003931	9.85576e-5	0.00042725
1800	1.308139	1.30804234	1.30823566	1.30771218	1.30856582	9.66615e-5	0.00042682
1950	1.41616992	1.41607653	1.4162633	1.41574383	1.416596	9.33877e-5	0.00042609
2100	1.52405859	1.52396868	1.5241485	1.52363325	1.52448393	8.99074e-5	0.00042534

Inverse Prediction

In our study, deflection took on the role of the independent variable, Y, load took on the role of our explanatory variable, X, and we established a quadratic prediction equation (equation 7.15). This prediction equation plays a critical role in how we will actually use this calibration curve in practice. However, there is a twist! Calibration curves are typically used in reverse; that is, what we observe is the *y* value and then we want to determine the *x* value that generated it. For our calibration study, this means that when we put a vehicle on the weighing platform it measures its Deflection (in), and then we determine its Load (lb) using an inverse prediction equation.

Inverse Prediction for a Simple Linear Regression Model

For a simple linear regression model, we can use the **Fit Model** platform in JMP by selecting **Estimates > Inverse Prediction** from within the Fit Model output window, Figure 7.43.

Figure 7.43 Inverse Prediction in JMP

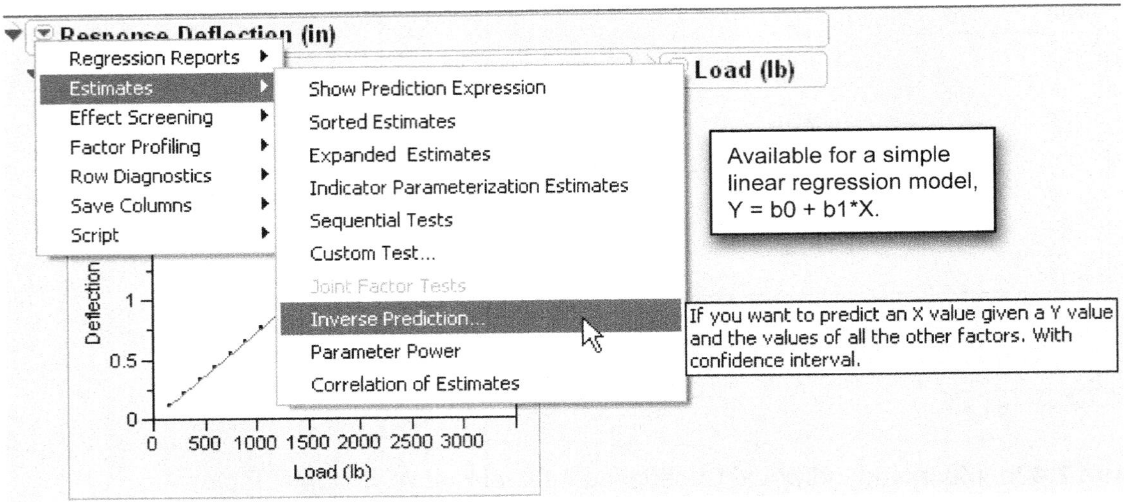

For example, for the first model that we fit to the calibration data, we can determine the Load (lb) for a measured deflection of 1 inch. When we select **Inverse Prediction**, additional fields are added to the end of the JMP output, which requires us to enter the deflection values for which we want to obtain the load. We can enter '1' in the first field. We also check the box to obtain confidence intervals with respect to an individual value, and then select **Run**. These steps, along with the results, are shown in Figure 7.44. If our load cell measured a deflection of 1 inch, then the predicted weight is 1376 lb, and it can range from 1370 lb to 1382 lb (95% confidence interval).

Figure 7.44 Inverse Prediction for Load Using Calibration Model

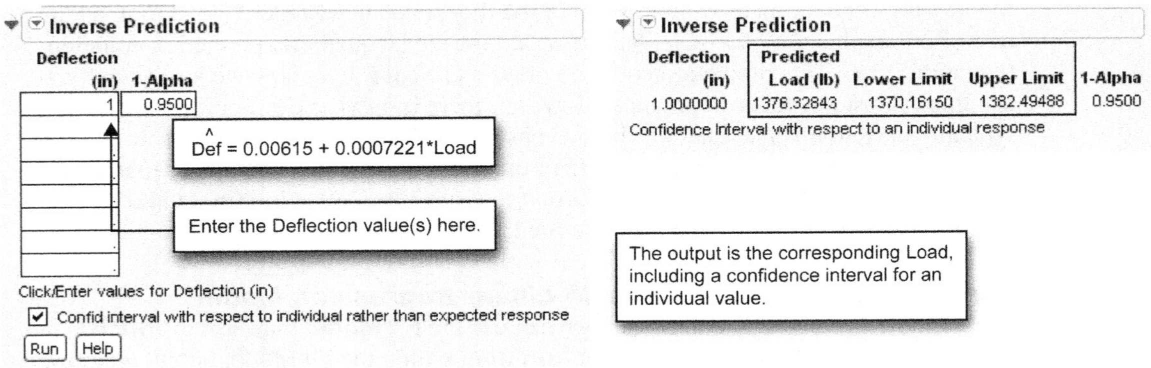

When we do an inverse prediction, we are back solving the estimated SLR model, using the given *y*, for *x*. That is, $x = (y - b_0)/ b_1$. For the example in Figure 7.44, the predicted load = (1 – 0.0061497) / 0.0007221 = 1376 lb.

How do we do inverse prediction for a quadratic model?

Unfortunately, in JMP this option is not available for more complex models, including the quadratic model that we ended up using for our calibration study. However, we can do inverse prediction by using our model, Equation 7.15, and solving the quadratic equation for load. A quadratic equation $b_0 + b_1\text{Load} + b_2\text{Load}^2$ has two solutions:

$$-b_1 - \sqrt{b_1^2 + 4b_2 Deflection - 4b_0 b_2}$$

and (7.16)

$$-b_1 + \sqrt{b_1^2 + 4b_2 Deflection - 4b_0 b_2}$$

Using the parameter estimates of the quadratic equation (equation 7.15), we get the following:

$$\text{Load} = 1.58178 \times 10^8 \left(0.0007221 - 0.000112446\sqrt{41.2477 - Deflection}\right) \quad (7.17a)$$

$$\text{Load} = 1.58178 \times 10^8 \left(0.0007221 + 0.000112446\sqrt{41.2477 - Deflection}\right) \quad (7.17b)$$

Which of these two roots, 7.17a or 7.17b, makes sense for our calibration problem? Let's create a JMP table with a deflection column containing values from 0 to 2.4 inches, using a step size of 0.1, and a total of 25 rows. This can be done using the formula editor and the **Row > Sequence** function. Again, using the formula editor, two more columns can be added to the JMP table, one for each of the two solutions for load. The JMP table, along with the plots for the two solutions 7.17a and 7.17b, is shown in Figure 7.45.

Figure 7.45 Inverse Prediction for Polynomial Calibration Model

	Deflection (in)	Load Solution 7.17a	Load Solution 7.17b
1	0	-12.239339	228452.907
2	0.1	126.316136	228314.351
3	0.2	265.040076	228175.628
4	0.3	403.933099	228036.735
5	0.4	542.995824	227897.672
6	0.5	682.228874	227758.439
7	0.6	821.632877	227619.035
8	0.7	961.208463	227479.459
9	0.8	1100.95627	227339.711
10	0.9	1240.87693	227199.791
11	1	1380.9711	227059.697
12	1.1	1521.23941	226919.428
13	1.2	1661.68253	226778.985
14	1.3	1802.3011	226638.367
15	1.4	1943.09578	226497.572
16	1.5	2084.06724	226356.6
17	1.6	2225.21614	226215.451
18	1.7	2366.54316	226074.124
19	1.8	2508.04897	225932.619
20	1.9	2649.73426	225790.933
21	2	2791.5997	225649.068
22	2.1	2933.64599	225507.022
23	2.2	3075.87382	225364.794
24	2.3	3218.28388	225222.384
25	2.4	3360.87689	225079.791

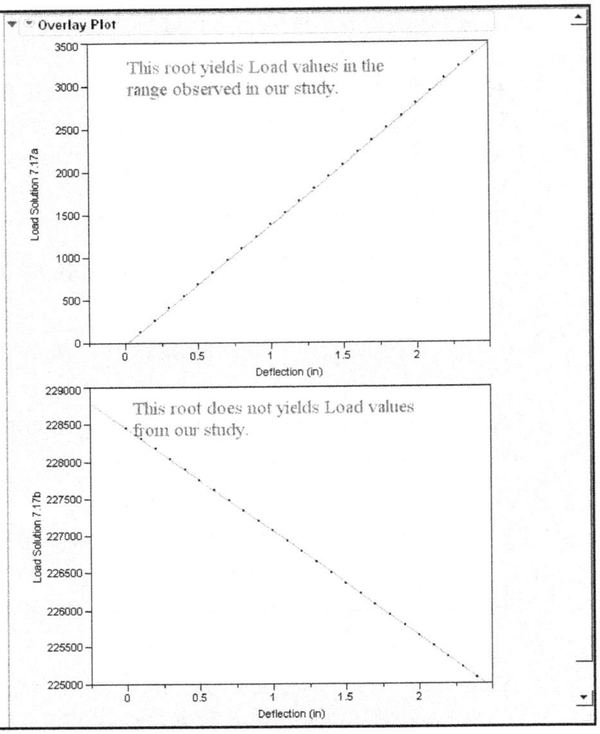

The first solution, equation 7.17a, yields values ranging from –12 to 3360 lb, while the second solution, equation 7.17b, yields values ranging from 225,079 to 228,452. In other words, deflection increases with load for the first solution, but decreases for the second solution. Therefore, the first solution should be used to develop the calibration curve for our study, because it is the one that agrees with reality.

Step 6: Summarize your results with key graphs and summary statistics

The main purpose of our study was to develop a calibration curve that can be used by customers who need to weigh large vehicles. In this case, or for any analysis for that matter, we must be able to relate our key findings in a way that can be understood by our audience. A plot of the inverse prediction, equation 7.17a, is the best way to show how the weight of an object (load) relates to deflection.

Load-Deflection Calibration curve

Figure 7.46 shows a plot of Load as a function of Deflection. This is the calibration curve that can be used to predict the load using the measured deflection of the load cell. This curve is given by Equation 7.17a, which is one of the two solutions to the quadratic model given in equation 7.15.

Figure 7.46 Load-Deflection Calibration Curve

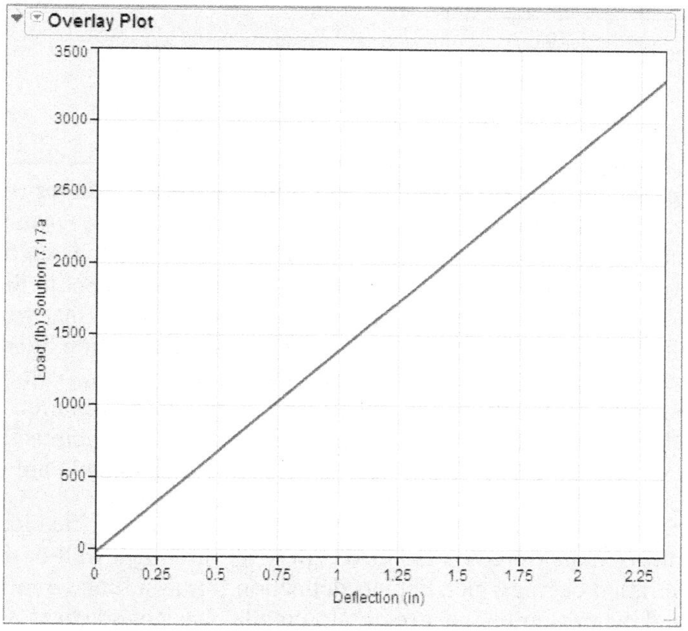

The calibration curve is

$$Load = 114220 - 17786.4 \times \sqrt{41.2477 - \text{Deflection}} \qquad (7.18)$$

For a vehicle placed in the load-cell, equation 7.18 gives the weight corresponding to the measured deflection. Note that for a deflection of 0 equation 7.18 gives a weight of –12.0376 lb. In theory we expect a 0 weight object to give a 0 deflection. This is an indication that the load cell might have a downward bias represented, in this case, by the intercept.

In addition to the plot we can also present a table with key deflection values and their corresponding predicted weight. Table 7.14 shows the predicted weights for a sample of deflections.

Table 7.14 Predicted Weights for Given Deflections According to Equation 7.18

Deflection (in)	Weight (lb)
0	−12.04
0.5	682.43
1	1,381.17
1.5	2,084.26
2	2,791.79
2.4	3,361.06

Step 7: Interpret the results and make recommendations

As we always recommend, we must translate our statistical findings in terms of the context of the situation at hand, so we can make decisions based on the outcomes of the analysis. Therefore, we should always go back to the problem statement in Step 1 and our scientific hypothesis to make sure that we answered the questions that we set out to answer with our study. Recall that you are preparing for the certification by a nationally recognized standards bureau for a new high-capacity canister style load cell entering this market. In order to come up to speed, you reproduced an existing calibration standard for equipment and vehicles up to 3,000 lb, based on a historical study circa 1975 by a National Institute of Standards and Technology (NIST) scientist named Paul Pontius.

This standard is based on establishing the relationship between a specified load and the corresponding deflection measured by the load cell. After running a well-designed study, you derived a calibration curve, which depicts deflection (in) as a function of the load (lb). You discovered that a simple linear regression model was not adequate for your calibration curve. While the impact to the predicted deflection was relatively small, a quadratic term was added to the model not only to get a better fit to the data, but also to improve the prediction ability. To see this, you calculated the average value of the absolute errors with $|Y - \hat{Y}|$ under both models. Under the linear model Average $(|Y - \hat{Y}|) = 0.00183$, while under the quadratic model, Average $(|Y - \hat{Y}|) = 0.00016$. This represents a one decimal place improvement in the absolute errors, and is smaller than the precision of the load cell, the pure error, $= \sqrt{4.61075E{-}8} = 0.00021$ (Figure 7.35), which is practically significant.

You also established an inverse prediction equation, equation 7.18, which can be hard coded into the load cell software to obtain the weight for any vehicle or heavy equipment appropriate for this class of load cell.

In conclusion, you have reproduced the calibration study and obtained useful results. You have also learned how to carry out a calibration study using a range of loads and replications in order to get the most accurate relationship between the variables of interest. This training will serve you well as you prepare for your upcoming certification and begin to establish new calibration standards.

7.5 Einstein's Data: The Rest of the Story

In Section 7.3 we used part of the data from Einstein's first published paper to illustrate many of the SLR concepts. We showed how to fit a simple linear regression model between Carbon and *c*. The data from which this example was taken actually contained two additional explanatory variables, Hydrogen and Oxygen, thus making it more suitable to a multiple linear regression. Although this chapter is aimed at simple linear regression, we will present the results of fitting a multiple linear regression model to Einstein's data.

The model that Einstein was interested in studying was:

$$c = \beta_1 \text{Carbon} + \beta_2 \text{Hydrogen} + \beta_3 \text{Oxygen} + \varepsilon \qquad (7.19)$$

The fit of equation 7.19 can be easily accomplished using the **Fit Model** platform in JMP by selecting c for the **Y** role and then Carbon, Hydrogen, and Oxygen for the **Model Effects**. Note that each regressor can be individually selected and added to the model window or all three can be selected at once and then added to the model window. When the number of Carbon, Hydrogen and Oxygen atoms is 0 we expect $c=0$; therefore, the intercept is not needed. To fit a model without an intercept, we select the **No Intercept** check box at the bottom of the window, as shown in Figure 7.47.

Figure 7.47 Fit Model Window for Equation 7.19 Model

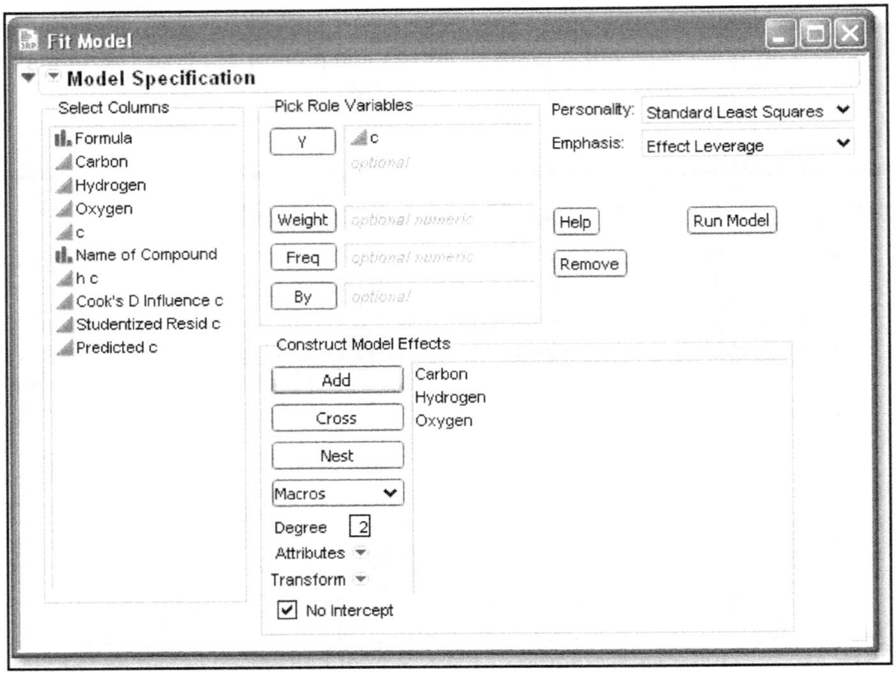

The resultant model, Figure 7.48, shows large effects of similar magnitude for Carbon and Oxygen, and a smaller effect for Hydrogen. The three *p*-values are less than our significance level $\alpha = 0.05$, and the RMSE is 13.01.

Figure 7.48 Multiple Regression Output for Einstein's Data

Response c

Whole Model

Summary of Fit

RSquare	.
RSquare Adj	.
Root Mean Square Error	13.00857
Mean of Response	402.3529
Observations (or Sum Wgts)	17

Analysis of Variance

Source	DF	Sum of Squares	Mean Square	F Ratio
Model	3	3016580.9	1005527	5942.024
Error	14	2369.1	169	Prob > F
C. Total	17	3018950.0		<.0001*

Tested against reduced model: Y=0

Parameter Estimates

Term	Estimate	Std Error	t Ratio	Prob>|t|
Carbon	48.05	2.213637	21.71	<.0001*
Hydrogen	3.63	1.607299	2.26	0.0402*
Oxygen	45.56	2.392354	19.04	<.0001*

Effect Tests

Source	Nparm	DF	Sum of Squares	F Ratio	Prob > F
Carbon	1	1	79728.454	471.1444	<.0001*
Hydrogen	1	1	865.476	5.1144	0.0402*
Oxygen	1	1	61364.877	362.6274	<.0001*

JMP Note 7.4: In Figure 7.48 both R^2 values are missing. This is because an R^2 calculated with a no-intercept model is misleading due to the fact that the Total Sum of Squares is not corrected for the mean. In other words, the no-intercept model is tested against the reduced model $y = 0 + \varepsilon$, which is indicated by the message **Tested against the reduced model Y=0** below the **Analysis of Variance** section of the output. For further details, see Section 2.4.4 of Freund et al. (2003).

It is interesting to think of a young Einstein using least squares estimation, and doing the calculations by hand, to estimate the coefficients in equation 7.19. Due to arithmetic and round off errors, his hand calculations were off, giving estimates for the coefficients for Carbon, Hydrogen, and Oxygen as $\beta_1 = 55.0$, $\beta_2 = -1.6$, and $\beta_3 = 46.8$, as compared with the ones obtained using JMP, $\beta_1 = 48.05$, $\beta_2 = 3.63$, and $\beta_3 = 45.56$.

Einstein's estimates enabled him to get fitted values and compare them to observed values. From this statistical analysis he stated that, "It can be seen that in almost all cases the deviations barely exceed the experimental errors and do not show any trend" (see also Iglewicz 2007). Residuals analysis, as we know it today, and the software tools to perform it, were not available when Einstein did his analysis. The studentized residual versus predicted plot, shown in Figure 7.49, identifies observation 1, Limonene, and observation 16, Valeraldehyde, as having large studentized residual values.

Figure 7.49 Einstein's Data: Studentized Residuals versus Predicted Plot

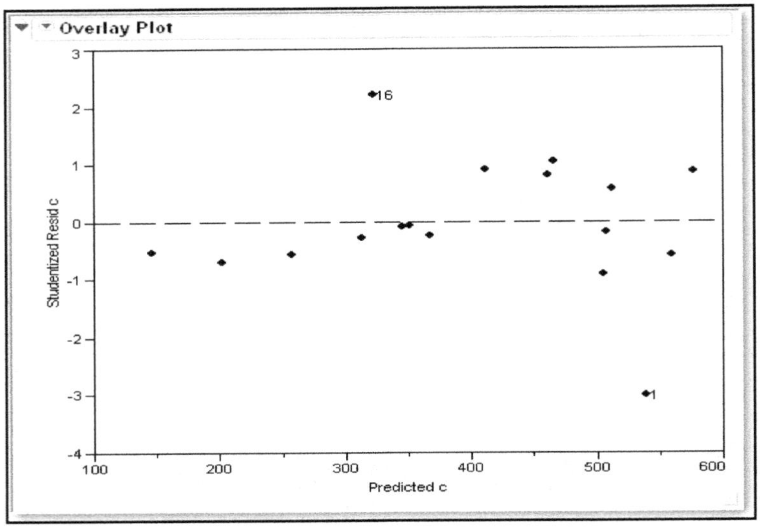

A bubble diagnostic plot, Figure 7.50, suggests that observation 1, Limonene, with values (Carbon=10, Hydrogen=16, Oxygen=0, c=510) is an influential observation, due to the large Cook's D value of 2.66. It is always a good practice to fit the model without these observations and compare the results to the fit with all of the data. Fitting the model excluding observations 1 and 16 shows a decrease in the slope for Oxygen, from 45.56 to 41.72, and a 49% reduction in the RMSE from 13.49 to 6.94. Of course, the decision to include or exclude influential observations is situation- and context-dependent.

As noted by Iglewicz (2007) "Einstein was capable of cleverly handling statistical tools and data in support of a useful scientific proposition." We hope that you use the statistical concepts and techniques we have shown you in this book to support your scientific and engineering investigations. We strongly believe that statistics, when combined with fundamental knowledge, good judgment, and an interactive and highly visual statistical software like JMP, is a powerful catalyst for new discoveries and improvements to your materials, products, and processes.

Figure 7.50 Einstein's Data: Diagnostic Bubble Plot

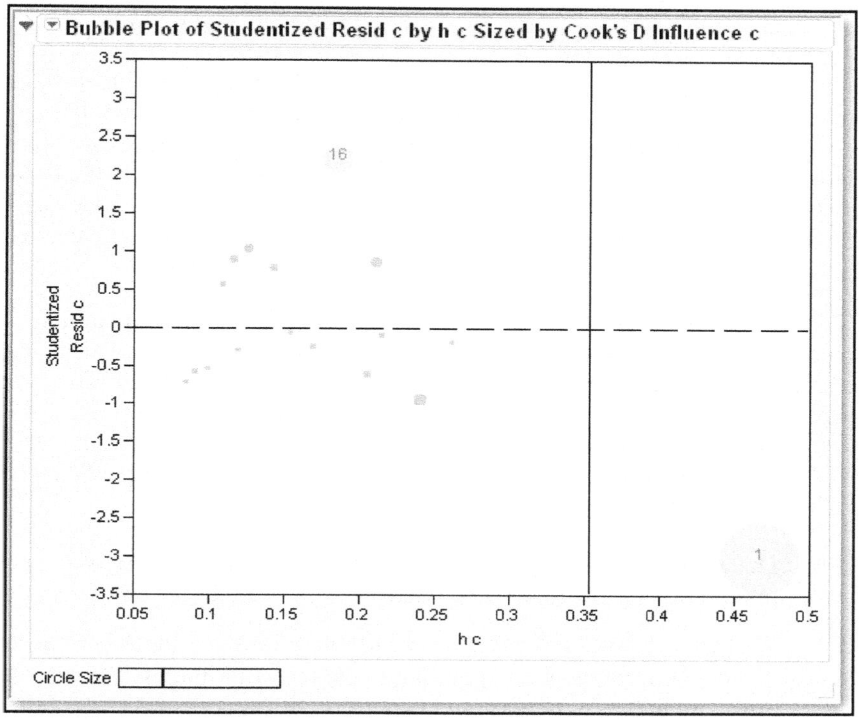

7.6 Summary

In this chapter we have learned how to model the linear relationship of a continuous factor on a continuous response. We discussed the difference between functional and statistical relationships between two variables and when simple linear regression is appropriate. We showed the form of the simple linear regression model (linear equation), $Y = \beta_0 + \beta_1 X + \varepsilon$ and reviewed how to interpret each term in the model, including the slope, β_1, which measures the change in response, Y, when the explanatory variable, X, changes by one unit.

Some concepts associated with analysis of variance (ANOVA) also apply to regression analysis. For example, we partition the variation from our data into the variation due to our model and that due to noise. This partition enables us to carry out a significance test for the estimate of the slope. If the *p*-value is less than our chosen value for α, we conclude that a significant linear relationship exists. In addition, we discussed how to check the quality of the fitted model using residuals analysis, R^2, and a lack-of-fit test and, in some instances, our model can be improved by the addition of polynomial terms.

The importance of designing a good study up front, which can enable us to check for model adequacy and develop a good predictive model, was discussed. If we have a choice, we should select a suitable range for our regressor variable that exploits the relationship between the two variables, and that will be of interest in the future. We also want to replicate some of our factor levels so we can carry out a lack-of-fit test to see whether we are missing any key terms in our model, and obtain an estimate of pure (replication) error.

We provided step-by-step instructions for establishing a calibration curve for load cells, including how to use JMP to help us conduct the corresponding statistical analysis, and make data-driven decisions. We also illustrated how to develop an inverse prediction equation to be able to use the calibration curve as is typically done.

Some of the key concepts and practices that we should remember from this and previous chapters are the following:

- Determine whether a simple linear regression model will suffice, or if additional polynomial terms might be needed.
- Write out the ANOVA table before running the study in order to account for the degrees of freedom corresponding to the factor of interest and the sources of noise, including pure error and lack of fit.
- Conduct the appropriate analysis using simple linear regression techniques.
- Fit the model without potentially influential observations and compare the results with the fit to the full data.
- Remember to clearly define the population of interest, the experimental unit, the observational unit, and how the study will be executed.
- Use the scientific method to make sure that the statistics are helping us support or refute our hypothesis.
- Select key graphs and analysis output to facilitate the presentation of results.
- *Always translate the statistical results back in the context of the original question.*

7.7 References

Beck, A., and P. Havas. 1989. *The Collected Papers of Albert Einstein, Volume 2, The Swiss Years: 1900–1909.* Princeton, NJ: Princeton University Press.

Draper, N. R., and H. Smith. 1998. *Applied Regression Analysis.* 3d ed. New York, NY: John Wiley & Sons.

Freund, R.J., R.C. Littell, and L. Creighton. 2003. *Regression Using JMP.* Cary, NC: SAS Institute, Inc.

Hahn, G.J. 1973. "The Coefficient of Determination Exposed!" *Chemical Technology* 3: 609–611.

Iglewicz, B. 2007. "Einstein's First Published Paper." *The American Statistician* 61, no. 4: 339–342.

Natrella, M. G. 1963. *Experimental Statistics, NBS Handbook 91*. Washington, D.C.: U.S. Department of Commerce, National Bureau of Standards. Reprinted 1966.

Weisberg, S. 1985. *Applied Linear Regression*. 2d ed. New York, NY: John Wiley & Sons.

Wheeler, D.J. 2005. *The Six Sigma Practitioner's Guide to Data Analysis*. Knoxville, TN: SPC Press, Inc.

Index

D

G

H

N

S

V

W

Y

Z

CPSIA information can be obtained
at www.ICGtesting.com
Printed in the USA
LVOW04s0601080417
530086LV00001B/1/P

9 781599 944883